Updates in Hypertension and Cardiovascular Protection

Series Editors

Giuseppe Mancia
Milano, Italy

Enrico Agabiti-Rosei
Brescia, Italy

The aim of this series is to provide informative updates on both the knowledge and the clinical management of a disease that, if uncontrolled, can very seriously damage the human body and is still among the leading causes of death worldwide. Although hypertension is associated mainly with cardiovascular, endocrine, and renal disorders, it is highly relevant to a wide range of medical specialties and fields – from family medicine to physiology, genetics, and pharmacology. The topics addressed by volumes in the series *Updates in Hypertension and Cardiovascular Protection* have been selected for their broad significance and will be of interest to all who are involved with this disease, whether residents, fellows, practitioners, or researchers.

More information about this series at http://www.springer.com/series/15049

Enrico Agabiti-Rosei
Anthony M. Heagerty • Damiano Rizzoni
Editors

Microcirculation in Cardiovascular Diseases

Editors
Enrico Agabiti-Rosei
Department of Clinical
and Experimental Sciences
University of Brescia
Brescia
Italy

Anthony M. Heagerty
Division of Cardiovascular Sciences
School of Medical Sciences
University of Manchester
Manchester
UK

Damiano Rizzoni
Department of Clinical
and Experimental Sciences
University of Brescia
Brescia
Italy

ISSN 2366-4606 ISSN 2366-4614 (electronic)
Updates in Hypertension and Cardiovascular Protection
ISBN 978-3-030-47803-2 ISBN 978-3-030-47801-8 (eBook)
https://doi.org/10.1007/978-3-030-47801-8

This Springer imprint is published by the registered company Springer Nature Switzerland AG
The registered company address is: Gewerbestrasse 11, 6330 Cham, Switzerland

Preface

The microcirculation represents a very important part of the vasculature, being directly responsible for the delivery of oxygen and nutrients to the peripheral tissues as well as for the increase of vascular resistance, which is recognized as the most frequent hemodynamic characteristic of established hypertension. Cardiovascular and metabolic diseases are commonly associated with alterations in the microcirculation, involving small arteries, arterioles, capillaries and postcapillary venules, i.e. vessels with a diameter from about 350 μm to those as small as 6–8 μm.

Alterations of the microcirculation in several cardiovascular diseases are diffuse and detectable in all vascular beds which may be assessed by current diagnostic methods. In fact, so far the evaluation of microcirculation in humans has been limited by the relative complexity of the techniques available and by the difficult access to several important vascular beds. New promising diagnostic methods have been recently proposed.

The presence of structural alterations in the microcirculation, in terms of remodelling of small arteries and rarefaction of distal microvessels and capillaries, is responsible for resistance to flow and for the maintenance and progressive worsening of hypertension. Moreover, they cause a reduced flow reserve in several important vascular districts, such as the coronary vascular bed. The clinical value of the assessment of alterations in the microcirculation is emphasized by the demonstration of their prognostic significance, by their improvement with some effective drugs and by the possible prognostic meaning of their changes during treatment.

This book was conceived to inform readers on the fundamental and updated advances in the pathophysiological mechanisms and the clinical aspects concerning the microcirculation, in cardiovascular and metabolic diseases, mainly in hypertension. The book has been written by a group of well-known and most respected European researchers and clinicians in this field. We hope to provide interesting information in order to improve the knowledge in this somewhat neglected part of the circulation, thus allowing a future better diagnosis and management of cardiovascular diseases.

Brescia, Italy	Enrico Agabiti-Rosei
Brescia, Italy	Damiano Rizzoni
Manchester, UK	Anthony M. Heagerty

Contents

Structure and Function of the Microcirculation

1

Christian Aalkjaer and Michael J. Mulvany

1.1 Location of Peripheral Resistance

1.1.1 Anatomy

The vasculature is usually divided into six categories: conduit arteries, small arteries, arterioles, capillaries, venules and veins [1, 2]. The division between conduit arteries and small arteries is arbitrary but commonly placed at the level of arteries with diameter 200 μm. The division between small arteries and arterioles rests on arterioles being defined as vessels with not more than 1–2 layers of smooth muscle [3]. Although previously the 'microcirculation' was considered alone to consist of the arterioles, the capillaries and the venules, it is now more often considered to include also the small arteries, at least to the extent that these contribute to the control of the peripheral resistance. 'Resistance vessels' are thus taken to include both arterioles and small arteries.

1.1.2 Contribution to Peripheral Resistance

A key to determining whether small arteries contribute to control of the peripheral resistance lies in measurement of the intravascular pressure along the vascular tree in order to determine which vessels are responsible. Clearly this is not a trivial task,

C. Aalkjaer (✉)
Department of Biomedicine, Aarhus University, Aarhus C, Denmark

Department of Biomedical Science, Copenhagen University, Copenhagen N, Denmark
e-mail: ca@biomed.au.dk

M. J. Mulvany
Department of Biomedicine, Aarhus University, Aarhus C, Denmark
e-mail: mjm@biomed.au.dk

E. Agabiti-Rosei et al. (eds.), *Microcirculation in Cardiovascular Diseases*, Updates in Hypertension and Cardiovascular Protection,
https://doi.org/10.1007/978-3-030-47801-8_1

since the microvessels are not normally accessible without anaesthesia-requiring surgery, and the surgery and the probes used to make the measurements may themselves disturb the haemodynamics. In humans it has been possible to make measurements of the pressure in the capillaries of the nailfold of fingers [4] and toes [5] without the use of anaesthesia. These measurements have indicated that nailfold capillary pressure at the apex of the capillary loop ranges from 10.5 to 22.5 mmHg [4] in normotensive individuals and higher in patients with essential hypertension [6] or with insulin-dependent diabetes mellitus [7]. Thus some 80% of the systemic blood pressure is dissipated proximal to the capillaries. The questions are: where in the precapillary vasculature does the pressure drop occur, and which vessels are involved in controlling the pressure profile under physiological and pathophysiological conditions? Another question is the extent to which the venules play a role in control of peripheral resistance; as the total venous resistance is around 15% of total peripheral resistance, the contribution may be small, although it could have physiologically important consequences for capillary pressure.

There have been few recent direct measurements of small artery pressure, but earlier studies have provided a consistent picture in exposed vascular beds of anaesthetized animals using servo-null measurement techniques [8]. Here a pipette is inserted into a vessel, and the pressure within the pipette is raised to a level required to just prevent outflow of blood; this then corresponds to the intravascular pressure. In the hamster cheek pouch, 7.5% of the precapillary pressure drop occurred in large arteries, 59% in small arteries (diameter ca. 70–300 μm) and 33% in arterioles [9]. These findings in hamster cheek pouch were confirmed by Joyner et al. [10] who found that 54% of the precapillary pressure drop occurred proximal to 1A arterioles (diameter ca. 100 μm) both in normotensive and renal hypertensive animals. Even greater pre-arteriole pressure drops were reported by Gore and Bohlen [9] in rat intestinal muscle (74% of precapillary pressure), consistent with previous measurements in arterioles of a variety of animals in a variety of vascular beds [8]. In the brain, 47% of total resistance in cerebrum and only 25% of total resistance in brain stem were found in vessels proximal to the arterioles [11]. Zweifach [12] found that ca. 40% of precapillary pressure drop occurred proximal to 60-μm arterioles in the cat mesentery. Meininger et al. [13] found that ca. 60% of precapillary pressure drop occurred proximal to 120-μm arterioles in the rat cremaster. Thus these studies indicate that the resistance of pre-arteriolar arteries is a significant portion of the precapillary pressure drop. In contrast, Delano and colleagues found that the pre-arteriolar pressure drop was as little as ca. 15% of precapillary pressure drop in rat tibialis muscle, although higher values were found in other skeletal muscles [14].

The general picture from these studies is that arteries proximal to those with diameter ca. 100 μm contribute substantially to the peripheral resistance, but since these are animal studies made under anaesthesia and substantial surgery, the relevance to the human situation is unclear. An alternative approach is to model the human vasculature as done by Blanco and colleagues [15] in order to make predictions of pressure in the small arteries. In particular, they have modelled the arterial network in the human brain. They predicted that, in the lenticulostriate arteriolar bed, about 30% of the precapillary pressure drop occurred proximal to 30/50-μm arteries. In the arterioles of posterior parietal arteriolar bed, the corresponding predictions were 60%. Thus although there are substantial differences between the two vascular beds, these theoretical findings are in general agreement with the experimental animal results, suggesting that in at least some vascular beds a substantial portion of the peripheral resistance lies in vessels proximal to the arterioles.

Despite the relevance of this question, it is surprising that there have been so few attempts to obtain information about the role of small arteries in the control of peripheral resistance under physiological conditions. The question was, however, addressed in our laboratory where Christensen and Fenger-Gron developed an ingenious method of measuring in conscious rat's blood pressure at the base of the mesenteric arcade with indwelling catheters [16, 17]. By the use of appropriate ligations, normal haemodynamics were maintained at the points of measurement. The main finding 5–17 hr. after surgery was that 31% of total pressure drop occurred between the superior mesenteric artery and vessels of diameter ca. 100 μm (R_{feed}). Immediately following surgery R_{feed} was only 16% indicating the importance of allowing animals to recover from the effects of surgery and anaesthesia. It was also found that the small arteries had specific responses to infusion of agonists (increase in R_{feed}: noradrenaline 151%, angiotensin II 0%, serotonin 414%). Spontaneous activity increased R_{feed} by 29% and environmental stress (loud noise) by 116%, the latter responses being blocked by prazosin. In our view, these findings provide good evidence for an important role for small arteries not only as part of but also in the control of peripheral resistance. The data, however, refer to a particular vascular bed in a particular animal. Given the large amount of work done on small arteries on the assumption that they are 'resistance vessels', it would clearly be useful if further studies be made to address this question.

1.2 Structure of Resistance Vessels

The role of resistance vessel structure in the aetiology of hypertension has been reviewed recently [18] and will be briefly summarized here.

1.2.1 Clinical Studies

Measurements of peripheral resistance using forearm plethysmography as pioneered by Folkow [19] have repeatedly indicated a structural basis for increased resistance of the resistance vasculature in essential hypertension. The results were consistent with a narrowing of the resistance vasculature and increase in the wall-to-lumen ratio. Such haemodynamic studies have been extended to other vascular beds, including the coronary circulation where essential hypertension has been shown to be associated with a reduced coronary reserve [20]. An alternative explanation for the increased peripheral resistance in hypertension is rarefaction with fewer parallel-connected vessels, and indeed evidence for rarefaction in essential hypertension precedes Folkow's studies [21]. The presence of rarefaction in hypertension has been widely confirmed [22, 23]. Microvascular rarefaction has two major consequences. Firstly, reduction of arteriolar density increases vascular resistance. Secondly, it disturbs the tissue delivery of oxygen and nutrients, thus contributing to target organ damage in hypertension. The clinical relevance of microvascular structure as a prognostically relevant end point has been recently reviewed [24]. There is also evidence that rarefaction precedes the onset of hypertension [25], suggesting that a primary defect in angiogenic mechanisms could be responsible for the development of hypertension.

1.2.2 Ex Vivo Evidence

The structure of resistance vessels has been much studied, in particular as regards changes associated with hypertension. Evidence concerning narrowing of the vessels has been obtained primarily from gluteal biopsies of small arteries [26], findings that have been widely confirmed [27]. The reason for the narrowing of resistance vessels in hypertension appears to be primarily due to changes in structure, and few functional changes have been observed [28]. The structural change is an inward eutrophic remodelling: a reorganization of the VSMC around a narrower lumen [29–31]. Indeed, Schiffrin and co-workers [32] concluded from their observations that small artery structure is one of the first manifestations of target organ damage, occurring before proteinuria or cardiac hypertrophy. Indeed, clinical evidence shows that, compared to controls, excessive microvascular structural abnormalities in the coronary and peripheral circulations are raised proportionally more than the blood pressure also suggesting that structural changes might precede the rise in blood pressure [33].

In addition to these changes in VSMC structure and function, the extracellular matrix is critically important for the altered properties of small arteries in hypertensive subjects. With chronic vasoconstriction, some degree of cell migration, secretion of fibrillar and nonfibrillar components, and rearrangement of extracellular matrix-cell interactions may occur [27]. Data from gluteal subcutaneous small arteries have indicated an age-dependent increase in ROS and in collagen in both EH and NT, greatest in EH [29, 34].

The prognostic value of abnormal small artery structure has been investigated in a number of studies. These studies have shown that an increased media-to-lumen ratio of gluteal small arteries is associated with increased cardiovascular risk [35, 36]. Furthermore it was shown that media-to-lumen ratio of small arteries on completion of 1 year of treatment was also predictive of increased cardiovascular risk [37]. There is also evidence that a hypertrophic response of the more proximal small arteries to essential hypertension is predictive of additional cardiovascular risk [38].

Izzard et al. [38] and Mulvany [39] have reviewed the mechanisms of inward eutrophic remodelling of small arteries from essential hypertensive individuals. Their conclusion was that chronic vasoconstriction is the stimulus for a structural reduction in lumen diameter, a conclusion supported by more recent experimental studies [40]. The nature of the contractile stimulus is still not fully resolved. Neural or humoral factors may be involved; although, Izzard et al. [38] favour the myogenic properties of the small arteries as the underlying mechanism. The myogenic vasoconstriction could serve to maintain wall stress at constant value.

1.2.3 Direct Observation

The retinal microcirculation has long been known to be abnormal in hypertension [41–43] and is an important tissue to study for hypertension-related organ damage [44]. More recently, major advances in the ability to analyse the retinal microcirculation have been developed using scanning-laser Doppler flowmetry (SLDF) [45] and adaptive optics [46]. The findings using these techniques are dealt with elsewhere in this volume (Chaps. 3 and 4).

1.3 Function of Resistance Vessels

1.3.1 Resistance Vessel Tone

As discussed above the microcirculation provides the major haemodynamic resistance in the vasculature and is therefore important for determining the blood pressure and the cardiac output. Through differential regulation of the tone of the microcirculation in different organs, the microcirculation furthermore controls the distribution of blood to the different organs. Finally changes in the tone of precapillary arteries and arterioles and the postcapillary venules control the capillary pressure which is important for lymph function, oedema formation, and in the kidney the filtration pressure. Collectively these functions are extremely important for the organism. This statement is underlined by the fact that abnormalities in the microcirculation are important for a large number of diseases with substantial individual and socio-economic impact.

The tone of the smooth muscle cells in the arterial wall determines the diameter of the arteries (on the background of the vessel structure) and therefore the hydrodynamic resistance. One might therefore say that the smooth muscle cell is the most important cell in the vascular wall. However, the tone of the smooth muscle cells is controlled by a host of factors released from the endothelium and by different types of nerves—predominantly of the sympathetic branch of the nervous system causing vasoconstriction—but also by vasodilator sensory nerves, sending antidromic signals to the smooth muscle cells. It is therefore evident that also endothelial cells and nerves are of central importance for the hydrodynamic resistance.

In the following sections, we will provide a short summary of the most important mechanisms controlling the smooth muscle tone and in particular the mechanisms responsible for oscillation of smooth muscle tone. Such oscillation is seen in many vascular beds and gives rise to vasomotion and flowmotion.

1.3.2 Contraction of Smooth Muscle Cells

The key parameter for contraction is the concentration of free calcium in the cytosol ($[Ca^{2+}]_i$). Ca^{2+} binds to calmodulin and that complex activates the myosin light chain kinase which phosphorylates the regulatory light chain of myosin to allow interaction between myosin and actin, hence contraction. An understanding of the control of smooth muscle contraction therefore involves an understanding of the control of $[Ca^{2+}]_i$. However, the sensitivity of the contractile machinery to $[Ca^{2+}]_i$ can be substantially modified, and it is therefore also important to understand how this $[Ca^{2+}]_i$ sensitivity is modified.

Flux of Ca^{2+} from the extracellular space into the smooth muscle cells is a main pathway leading to increase of $[Ca^{2+}]_i$. The main influx pathway is the L-type Ca^{2+} channel, which opens in response to a depolarization of the cells. When the channel opens Ca^{2+} runs into the cell down a steep electrochemical gradient and $[Ca^{2+}]_i$ increases. There is an associated release of Ca^{2+} from the sarcoplasmic reticulum which contributes to the $[Ca^{2+}]_i$ increase. The balance between the two sources of Ca^{2+} is not well studied and may depend on the agonist. Ca^{2+} is pumped out of the cells by a Ca^{2+} ATPase situated in the sarcolemma, while the Ca^{2+} released from the sarcoplasmic reticulum is pumped back into the sarcoplasmic reticulum by another Ca^{2+} ATPase sometimes called SERCA. Other channels of the TRP-channel family also contribute to the influx of Ca^{2+}, but it seems that the L-type Ca^{2+} channels are the functionally most significant, but this may also depend on the agonist used.

The main regulation of the Ca^{2+} sensitivity occurs via inhibition of the myosin light chain phosphatase, which dephosphorylates the myosin light chain [47]. The phosphatase is inhibited after activation of G-coupled vasoconstrictor receptors. A main pathway is via the monomeric G protein RhoA and its associated effector protein Rho-kinase. Rho-kinase dephosphorylates the myosin light chain either directly or via the phosphatase inhibitor CPI-17. The role of Ca^{2+} sensitivity for resistance vessel tone is substantial and may easily account for 50% or more of the smooth muscle tone in these arteries, indicating that Ca^{2+}-induced activation of the myosin light chain kinase is far from sufficient to release the full potential of the muscle cells.

1.3.3 Vasomotion

Vasomotion—the oscillation of vascular tone—is a prominent feature in the microcirculation [48, 49]. Vasomotion gives rise to oscillations of flow in microcirculatory districts, i.e. flowmotion. Vasomotion is prevalent both in vivo and in vitro and is intrinsic to the vascular wall, although the environment modifies the prevalence, amplitude and frequency. Although vasomotion is predominantly seen in the microvasculature, also large arteries [50, 51] exhibit vasomotion, and the first report on vasomotion was in the bat vein [52]. Most if not all microvascular areas may exhibit vasomotion.

The occurrence of vasomotion leads to two questions: (1) What are the cellular mechanisms that ensure a synchronized oscillation of the smooth muscle cell activity in the vascular wall and (2) what are the consequences of vasomotion, i.e. does vasomotion ensure a better or worse delivery of oxygen and removal of CO_2 and other metabolites from tissues? Both questions have at best only partial answers.

1.3.4 What Are the Cellular Mechanisms Responsible for Vasomotion?

The cellular mechanisms responsible for vasomotion involve both an oscillator, which sets up an oscillation of tone in the individual smooth muscle cell, and a mechanism that ensures synchronization of the smooth muscle cells oscillatory activity in the vascular wall. Perhaps the only feature related to vasomotion which is agreed by everybody is that an oscillation of the smooth muscle cell membrane potential is the background for the oscillation of the individual smooth muscle cell tone and also for the synchronization of the smooth muscle cells. When the smooth muscle cell membrane potential has been measured during vasomotion, an oscillation has been observed [53–61]. This strongly suggests that $[Ca^{2+}]_i$ in the smooth muscle cells is also oscillating in a synchronized manner—and this is indeed the case [62, 63]. This coupling between membrane potential oscillations and oscillations of $[Ca^{2+}]_i$ immediately explains why vasomotion is sensitive to L-type Ca^{2+} channel blockers [64]. The coupling between membrane potential oscillations and oscillations of $[Ca^{2+}]_i$ furthermore means that vasomotion only occurs in a situation where the membrane potential oscillation will result in significant changes in the flux of Ca^{2+} through the L-type Ca^{2+} channels. This again dictates that vasomotion cannot occur at low and high levels of activation, where the membrane potential may be in a range where it has little effect on $[Ca^{2+}]_i$.

The important question is therefore how the oscillation of the membrane potential is set up. This may vary between different vascular beds. Rat mesenteric small arteries exhibit a very prevalent and regular vasomotion both in vivo and in vitro when the arteries are submaximally activated by noradrenaline or another vasoconstrictor. Submaximal activation ensures partial depolarization and therefore brings the smooth muscle cells into a state where membrane potential oscillations will lead to oscillations of $[Ca^{2+}]_i$ and consequently tone [62]. In these arteries, we demonstrated how an oscillatory release of Ca^{2+} from the sarcoplasmic reticulum followed by an oscillatory uptake into the sarcoplasmic reticulum—a so-called cytosolic oscillator—leads to an oscillatory activity of a Ca^{2+}-sensitive conductance in the membrane [62]. The resulting oscillatory current feeds into the neighbouring cells via gap junctions. This causes changes in the membrane potential in these neighbours. The changes in membrane potential facilitate the release and uptake of Ca^{2+} in the neighbours and in this way entrain the cytosolic oscillators in the smooth muscle cells within the vascular wall [62, 65, 66]. This causes synchronized oscillations in the membrane potential of the smooth muscle cells and vasomotion. The model thus combines the cytosolic oscillator with a membrane oscillator, i.e. the oscillation of the Ca^{2+}-activated Cl^- conductance to set up vasomotion.

Vasomotion is influenced by the endothelium. But while some find that endothelium removal inhibits vasomotion [54, 67–74], others find that endothelial removal has little effect on vasomotion [61, 75–79]. The reason for this ambiguity is not clear. Although differences in arterial preparations (isobaric or isometric, strips or segments of arteries) and animal species investigation may explain some of the ambiguity, since, e.g. vasomotion of rat mesenteric small arteries has been shown to be either endothelial dependent [54] or endothelial independent [77, 79] under seemingly similar conditions. It has been suggested that part of the explanation may be that the absence of endothelium could lead to a contractile state where vasomotion is not feasible [78]. It is possible that removal of the endothelium will depolarize the membrane to a level where oscillations in membrane potential would not lead to oscillations of $[Ca^{2+}]_I$ and hence vasomotion. Although this may explain some of the discrepancy, in our hands, even in experiments based on concentration-response curves to the vasoconstrictor agonist (thus covering the full range of membrane potentials and secondary messenger activation), vasomotion is still not seen after endothelial removal [68]. Perhaps another explanation may be that vasomotion is critically dependent on minor changes in experimental conditions, e.g. content of physiological solution, stretch of the smooth muscle cells and vasoconstrictor agonist. Thus we have found that vasomotion that occurs in a particular vessel may disappear later during the experiment even though no intervention has been introduced. Also it may be present in blood vessels from one animal but not in the same blood vessel taken from a new animal the next day, or for unknown reason vasomotion may not be present for some months and then suddenly reappear.

Some evidence for the type of conditions needed for vasomotion comes from our observation [68] that in rat mesenteric small arteries with the endothelium removed no vasomotion was seen under isometric conditions. However, the presence of 1 mM of the anion transport inhibitor DIDS induced vasomotion which was SCN^- sensitive and therefore likely Cl^- channel dependent, which is unexpected in the presence of an anion transport inhibitor [68]. Also 10 μM Zn^{2+} induced vasomotion in the rat mesenteric small arteries without endothelium which was partly SCN^- sensitive [68]. Although the mechanism responsible for the DIDS- and Zn^{2+}-induced vasomotion is not known, these observations indicate that even relatively minor changes in experimental conditions may lead to vasomotion reflecting that several pathways may lead to vasomotion.

In the rat mesenteric small arteries [54] and also in other blood vessels such as the hamster aorta [50], vasomotion is dependent on a steady supply of cGMP. Importantly this is not the case in all arteries, e.g. the prevalence of vasomotion is enhanced in rat cerebral arteries when cGMP is reduced [80, 81]. The cGMP dependency was used as a means to find the membrane conductance important for vasomotion in rat mesenteric small arteries. It turned out that the smooth muscle cells in rat mesenteric small arteries, and in many other vascular sections, with the exception of pulmonary arteries, have a cGMP-dependent Ca^{2+}-activated Cl^- conductance [82, 83]. This conductance is also present in rat colonic smooth muscle cells, i.e. not confined to vascular smooth muscle cells [83]. Based on knock-down experiments, it was demonstrated that the cGMP-dependent Ca^{2+}-activated Cl^- conductance required the presence of two proteins TMEM16A and bestrophins [84, 85]. TMEM16A is a membrane-bound protein now known to be responsible for Ca^{2+}-activated Cl^- conductances in many tissues including vascular smooth muscle cells [86, 87], while the role of bestrophins is still uncertain. However, the presence of bestrophin 3 (and possibly also bestrophins 1 and 2) changes the biophysical and pharmacological profile of the Ca^{2+}-activated Cl^- conductance in vascular smooth muscle and makes the current cGMP dependent [84, 88]. Knock-down of TMEM16A reduces the expression of the bestrophins [85]. It is therefore difficult to know whether bestrophins provide a Ca^{2+}-activated Cl^- conductance in their own right or are subunits to TMEM16A. The requirement of both of these membrane proteins for vasomotion in rat mesenteric small arteries was demonstrated by the finding that knock-down of the proteins inhibited vasomotion [84, 85].

Using the data obtained in these experiments, it was possible to model vasomotion. This further substantiates the idea that a mixed cytosolic and membrane oscillator with an obligatory role of a cGMP-dependent Ca^{2+}-activated Cl^- conductance is the basis for vasomotion at least in rat mesenteric small arteries [65, 66]. It will be important to find out to what extent this model has applicability in other vascular beds.

Other models of vasomotion have been suggested, showing that other ion channels, e.g. K^+ channels may be important for setting up vasomotion. There is still, however, relatively little experimental support for these alternative models.

1.3.5 Is Vasomotion Good or Bad?

Theoretical considerations indicate that the hydraulic conductance is larger in arteries with an oscillating diameter compared with an artery with the same mean diameter where the diameter is constant. This can easily be derived from Poiseuille's equation and is a reflection of the non-linear relationship between artery diameter and hydraulic conductance. This may not though constitute an important advantage because the mean diameter and thus hydraulic conductance can easily be modified through many pathways. Another potential advantage is that it may be easier to obtain a precise control of haemodynamic resistance in a system with oscillations compared to a steady system. The importance of this is though also difficult to evaluate. So the question of whether vasomotion enhances the delivery of oxygen and substrates to the tissues and enhances the washout of CO_2 or vasomotion has proven very difficult to answer unambiguously.

Oscillations of $[Ca^{2+}]_i$ have, in other cells systems, been shown to encode for transcriptional regulation, and, depending on amplitude and frequency, different transcriptional patterns are activated [89, 90]. The importance of this in vascular smooth muscles is not known, but, in vascular endothelial cells, the expression of vascular cell adhesion molecule 1 (VCAM1) is modulated by the frequency of Ca^{2+} oscillations for a constant Ca^{2+} concentration [91]. In smooth muscle cells, the amplitude and frequency of the oscillations of $[Ca^{2+}]_i$ were decoded by the myosin light chain phosphatase, which will modify the Ca^{2+}-independent contraction of the smooth muscle cells [92]. Also in other cell systems, the frequency of oscillations is decoded by enzymes [93], e.g. oscillating calcium activates both calcineurin and protein kinase II (CaMKII), but increased frequencies shift the magnitude of activation from calcineurin to CaMKII [94]. Furthermore mitochondrial function may also be dependent on the amplitude or frequency of $[Ca^{2+}]_i$ oscillations [95]. Whether these effects of $[Ca^{2+}]_i$ oscillations are prevalent in vascular smooth muscle cells is unknown.

1.4 Conclusion

The small arteries play a crucial role in the determination of, and control of, peripheral resistance. The resistance they present is the result of interplay between their structural and functional properties. The functional properties are determined by the smooth muscle cells, the endothelial cells and the sympathetic and sensory nerves in the vessel wall predominantly. The interaction between these cell types is dynamic, and often results in the development of an oscillation of vascular tone (vasomotion), which consequently leads to oscillations in flow (flowmotion). It will be one of the tasks of future studies to elucidate the consequences of these dynamic responses of small arteries.

References

1. Davis MJ, Ferrer PN, Gore RW. Vascular anatomy and hydrostatic pressure profile in the hamster cheek pouch. Am J Physiol. 1986;250:H291–303.
2. Mulvany MJ, Aalkjaer C. Structure and function of small arteries. Physiol Rev. 1990;70:921–61.
3. Rhodin JAG. Architecture of the vessel wall. In: Bohr DF, Somlyo AP, Sparks HV, editors. Handbook of physiology, section 2, the cardiovascular system, volume 2, vascular smooth muscle. American Physiological Society: Bethesda, MD; 1980.
4. Shore AC. Capillaroscopy and the measurement of capillary pressure. Br J Clin Pharmacol. 2000;50:501–13.
5. De Graaff JC, Ubbink DT, Lagarde SM, Jacobs MJ. Postural changes in capillary pressure in the hallux of healthy volunteers. J Appl Physiol. 2003;95:2223–8.
6. Williams SA, Boolell M, Macgregor GA, Smaje LH, Wasserman SM, Tooke JE. Capillary hypertension and abnormal pressure dynamics in patients with essential hypertension. Clin Sci (Lond). 1990;79:5–8.
7. Sandeman DD, Shore AC, Tooke JE. Relation of skin capillary pressure in patients with insulin-dependent diabetes mellitus to complications and metabolic control. N Engl J Med. 1992;327:760–4.
8. Intaglietta M, Pawula RF, Tompkins WR. Pressure measurements in the mammalian microvasculature. Microvasc Res. 1970;2:212–20.
9. Gore RW, Bohlen HG. Microvascular pressures in rat intestinal muscle and mucosal villi. Am J Physiol. 1977;233:H685–93.
10. Joyner WL, Davis MJ, Gilmore JP. Intravascular pressure distribution and dimensional analysis of microvessels in hamsters with renovascular hypertension. Microvasc Res. 1981;22:190–8.

11. Faraci FM, Mayhan WG, Heistad DD. Segmental vascular responses to acute hypertension in cerebrum and brain stem. Am J Phys. 1987;252:H738–42.
12. Zweifach BW. Quantitative studies of microcirculatory structure and function. I. Analysis of pressure distribution in the terminal vascular bed in cat mesentery. Circ Res. 1974;34:843–57.
13. Meininger GA, Fehr KL, Yates MB. Anatomic and hemodynamic characteristics of the blood vessels feeding the cremaster skeletal muscle in the rat. Microvasc Res. 1987;33:81–97.
14. Delano FA, Schmid Schonbein GW, Skalak TC, Zweifach BW. Penetration of the systemic blood pressure into the microvasculature of rat skeletal muscle. Microvasc Res. 1991;41:92–110.
15. Blanco PJ, Muller LO, Spence JD. Blood pressure gradients in cerebral arteries: a clue to pathogenesis of cerebral small vessel disease. Stroke Vasc Neurol. 2017;2:108–17.
16. Fenger Gron J, Mulvany MJ, Christensen KL. Mesenteric blood pressure profile of conscious, freely moving rats. J Physiol Lond. 1995;488:753–60.
17. Fenger Gron J, Mulvany MJ, Christensen KL. Intestinal blood flow is controlled by both feed arteries and microcirculatory resistance vessels in freely moving rats. J Physiol Lond. 1997;498:215–24.
18. Mulvany M, Agabiti-Rosei E, Struijker-Boudier H. Structural cardiovascular changes in hypertension. In: Mancia G, Grassi G, Tsioufis K, Dominiczak A, Agabiti-Rosei E, editors. Manual of hypertension: European Society of Hypertension. 2019;81–8.
19. Folkow B. Structural, myogenic, humoral and nervous factors controlling peripheral resistance. In: Harington M, editor. Hypotensive Drugs. London: Pergamon; 1956.
20. Schafer S, Kelm M, Mingers S, Strauer BE. Left ventricular remodeling impairs coronary flow reserve in hypertensive patients. J Hypertens. 2002;20:1431–7.
21. Ruedemann AD. Conjunctival vessels. J Am Med Assoc. 1933;101:1427–81.
22. Antonios TF, Singer DR, Markandu ND, Mortimer PS, Macgregor GA. Structural skin capillary rarefaction in essential hypertension. Hypertension. 1999;33:998–1001.
23. Struijker Boudier HA, Le Noble JL, Messing MW, Huijberts MS, Le Noble FA, Van Essen H. The microcirculation and hypertension. J Hypertens Suppl. 1992;10:S147–56.
24. Agabiti-Rosei E, Rizzoni D. Microvascular structure as a prognostically relevant endpoint. J Hypertens. 2017;35:914–21.
25. Antonios TF, Rattray FM, Singer DR, Markandu ND, Mortimer PS, Macgregor GA. Rarefaction of skin capillaries in normotensive offspring of individuals with essential hypertension. Heart. 2003;89:175–8.
26. Aalkjaer C, Heagerty AM, Petersen KK, Swales JD, Mulvany MJ. Evidence for increased media thickness, increased neuronal amine uptake, and depressed excitation-contraction coupling in isolated resistance vessels from essential hypertensives. Circ Res. 1987;61:181–6.
27. Schiffrin EL. Remodeling of resistance arteries in essential hypertension and effects of antihypertensive treatment. Am J Hypertens. 2004;17:1192–200.
28. Aalkjaer C, Heagerty AM, Bailey I, Mulvany MJ, Swales JD. Studies of isolated resistance vessels from offspring of essential hypertensive patients. Hypertension. 1987;9(Suppl III):III-155–8.
29. Bruno RM, Duranti E, Ippolito C, Segnani C, Bernardini N, Di Candio G, Chiarugi M, Taddei S, Virdis A. Different impact of essential hypertension on structural and functional age-related vascular changes. Hypertension. 2017;69:71–8.
30. Heagerty AM, Aalkjaer C, Bund SJ, Korsgaard N, Mulvany MJ. Small artery structure in hypertension—dual processes of remodeling and growth. Hypertension. 1993;21:391–7.
31. Mulvany MJ, Baumbach GL, Aalkjaer C, Heagerty AM, Korsgaard N, Schiffrin EL, Heistad DD. Vascular remodelling (letter to editor). Hypertension. 1996;28:505–6.
32. Park JB, Schiffrin EL. Small artery remodeling is the most prevalent (earliest?) form of target organ damage in mild essential hypertension. J Hypertens. 2001;19:921–30.
33. Eftekhari A, Mathiassen ON, Buus NH, Gotzsche O, Mulvany MJ, Christensen KL. Disproportionally impaired microvascular structure in essential hypertension. J Hypertens. 2011;29:896–905.

34. Intengan HD, Deng LY, Li JS, Schiffrin EL. Mechanics and composition of human subcutaneous resistance arteries in essential hypertension. Hypertension. 1999;33:569–74.
35. Mathiassen ON, Buus NH, Larsen ML, Mulvany MJ, Christensen KL. Small artery structure adapts to vasodilatation rather than to blood pressure during antihypertensive treatment. J Hypertens. 2007;25:1027–34.
36. Rizzoni D, Porteri E, Boari GE, De Ciuceis C, Sleiman I, Muiesan ML, Castellano M, Miclini M, Agabiti-Rosei E. Prognostic significance of small-artery structure in hypertension. Circulation. 2003;108:2230–5.
37. Buus NH, Mathiassen ON, Fenger-Gron M, Praestholm MN, Sihm I, Thybo NK, Schroeder AP, Thygesen K, Aalkjaer C, Pedersen OL, Mulvany MJ, Christensen KL. Small artery structure during antihypertensive therapy is an independent predictor of cardiovascular events in essential hypertension. J Hypertens. 2013;31:791–7.
38. Izzard AS, Rizzoni D, Agabiti-Rosei E, Heagerty AM. Small artery structure and hypertension: adaptive changes and target organ damage. J Hypertens. 2005;23:247–50.
39. Mulvany MJ. Small artery remodeling and significance in the development of hypertension. News Physiol Sci. 2002;17:105–9.
40. Castorena-Gonzalez JA, Staiculescu MC, Foote C, Martinez-Lemus LA. Mechanisms of the inward remodeling process in resistance vessels: is the actin cytoskeleton involved? Microcirculation. 2014;21:219–29.
41. Rizzoni D, Muiesan ML. Retinal vascular caliber and the development of hypertension: a meta-analysis of individual participant data. J Hypertens. 2014;32:225–7.
42. Witt N, Wong TY, Hughes AD, Chaturvedi N, Klein BE, Evans R, Mcnamara M, Thom SA, Klein R. Abnormalities of retinal microvascular structure and risk of mortality from ischemic heart disease and stroke. Hypertension. 2006;47:975–81.
43. Wong TY, Mitchell P. Hypertensive retinopathy. N Engl J Med. 2004;351:2310–7.
44. Cuspidi C, Sala C. Do microvascular retinal changes improve cardiovascular risk estimation? J Hypertens. 2012;30:682–4.
45. Ritt M, Harazny JM, Ott C, Schlaich MP, Schneider MP, Michelson G, Schmieder RE. Analysis of retinal arteriolar structure in never-treated patients with essential hypertension. J Hypertens. 2008;26:1427–34.
46. Gallo A, Mattina A, Rosenbaum D, Koch E, Paques M, Girerd X. Retinal arteriolar remodeling evaluated with adaptive optics camera: relationship with blood pressure levels. Annales de Cardiologie et d'Angeiologie. 2016;65:203–7.
47. Somlyo AP, Somlyo AV. Ca^{2+} sensitivity of smooth muscle and nonmuscle myosin II: modulated by G proteins, kinases, and myosin phosphatase. Physiol Rev. 2003;83:1325–58.
48. Schmidt JA. Periodic hemodynamics in health and disease. Heidelberg, Berlin: Springer; 1996.
49. Aalkjaer C, Nilsson H. Vasomotion: cellular background for the oscillator and for the synchronization of smooth muscle cells. Br J Pharmacol. 2005;144:605–16.
50. Jackson WF, Mulsch A, Busse R. Rhythmic smooth muscle activity in hamster aortas is mediated by continuous release of NO from the endothelium. Am J Phys. 1991;260:H248–53.
51. Palacios J, Vega JL, Paredes A, Cifuentes F. Effect of phenylephrine and endothelium on vasomotion in rat aorta involves potassium uptake. J Physiol Sci. 2013;63:103–11.
52. Jones TW. Discovery that the veins of the bat's wing are endowed with rhythmical contractility and that onward flow of blood is accelerated by each contraction. Phil Trans Roy Soc Lond. 1852;142:131–6.
53. Garland CJ. Influence of the endothelium and alpha-adrenoreceptor antagonists on responses to noradrenaline in the rabbit basilar artery. J Physiol. 1989;418:205–17.
54. Gustafsson H, Mulvany MJ, Nilsson H. Rhythmic contractions of isolated small arteries from rat: influence of the endothelium. Acta Physiol Scand. 1993;148:153–63.
55. Hayashida N, Okui K, Fukuda Y. Mechanism of spontaneous rhythmic contraction in isolated rat large artery. Jpn J Physiol. 1986;36:783–94.

56. Matchkov VV, Rahman A, Peng H, Nilsson H, Aalkjaer C. Junctional and nonjunctional effects of heptanol and glycyrrhetinic acid derivates in rat mesenteric small arteries. Br J Pharmacol. 2004;142:961–72.
57. Mulvany MJ, Nilsson H, Flatman JA. Role of membrane potential in the response of rat small mesenteric arteries to exogenous noradrenaline stimulation. J Physiol. 1982;332:363–73.
58. Rahman A, Hughes A, Matchkov V, Nilsson H, Aalkjaer C. Antiphase oscillations of endothelium and smooth muscle [Ca^{2+}] in vasomotion of rat mesenteric small arteries. Cell Calcium. 2007;42:536–47.
59. Rahman A, Matchkov V, Nilsson H, Aalkjaer C. Effects of cGMP on coordination of vascular smooth muscle cells of rat mesenteric small arteries. J Vasc Res. 2005;42:301–11.
60. Segal SS, Beny JL. Intracellular recording and dye transfer in arterioles during blood flow control. Am J Phys. 1992;263:H1–7.
61. Von Der Weid PY, Beny JL. Simultaneous oscillations in the membrane potential of pig coronary artery endothelial and smooth muscle cells. J Physiol. 1993;471:13–24.
62. Peng H, Matchkov V, Ivarsen A, Aalkjaer C, Nilsson H. Hypothesis for the initiation of vasomotion. Circ Res. 2001;88:810–5.
63. Schuster A, Oishi H, Beny JL, Stergiopulos N, Meister JJ. Simultaneous arterial calcium dynamics and diameter measurements: application to myoendothelial communication. Am J Physiol Heart Circ Physiol. 2001;280:H1088–96.
64. Gustafsson H, Nilsson H. Rhythmic contractions of isolated small arteries from rat: role of calcium. Acta Physiol Scand. 1993;149:283–91.
65. Jacobsen JC, Aalkjaer C, Nilsson H, Matchkov VV, Freiberg J, Holstein-Rathlou NH. Activation of a cGMP-sensitive calcium-dependent chloride channel may cause transition from calcium waves to whole cell oscillations in smooth muscle cells. Am J Physiol Heart Circ Physiol. 2007;293:H215–28.
66. Jacobsen JC, Aalkjaer C, Nilsson H, Matchkov VV, Freiberg J, Holstein-Rathlou NH. A model of smooth muscle cell synchronization in the arterial wall. Am J Physiol Heart Circ Physiol. 2007;293:H229–37.
67. Akata T, Kodama K, Takahashi S. Role of endothelium in oscillatory contractile responses to various receptor agonists in isolated small mesenteric and epicardial coronary arteries. Jpn J Pharmacol. 1995;68:331–43.
68. Boedtkjer DM, Matchkov VV, Boedtkjer E, Nilsson H, Aalkjaer C. Vasomotion has chloride-dependency in rat mesenteric small arteries. Pflugers Arch. 2008;457:389–404.
69. Haddock RE, Grayson TH, Brackenbury TD, Meaney KR, Neylon CB, Sandow SL, Hill CE. Endothelial coordination of cerebral vasomotion via myoendothelial gap junctions containing connexins 37 and 40. Am J Physiol Heart Circ Physiol. 2006;291:H2047–56.
70. Huang Y, Cheung KK. Endothelium-dependent rhythmic contractions induced by cyclopiazonic acid in rat mesenteric artery. Eur J Pharmacol. 1997;332:167–72.
71. Jackson WF. Arteriolar vasomotion: a role for NO in the microcirculation? Resistance arteries, structure and function. In: Proceedings of the Third International Symposium on Resistance Arteries; 21–25 May 1991; Rebild, Skorping, Denmark. 1991. p. 199–203.
72. Jackson WF. Role of endothelium-derived nitric oxide in vasomotion. In: Rubanyi GM, editor. Mechanoreception by the Vascular Wall: Futura Publishing Company; 1993. p. 173–95.
73. Mauban JR, Wier WG. Essential role of EDHF in the initiation and maintenance of adrenergic vasomotion in rat mesenteric arteries. Am J Physiol Heart Circ Physiol. 2004;287:H608–16.
74. Okazaki K, Seki S, Kanaya N, Hattori J, Tohse N, Namiki A. Role of endothelium-derived hyperpolarizing factor in phenylephrine-induced oscillatory vasomotion in rat small mesenteric artery. Anesthesiology. 2003;98:1164–71.
75. Freeman KA, Mao A, Nordberg LO, Pak J, Tallarida RJ. The relationship between vessel wall tension and the magnitude and frequency of oscillation in rat aorta. Life Sci. 1995;56:Pl129–34.
76. Omote M, Mizusawa H. The role of sarcoplasmic reticulum in endothelium-dependent and endothelium-independent rhythmic contractions in the rabbit mesenteric artery. Acta Physiol Scand. 1993;149:15–21.

77. Sell M, Boldt W, Markwardt F. Desynchronising effect of the endothelium on intracellular Ca^{2+} concentration dynamics in vascular smooth muscle cells of rat mesenteric arteries. Cell Calcium. 2002;32:105–20.
78. Seppey D, Sauser R, Koenigsberger M, Beny JL, Meister JJ. Does the endothelium abolish or promote arterial vasomotion in rat mesenteric arteries? Explanations for the seemingly contradictory effects. J Vasc Res. 2008;45:416–26.
79. Seppey D, Sauser R, Koenigsberger M, Beny JL, Meister JJ. Intercellular calcium waves are associated with the propagation of vasomotion along arterial strips. Am J Physiol Heart Circ Physiol. 2010;298:H488–96.
80. Dirnagl U, Lindauer U, Villringer A. Nitric oxide synthase blockade enhances vasomotion in the cerebral microcirculation of anesthetized rats. Microvasc Res. 1993;45:318–23.
81. Lacza Z, Herman P, Gorlach C, Hortobagyi T, Sandor P, Wahl M, Benyo Z. NO synthase blockade induces chaotic cerebral vasomotion via activation of thromboxane receptors. Stroke. 2001;32:2609–14.
82. Matchkov VV, Aalkjaer C, Nilsson H. A cyclic GMP-dependent calcium-activated chloride current in smooth-muscle cells from rat mesenteric resistance arteries. J Gen Physiol. 2004;123:121–34.
83. Matchkov VV, Aalkjaer C, Nilsson H. Distribution of cGMP-dependent and cGMP-independent Ca^{2+} -activated Cl^- conductances in smooth muscle cells from different vascular beds and colon. Pflugers Arch. 2005;451:371–9.
84. Broegger T, Jacobsen JC, Secher Dam V, Boedtkjer DM, Kold-Petersen H, Pedersen FS, Aalkjaer C, Matchkov VV. Bestrophin is important for the rhythmic but not the tonic contraction in rat mesenteric small arteries. Cardiovasc Res. 2011;91:685–93.
85. Dam VS, Boedtkjer DM, Nyvad J, Aalkjaer C, Matchkov V. TMEM16A knockdown abrogates two different Ca^{2+} -activated Cl^- currents and contractility of smooth muscle in rat mesenteric small arteries. Pflugers Arch. 2014a;466:1391–409.
86. Dam VS, Boedtkjer DM, Aalkjaer C, Matchkov V. The bestrophin- and TMEM16A-associated Ca^{2+} – activated Cl^- channels in vascular smooth muscles. Channels (Austin). 2014b;8:361–9.
87. Oh U, Jung J. Cellular functions of TMEM16/anoctamin. Pflugers Arch. 2016;468:443–53.
88. Matchkov VV, Larsen P, Bouzinova EV, Rojek A, Boedtkjer DM, Golubinskaya V, Pedersen FS, Aalkjaer C, Nilsson H. Bestrophin-3 (vitelliform macular dystrophy 2-like 3 protein) is essential for the cGMP-dependent calcium-activated chloride conductance in vascular smooth muscle cells. Circ Res. 2008;103:864–72.
89. Dolmetsch RE, Xu K, Lewis RS. Calcium oscillations increase the efficiency and specificity of gene expression. Nature. 1998;392:933.
90. Li W-H, Llopis J, Whitney M, Zlokarnik G, Tsien RY. Cell-permeant caged InsP3 ester shows that Ca^{2+} spike frequency can optimize gene expression. Nature. 1998;392:936.
91. Zhu L, Luo Y, Chen T, Chen F, Wang T, Hu Q. Ca^{2+} oscillation frequency regulates agonist-stimulated gene expression in vascular endothelial cells. J Cell Sci. 2008;121:2511–8.
92. Sward K, Dreja K, Lindqvist A, Persson E, Hellstrand P. Influence of mitochondrial inhibition on global and local $[Ca^{2+}]_i$ in rat tail artery. Circ Res. 2002;90:792–9.
93. Smedler E, Uhlen P. Frequency decoding of calcium oscillations. Biochim Biophys Acta. 2014;1840:964–9.
94. Li L, Stefan MI, Le Novere N. Calcium input frequency, duration and amplitude differentially modulate the relative activation of calcineurin and CaMKII. PLoS One. 2012;7:e43810.
95. Collins TJ, Lipp P, Berridge MJ, Bootman MD. Mitochondrial Ca^{2+} uptake depends on the spatial and temporal profile of cytosolic Ca^{2+} signals. J Biol Chem. 2001;276:26411–20.

Assessment of Small Artery Structure and Function by Micromyography

2

Michael J. Mulvany and Christian Aalkjaer

2.1 Introduction

Since Harvey's discovery of the circulation of the blood and Hales' first measurements of blood pressure, the basis for the determination and control of peripheral resistance has been an object of intense interest. The classic Langendorff heart preparation was the first to provide information about the resistance of the cardiac vasculature and its response to drugs [1] although the precise location of the responses could not be determined. To obtain information about the location of the resistance, Zimmermann made in vivo measurements in cats of pressure responses in the aorta, the dorsalis pedis artery and a metatarsal vein [2]. This allowed him to calculate total resistance, arterial resistance, small vessel resistance, and venous resistance. To obtain more specific information about the contribution of small arteries, Uchida and colleagues [3] developed a technique for measurement of small artery resistance by dissecting branches of the middle cerebral artery and of mesenteric artery from the mesojejunum. By leaving them attached at one end to the artery from which they branch, it was possible to perfuse them and study changes in their responsiveness. Since the majority of the resistance came from the small arteries diameter 50–250 μm at the end of the preparation, the responses of these to various drugs could be assessed. All these measurements were, however, indirect, and direct measurements of the mechanical responses to drugs were clearly to be preferred.

M. J. Mulvany (✉)
Department of Biomedicine, Aarhus University, Aarhus C, Denmark
e-mail: mjm@biomed.au.dk

C. Aalkjaer
Department of Biomedicine, Aarhus University, Aarhus C, Denmark

Department of Biomedical Science, Copenhagen University, Copenhagen N, Denmark
e-mail: ca@biomed.au.dk

E. Agabiti-Rosei et al. (eds.), *Microcirculation in Cardiovascular Diseases*, Updates in Hypertension and Cardiovascular Protection,
https://doi.org/10.1007/978-3-030-47801-8_2

2.2 Historical Background for Development of Micromyography

Direct mechanical measurements on cylindrical smooth muscle tissues were first based on strips taken from large specimens such as bovine and swine bronchi where shortening in response to drugs was measured using a cantilever and a kymograph [4, 5]. Similar isotonic experiments with material from smaller animals were made much later by Castillo and De Beer using rings of guinea pig bronchi connected in chains to amplify the drug-induced isotonic shortening [6]. Furchgott developed a method for experiments with aortae [7] in which the aortic wall was cut in a spiral strip (oriented at about 15 degrees relative to its long axis) to produce a specimen several cm long which could be strung between a fixed point and a suitably loaded cantilever allowing kymograph recordings. This delicate technique was later applied—remarkably—to small arteries diameter 200–300 μm by Bohr and colleagues [8] in a setup allowing isometric measurement of tension.

The earliest reference to the use of a ring preparation of an artery we have found is that of Nielsen and Owman who dissected rings of cat middle cerebral arteries which were then slipped onto two prongs of which one was fixed and the other attached to a Statham FT 03C force transducer for measurement of isometric tension [9]. This technique was extended to small vessels by Bevan and Osher [10]. They threaded a rabbit posterior inferior cerebellar small artery (diameter 200 μm) onto two wires, each of which was connected at both ends to one of two plates. One of the plates was fixed, and the other plate was mounted on a Statham G10B strain gauge. The wires were tightened using screws allowing isometric measurement of drug-induced changes in force. The records shown suggest that there may have been considerable noise in the transducer output, but this does not seem to have been investigated in any detail. In any event, these authors did not themselves take their elegant technique further, but the concept was taken up by Mulvany and Halpern [11], who developed (with the important input of toolmaker Mr. Jo Trono, University of Vermont workshop) a stable myograph based on the Bevan and Osher configuration using a sensitive temperature-compensated semiconductor strain gauge (Kistler-Morse DSC6). The myograph was built around a water immersion lens (Zeiss 40x, NA 0.75) which allowed visualization of smooth muscle cells within the vessel wall using Nomarski interference contrast optics, as well as measurements of wall and media thickness of the portion of the vessel that wrapped around the 32-μm tungsten mounting wires. These optics allowed smooth muscle cells in the vascular wall to be visualized (Fig. 2.1). In subsequent years Mulvany and Halpern in collaboration and then separately further developed the myograph (known as the "wire myograph") and two versions became (and are) commercially available (Danish Myo Technology, Aarhus, Denmark; Living Systems Instrumentation, St. Alban's, Vermont, USA). A video showing the mounting technique is available: https://www.youtube.com/watch?v=fSD1Ee4G6_U. Other companies are also marketing wire myographs (e.g., Radnoti LLC, Covina, CA, USA), but these seem to have been used primarily for investigating larger arteries [13].

The Journal of Physiology, Vol. 275 **Plate 1**

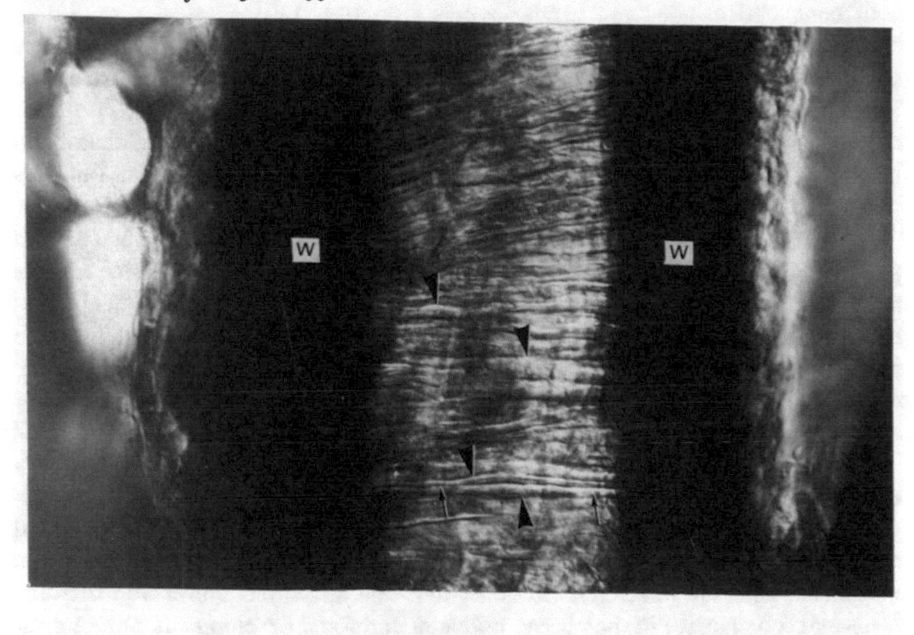

Fig. 2.1 Rat mesenteric small artery mounted on wire myograph under microscope using Nomarski differential contrast optics and Zeiss 40x water immersion lens. The microscope is focused within the upper media. Arrow heads show individual smooth muscle cells. The bar shows 50 μm. Reproduced with permission from [12]

2.3 Wire Myograph

The development of the myograph technique allowed detailed investigations of the structure and function of small arteries [14] in particular as related to abnormalities associated with hypertension. Here small arteries were defined as pre-arteriolar vessels with lumen diameter less than approximately 500 μm. A useful procedure was the use of the passive internal circumference—wall tension relation to estimate—using the Laplace relation, the lumen diameter that the vessel would have had in situ when relaxed and subjected to a specific internal pressure [15], a procedure that has been confirmed [16, 17] and widely cited (1930 citations pr. 27-11-2018, Web of Science). From measurements on the myograph of wall and media thickness, estimates could thus also be made of the physiologically important wall-to-lumen and media-to-lumen ratios [18]. Thus the media-to-lumen ratio of small arteries was found to be increased in animal models of hypertension and that this to some extent could be decreased by antihypertensive treatment [14].

An important extension of the animal studies concerning the remodeling [19] of small arteries came with the development of the human gluteal biopsy which allowed measurements of gluteal small arteries in patients with essential hypertension compared to age- and gender-matched controls [20]. Similarly, studies in hypertensive individuals have also been made of small arteries before and after treatment, of which many studies have now been made in different laboratories [21]. These studies have confirmed the animal studies showing that the media-to-lumen ratio of small arteries in essential hypertensive patients is increased and that it can be decreased by blockers of the renin-angiotensin system and by calcium blockers and (to a certain extent) diuretics, but not by beta blockers.

The technique has been extensively used to make pharmacological studies. With practice the time taken to dissect and mount vessels is minimal and comparable to traditional aorta ring preparations. Myograph configurations are available that allow simultaneous measurement of four vessels thus allowing large numbers of experiments to be made on tissues relevant for determination and control of the peripheral circulation. The endothelial dependence of responses may be determined by removal or damage of the endothelium by rubbing the lumen either with the end of a wire or a human hair [22].

The geometric arrangement of the myograph makes it suitable for making simultaneous measurements of membrane potential and tension where glass micropipettes can be inserted from above into smooth muscle cells lying in the upper wall of the vessel between the mounting wires [23]. This allows measurement of in situ membrane potential under relaxed conditions and also direct correlation between drug-induced changes in membrane potential and force development. Direct measurement of endothelial membrane potential with microelectrodes is also possible [24]. The development of fluorescent dyes allowed simultaneous measurements of force and of intracellular pH [25] or of free cytosolic calcium [26]. Indeed in some instances, simultaneous measurements of force, cytosolic calcium and membrane potential could be made [27], the measurements showing that, during noradrenaline-induced phasic activity, changes in membrane potential preceded changes in cytosolic calcium which in turn preceded changes in tension. Peng and colleagues [28] imaged the calcium using confocal microscopy, which allowed subcellular resolution of calcium transients to be assessed. With this technology calcium waves (perhaps following intermittent release of Ca^{2+} from the sarcoplasmic reticulum) within the smooth muscle cells of mounted small arteries could be observed. These were initially unsynchronized between the vascular smooth muscle cells but later became synchronized to initiate vasomotion. The development of nitric oxide electrodes [29] was further developed to allow simultaneous measurements of nitric oxide release and tension showing that in rat vessels the nitric oxide release did indeed precede the relaxation. On the other hand, in human subcutaneous small arteries [30], the results showed that acetylcholine relaxation is dependent on a non-NO, non-prostanoid endothelium-dependent hyperpolarization.

Other parameters can also be measured. Thus, histomorphometric studies using the "disector" method allowed measurement of cell number and size [31]. Mechanical properties of the vessel wall have been evaluated on the basis of the passive wall tension—internal circumference relation and the wall thickness [32]. Developments in biochemical techniques have allowed precision biochemical measurements to be made in 2-mm segments of small arteries under clearly defined mechanical conditions. For example, simultaneous measurements of phosphorylation of myosin phosphatase targeting subunit 1 (MYPT1-Thr855), phosphorylation of regulatory myosin light chain (MLC2-Ser19), and tension in rat mesenteric small arteries [33]. It was also possible to detect transglutaminases (TG1–TG7) in these vessels by RT-PCR and immunoblotting.

2.4 Pressure Myograph

While the wire myograph technique is relatively simple to use, it is clear that the configuration is far from physiological. Furthermore the contractile mode is (approximately) isometric and unlike the perhaps more physiological isobaric mode; this may be the cause of the apparent difference in sensitivity to drugs between wire myographs and pressure myographs [34]. A closer approximation to the in vivo situation is obtained using the pressure myograph technique. Here vessels are cannulated at each end and held under pressure while being perfused and immersed in the solution contained in the myograph chamber. First developed by Duling for microvessels of diameter 12–112 μm [35], the technique was further developed by Halpern and colleagues to enable continuous electronic measurement of lumen diameter (Fig. 2.2 [36]) so that the vessel response to drugs or to changes

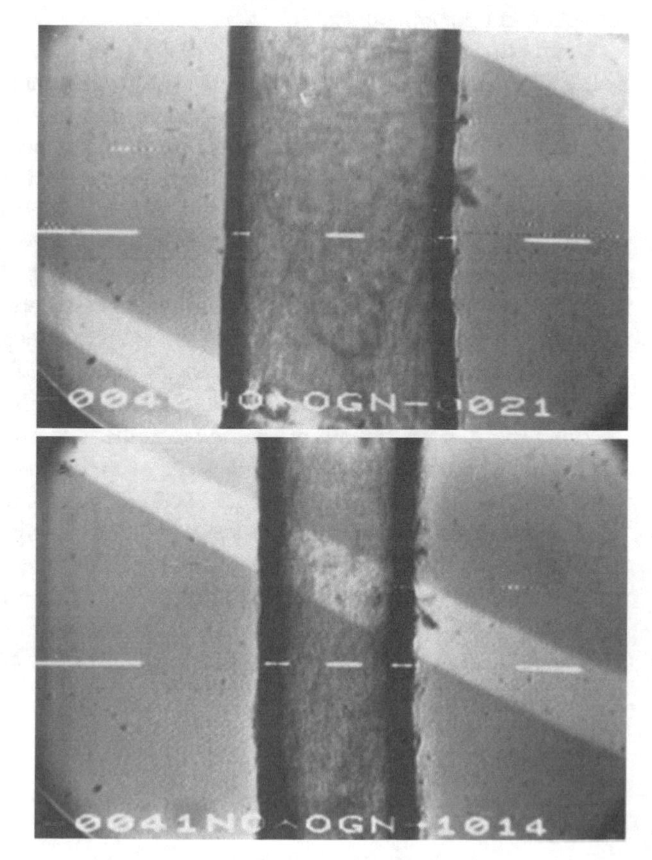

Fig. 2.2 Video-imaged mesenteric artery before (upper) and during (lower) a contraction elicited by transmural electrical stimulation. The photographs were taken approximately 10 s apart. Note the selected scan line, windows, slightly offset highlighted portions from which wall thickness and diameter measurements were calculated and recorded. Reproduced with permission from [36]

in intraluminal pressure could be determined. One advantage of this technique was investigation of the myogenic response of small arteries that is the contractile response produced by a rise in intraluminal pressure [37, 38]. Another advantage is measurement of the response to changes in intraluminal flow [39]. This technique was further developed by VanBavel [40] and further refined by Danish Myo Technology (Aarhus, Denmark) to allow continuous sampling of internal diameter and pressure at 10 Hz [41].

A difficulty with the pressure myograph technique is obtaining precise measurements of the luminal diameter since upon contraction the inner layers become concertinaed. Probably a more precise measurement could be based on measurements of inner and outer diameters while relaxed (these are clearly defined) and then monitoring the outer diameter, that continues to be clearly defined, and calculating the mean inner diameter on the basis that the volume of the wall remains constant; this approach seems, however, not to have been used. For measurements of responses of the endothelium to drugs, it is an advantage if the drugs are applied through the micropipettes perfusing the lumen [42]. A neat method of removing endothelial function is to pass an air bubble through the lumen [43]. Experiments to elucidate the mechanisms of small artery remodeling were performed by subjecting small arteries mounted in a pressure myograph to long-term culture [44]. The measurements showed that inward remodeling of small arteries is related to persistent active reduction in lumen diameter.

Comparisons of characteristics obtained using the wire- and pressure-myograph approaches showed that while the passive mechanical characteristics were similar (and that the wire myograph estimates of lumen diameter for a given pressure were similar to those obtained on the pressure-myograph), there were substantial differences in sensitivity to drugs (e.g., the threshold concentration to noradrenaline was an order of magnitude lower on the pressure myograph compared to the wire myograph) [16]. The principle of these pressure myographs has also been used as the basis for instruments developed elsewhere [45]. Other authors have developed an isovolumic myograph, where responses are measured in terms of pressure responses [46]. Günther and colleagues described a "microfluidic platform" for probing small artery structure and function [47]. The mounting procedure is described as semiautomatic, and the authors provided data showing that the instrument allows measurement of contractile responses under conditions where there were dynamic changes in the microenvironment. However, there are few if any reports where this seemingly promising technique has been used. A more recent development for a pressure myograph was division of the superfusion chamber halfway along the vessel into two compartments, allowing an independent superfusion of the arterial segment in each compartment [48]. This study provided support for maintained conduction of vasoactive responses to physiological agonists in rat mesenteric small arteries likely via gap junctions and endothelial cells.

2.5 Assessment of Vascular Structure and Function in Vivo

The in vitro analyses of isolated small arteries and veins have provided important information on small artery function. It is nevertheless the goal to understand how these vessels function in vivo, and in vivo techniques for evaluation of small artery structure and function ought therefore to be developed.

In other chapters of this book, the assessment of retinal arteriolar morphology and function with scanning laser Doppler flowmetry (SLDF) and adaptive optics is described. In this section we will discuss how these parameters are measured with intravital microscopy and laser speckle analysis. Intravital microscopy involves visualization of the vasculature using microscopy techniques and is applicable to experimental animals mainly during anesthesia. Laser speckle imaging analyses the speckle pattern obtained when a laser beam is focused on a moving target such as the erythrocytes in the vascular compartment and will not be considered further here.

Intravital microscopy has been used systematically to study the microcirculation in vivo for the last ca. 70 years. Assessment of the exteriorized mesenteric circulation from anaesthetized dogs and rodents was pioneered by Zweifach's laboratory [49–51] and has been used in Zweifach's laboratory for 50 years. This preparation has also been used extensively by Altura and his associates [52, 53] and several other groups. In this connection it is worth mentioning that Lombard's laboratory [54] was able to measure membrane potentials of the smooth muscle cells under in vivo conditions using this preparation. Later other microvascular preparations were developed where small arteries and arterioles could be visualized in vivo, e.g., the hamster cheek pouch [55, 56], the mouse cremaster muscle [57, 58], and the gracilis artery [59, 60]. These studies have provided substantial insight into vascular physiology and pharmacology, but it is not the intention in this review to discuss these techniques in detail.

Recently we [61] developed a hybrid between the in vitro isolated vessel myographs and the intravital microscopy preparations mentioned above. In this setup, the rats are anesthetized and the intestine exteriorized. A ca. 6 mm long first or second order branch of the mesenteric artery is isolated in a tissue chamber containing ca. 200 μl solution. The entrance and exit from the chamber is sealed with high vacuum grease, so that the artery or vein segment under investigation can be bathed in solutions which are different from the solution bathing the blood vessels outside the chamber and the intestinal wall. It is also possible to expose the segment under investigation to drugs without the drugs affecting the blood vessels outside the chamber or the intestine. In this way the preparation maintains some of the control which is the virtue of the in vitro myograph experiments and at the same time does this with the blood vessels in situ in the living rodent. In this way the effect of various interventions can be assessed under controlled conditions in vivo without the responses being affected by major changes in hemodynamics or metabolism of the tissue supplied by the arteries. Thus, for example, the effect of VEGF [62] or ouabain [63] or the effect of knockout of proteins relevant for vascular tone [64] has been investigated.

2.6 Perspectives

Much of the work elucidating the physiology and pharmacology of the resistance vasculature over the past 40 years worldwide has been performed using the techniques described above. Many laboratories have been involved in developing the techniques as described at the regular meetings of the International Symposia on Resistance Arteries, the proceedings of some of which are referenced here [65–67]. The work has permitted detailed understanding of many of the mechanisms involved in the determination and control of the peripheral resistance and has been extensively used in the development of relevant drugs. This wealth of knowledge now needs to be synthesized to allow better understanding of how the small arteries operate and interact in the intact animal. In vivo techniques need to be developed to allow this. Combination of the in vitro myograph techniques with such in vivo studies in unanesthetized individuals will be the key to further development.

References

1. Langendorff O. Untersuchungen am überlebenden Säugetierherzen. Pflugers Archiv. 1895;61:291–332.
2. Zimmermann BG. Measurement of total resistance, proximal arterial resistance, small vessel resistance and venous resistance. J Pharmacol Exp Ther. 1964;146:200–8.
3. Uchida E, Bohr DF, Hoobler SW. A method for studying isolated resistance vessels from rabbit mesentery and brain and their responses to drugs. Circ Res. 1967;21:525–36.
4. Macht DI, Ting GC. A study of antispasmodic drugs on the bronchus. J Pharmacol Exp Ther. 1921;18:373–98.
5. Trendelenburg P. Physiologische und pharmakologische Untersuchungen an tier isolierten Bronchialmuskulatur. Archiv für Experimentelle Pathologie Und Pharmakologie. 1912;69:79–107.
6. Castillo JC, De Beer EJ. The tracheal chain; a preparation for the study of antispasmodics with particular reference to bronchodilator drugs. J Pharmacol Exp Ther. 1947;90:104–9.
7. Furchgott RF, Bhadrakom S. Reactions of strips of rabbit aorta to epinephrine, isopropylarterenol, sodium nitrite and other drugs. J Pharmacol Exp Ther. 1953;108:129–43.
8. Bohr DF, Goulet PL, Taquini AC Jr. Direct tension recording from smooth muscle of resistance vessels from various organs. Angiology. 1961;12:478–85.
9. Nielsen KC, Owman C. Contractile response and amine receptor mechanisms in isolated middle cerebral artery of the cat. Brain Res. 1971;27:33–42.
10. Bevan JA, Osher JV. A direct method for recording tension changes in the wall of small blood vessels in vitro. Agents Actions. 1972;2:257–60.
11. Mulvany MJ, Halpern W. Mechanical properties of vascular smooth muscle cells in situ. Nature. 1976;260:617–9.
12. Mulvany MJ, Warshaw D, Halpern W. Mechanical properties of smooth muscle cells in the wall of arterial resistance vessels. J Physiol. 1978;275:85–101.
13. Pelham CJ, Drews EM, Agrawal DK. Vitamin D controls resistance artery function through regulation of perivascular adipose tissue hypoxia and inflammation. J Mol Cell Cardiol. 2016;98:1–10.
14. Mulvany MJ, Aalkjaer C. Structure and function of small arteries. Physiol Rev. 1990;70:921–61.
15. Mulvany MJ, Halpern W. Contractile properties of small arterial resistance vessels in spontaneously hypertensive and normotensive rats. Circ Res. 1977;41:19–26.

16. Buus NH, Vanbavel E, Mulvany MJ. Differences in sensitivity of rat mesenteric small arteries to agonists when studied under isobaric and isometric conditions. Br J Pharmacol. 1994;112:579–89.
17. Falloon BJ, Stephens N, Tulip JR, Heagerty AM. Comparison of small artery sensitivity and morphology in pressurized and wire-mounted preparations. Am J Phys Heart Circ Phys. 1995;268:H670–8.
18. Mulvany MJ, Hansen PK, Aalkjaer C. Direct evidence that the greater contractility of resistance vessels in spontaneously hypertensive rats is associated with a narrowed lumen, a thickened media, and an increased number of smooth muscle cell layers. Circ Res. 1978a;43:854–64.
19. Mulvany MJ, Baumbach GL, Aalkjaer C, Heagerty AM, Korsgaard N, Schiffrin EL, Heistad DD. Vascular remodelling (letter to editor). Hypertension. 1996;28:505–6.
20. Aalkjaer C, Heagerty AM, Petersen KK, Swales JD, Mulvany MJ. Evidence for increased media thickness, increased neuronal amine uptake, and depressed excitation-contraction coupling in isolated resistance vessels from essential hypertensives. Circ Res. 1987;61:181–6.
21. Agabiti-Rosei E, Heagerty AM, Rizzoni D. Effects of antihypertensive treatment on small artery remodelling. J Hypertens. 2009;27:1107–14.
22. Osol G, Cipolla M, Knutson S. A new method for mechanically denuding the endothelium of small (50-150mym) arteries with a human hair. Blood Vessels. 1989;26:320–4.
23. Mulvany MJ, Nilsson H, Flatman JA. Role of membrane potential in the response of rat small mesenteric arteries to exogenous noradrenaline stimulation. J Physiol. 1982;332:363–73.
24. Dora KA, Xia J, Duling BR. Endothelial cell signaling during conducted vasomotor responses. Am J Physiol Heart Circ Physiol. 2003;285:H119–26.
25. Aalkjaer C, Cragoe EJ. Intracellular pH regulation in resting and contracting segments of rat mesenteric resistance vessels. J Physiol. 1988;402:391–410.
26. Jensen PE, Mulvany MJ, Aalkjaer C, Nilsson H, Yamaguchi H. Free cytosolic Ca(2+) measured with Ca(2+)−selective electrodes and fura 2 in rat mesenteric resistance arteries. *Am. J Physiol*. 1993;265:H741–6.
27. Nilsson H, Jensen PE, Mulvany MJ. Minor role for direct adrenoceptor-mediated calcium entry in rat mesenteric small arteries. J Vasc Res. 1994;31:314–21.
28. Peng H, Matchkov V, Ivarsen A, Aalkjaer C, Nilsson H. Hypothesis for the initiation of vasomotion. Circ Res. 2001;88:810–5.
29. Simonsen U, Wadsworth RM, Buus NH, Mulvany MJ. In vitro simultaneous measurements of relaxation and nitric oxide concentration in rat superior mesenteric artery. J Physiol. 1999;516:271–82.
30. Buus NH, Simonsen U, Pilegaard HK, Mulvany MJ. Nitric oxide, prostanoid and non-NO, non-prostanoid involvement in acetylcholine relaxation of isolated human small arteries. Br J Pharmacol. 2002;129:184–92.
31. Mulvany MJ, Baandrup U, Gundersen HJG. Evidence for hyperplasia in mesenteric resistance vessels of spontaneously hypertensive rats using a 3-dimensional disector. Circ Res. 1985a;57:794–800.
32. Mulvany MJ. Biophysical aspects of resistance vessels studied in spontaneous and renal hypertensive rats. Acta Physiol Scand. 1988;133(suppl 571):129–38.
33. Engholm M, Pinilla E, Mogensen S, Matchkov V, Hedegaard ER, Chen H, Mulvany MJ, Simonsen U. Involvement of transglutaminase 2 and voltage-gated potassium channels in cystamine vasodilatation in rat mesenteric small arteries. Br J Pharmacol. 2016;173:839–55.
34. Koenigsberger M, Sauser R, Seppey D, Beny JL, Meister JJ. Calcium dynamics and vasomotion in arteries subject to isometric, isobaric, and isotonic conditions. Biophys J. 2008;95:2728–38.
35. Duling BR, Gore RW, Dacey RG, Damon DN. Methods for isolation, cannulation, and in vitro study of single microvessels. Am J Physiol. 1981;241:H108–16.
36. Halpern W, Osol G, Coy GS. Mechanical behavior of pressurized in-vitro pre- arteriolar vessels determined with a video system. Ann Biomed Eng. 1984;12:463–79.
37. Osol G, Halpern W. Myogenic properties of cerebral blood vessels from normotensive and hypertensive rats. Am J Physiol. 1985;249:H914–21.

38. Wallis SJ, Firth J, Dunn WR. Pressure-induced myogenic responses in human isolated cerebral resistance arteries. Stroke. 1996;27:2287–90.
39. Bevan JA, Joyce EH. Flow-induced resistance artery tone: balance between constrictor and dilator mechanisms. Am J Physiol. 1990;258:H663–8.
40. Vanbavel E, Mooij T, Giezeman MJ, Spaan JA. Cannulation and continuous cross-sectional area measurement of small blood vessels. J Pharmacol Methods. 1990;24:219–27.
41. Christensen FH, Hansen T, Stankevicius E, Buus NH, Simonsen U. Elevated pressure selectively blunts flow-evoked vasodilatation in rat mesenteric small arteries. Br J Pharmacol. 2007;150:80–7.
42. Falloon BJ, Bund SJ, Tulip JR, Heagerty AM. In vitro perfusion studies of resistance artery function in genetic hypertension. Hypertension. 1993;22:486–95.
43. Mcneish AJ, Dora KA, Garland CJ. Possible role for K+ in endothelium-derived hyperpolarizing factor-linked dilatation in rat middle cerebral artery. Stroke. 2005;36:1526–32.
44. Bakker EN, Van Der Meulen ET, Van Den Berg BM, Everts V, Spaan JA, Vanbavel E. Inward remodeling follows chronic vasoconstriction in isolated resistance arteries. J Vasc Res. 2002;39:12–20.
45. Bell JS, Adio AO, Pitt A, Hayman L, Thorn CE, Shore AC, Whatmore JL, Winlove CP. Microstructure and mechanics of human resistance arteries. Am J Physiol Heart Circ Physiol. 2016;311:H1560–8.
46. Lu X, Kassab GS. Assessment of endothelial function of large, medium, and small vessels: a unified myograph. Am J Physiol Heart Circ Physiol. 2011;300:H94–H100.
47. Gunther A, Yasotharan S, Vagaon A, Lochovsky C, Pinto S, Yang J, Lau C, Voigtlaender-Bolz J, Bolz SS. A microfluidic platform for probing small artery structure and function. Lab Chip. 2010;10:2341–9.
48. Palao T, Van Weert A, De Leeuw A, De Vos J, Bakker E, Van Bavel E. Sustained conduction of vasomotor responses in rat mesenteric arteries in a two-compartment in vitro set-up. Acta Physiol (Oxf). 2018;224:e13099.
49. Suzuki H, Schmid-Schonbein GW, Suematsu M, Delano FA, Forrest MJ, Miyasaka M, Zweifach BW. Impaired leukocyte-endothelial cell interaction in spontaneously hypertensive rats. Hypertension. 1994;24:719–27.
50. Zweifach BW. Quantitative studies of microcirculatory structure and function. I. Analysis of pressure distribution in the terminal vascular bed in cat mesentery. Circ Res. 1974;34:843–57.
51. Zweifach BW, Lee RE, Hyman C, Chambers R. Omental circulation in morphinized dogs subjected to graded hemorrhage. Ann Surg. 1944;120:232–50.
52. Altura BM. Selective microvascular constrictor actions of some neurohypophyseal peptides. Eur J Pharmacol. 1973;24:49–60.
53. Altura BM, Hershey SG. Pharmacology of neurohypophyseal hormones and their synthetic analogues in the terminal vascular bed. Structure-activity relationships. Angiology. 1967;18:428–39.
54. Lombard JH, Burke MJ, Contney SJ, Willems WJ, Stekiel WJ. Effect of tetrodotoxin on membrane potentials and active tone in vascular smooth muscle. Am J Phys. 1982;242:H967–72.
55. Duling BR. The preparation and use of the hamster cheek pouch for studies of the microcirculation. Microvasc Res. 1973;5:423–9.
56. Fulton GP, Jackson RG, Lutz BR. Cinephotomicroscopy of normal blood circulation in the cheek pouch of the hamster. Science. 1947;105:361–2.
57. Baez S. An open cremaster muscle preparation for the study of blood vessels by in vivo microscopy. Microvasc Res. 1973;5:384–94.
58. Grant RT. Direct observation of skeletal muscle blood vessels (rat cremaster). J Physiol. 1964;172:123–37.
59. Boettcher M, De Wit C. Distinct endothelium-derived hyperpolarizing factors emerge in vitro and in vivo and are mediated in part via connexin 40-dependent myoendothelial coupling. Hypertension. 2011;57:802–8.
60. Henrich HN, Hecke A. A gracilis muscle preparation for quantitative microcirculatory studies in the rat. Microvasc Res. 1978;15:349–56.

61. Nyvad J, Mazur A, Postnov DD, Straarup MS, Soendergaard AM, Staehr C, Brondum E, Aalkjaer C, Matchkov VV. Intravital investigation of rat mesenteric small artery tone and blood flow. J Physiol. 2017;595:5037–53.
62. Egholm C, Khammy MM, Dalsgaard T, Mazur A, Tritsaris K, Hansen AJ, Aalkjaer C, Dissing S. GLP-1 inhibits VEGFA-mediated signaling in isolated human endothelial cells and VEGFA-induced dilation of rat mesenteric arteries. Am J Physiol Heart Circ Physiol. 2016;311:H1214–h1224.
63. Bouzinova EV, Hangaard L, Staehr C, Mazur A, Ferreira A, Chibalin AV, Sandow SL, Xie Z, Aalkjaer C, Matchkov VV. The alpha2 isoform Na,K-ATPase modulates contraction of rat mesenteric small artery via cSrc-dependent Ca(2+) sensitization. Acta Physiol (Oxf). 2018;224:e13059.
64. Dam VS, Boedtkjer DM, Nyvad J, Aalkjaer C, Matchkov V. TMEM16A knockdown abrogates two different Ca(2+)-activated Cl(−) currents and contractility of smooth muscle in rat mesenteric small arteries. Pflugers Arch. 2014;466:1391–409.
65. Mulvany MJ, Strandgaard S, Hammersen F. EDITORSResistance vessels: physiology, pharmacology and hypertensive pathology. Adv Appl Microcirc. 1985;8:1–236.
66. Abstracts from the 11th International Symposium on Resistance Arteries: From Molecular Machinery to Clinical Challenges, September 7–11, 2014, Banff, Alberta, Canada. J Vasc Res, 2014 Banff. 2014 1–156.
67. Withers S, Greenwood I, Mcneish A, Heagerty AM. Abstracts from 12th International Symposium on Resistance Arteries (ISRA 2017), September 3-6, 2017, Manchester, UK. J Vasc Res. 2017;54(Suppl 2):1–62.

Assessment of Retinal Arteriolar Morphology by SLDF

3

Christian Ott and Roland E. Schmieder

3.1 Introduction

The clinical importance of alterations in the microcirculation, vascular remodeling, are well clinically established [1]. Unfortunately, the gold standard for evaluation of small artery and arteriolar structure of isolated subcutaneous small vessels requires an invasive procedure, namely, the performance of a biopsy of subcutaneous tissue. Hence, this methodology is not suitable for routine patient management and its use is limited to scientific purposes. However, retinal arterioles abnormalities seem to mirror structural changes seen in other end organ tissues, including subcutaneous tissue [2]. Already in 1939, the exceptional role of retinal arterioles were recognized by Keith et al., "because the arterioles are small and are difficult to visualize in the peripheral organs, for example, in the skin, mucous membranes, and voluntary muscle, the retina, as seen through the ophthalmoscope, offers a unique opportunity for observing these small vessels from time to time. Therefore, we think that certain visible changes of the retinal arterioles have been of exceptional value in affording a clearer clinical conception of altered arteriolar function throughout the body" [3].

C. Ott
Department of Nephrology and Hypertension, Friedrich-Alexander University Erlangen-Nürnberg, Erlangen, Germany

Department of Nephrology and Hypertension, Paracelsus Medical University, Nürnberg, Germany
e-mail: christian.ott@uk-erlangen.de

R. E. Schmieder (✉)
Department of Nephrology and Hypertension, Friedrich-Alexander University Erlangen-Nürnberg, Erlangen, Germany
e-mail: roland.schmieder@uk-erlangen.de

E. Agabiti-Rosei et al. (eds.), *Microcirculation in Cardiovascular Diseases*, Updates in Hypertension and Cardiovascular Protection,
https://doi.org/10.1007/978-3-030-47801-8_3

In the last decades, new approaches were developed to detect reliable (early changes of) retinal arteriolar morphology. One of these promising approaches, introduced about two decades ago by our study group, is scanning laser Doppler flowmetry (SLDF) with automated full field perfusion image analysis (AFFPIA). With this technique, assessment of various parameters of retinal arteriolar vascular morphology and function becomes available.

3.2 Assessment of Retinal Arteriolar Morphology

For the assessment of retinal arteriolar morphology, Heidelberg Retina flowmetry (HRF) based on SLDF with a first-class laser light of 670 nm wavelength (Heidelberg Flowmeter, Heidelberg Engineering) is used. An arteriole sized 80–140 μm of the superficial layer in a retinal sample of 2.56 × 0.64 × 0.30 mm is scanned within 2 s at a resolution of 256 points × 64 lines × 128 lines. These scans are performed in the juxtapapillary area 2–3 mm superior to the optic nerve (normally standardized on the right eye) (Fig. 3.1). A distinct length of the arteriole reflecting arteriolar morphology during one heartbeat (systole and diastole) is used, and diameters are assessed every 10 μm. If there is no exact straight line of the specific chosen arteriole length, the software automatically adjusts the cross-sections perpendicular to this line against each other, referring to the cross-section with the lowest y coordinate. Finally, mean of measured diameters are calculated, and standardized average from three singular measurements are used for further analyses. Analyses of diameters are done offline with AFFPIA (current SLDF version 4.0) (Fig. 3.2). Historically, the assessments of outer (vessel) and inner (lumen) retinal arteriolar diameter are possible since SLDF version 3.6/3.7, and, over the years, the software

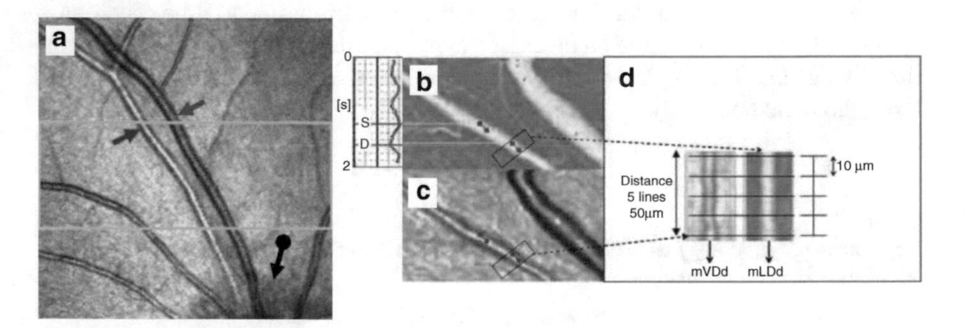

Fig. 3.1 (**a**) Exemplary retinal image (temporal superior of the optic nerve in the right eye) with marked arteriole (red arrow), venule (blue arrow), and optic nerve (black arrow); green rectangle marks the scanned area. (**b**) Fragment selection of the arteriole (perfusion image) for assessment of systolic and diastolic structural parameter. (**c**) Automatically on reflexion image the same vessel fragment (see **b**) is selected by AFFPIA. (**d**) Structural parameter analysis (here diastolic phase) of the selected part of the arteriole—mean diastolic vessel (mVD[=OD]d) and lumen diameter (mLD[=ID]d) averaged from every 10 μm distance measurement of the analyzed arteriole fragment (adapted from [4, 40])

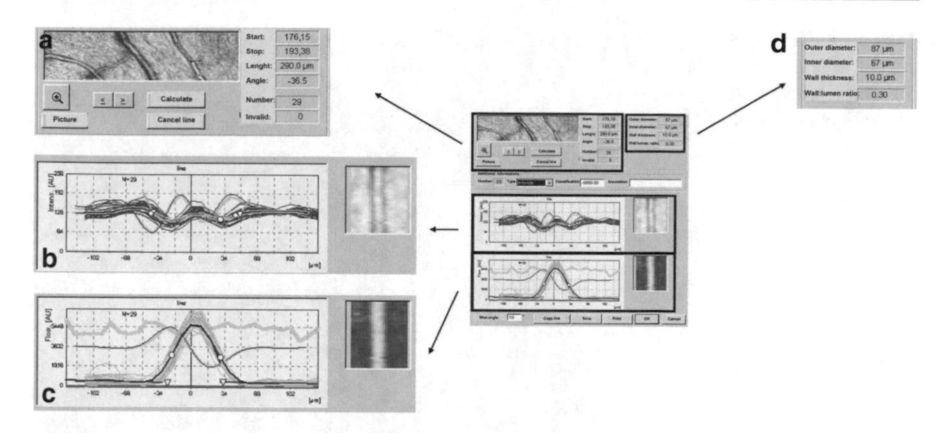

Fig. 3.2 Screenshot of the current AFFPIA version 4.0 program. (**a**) Reflection image of a retinal sample with a retinal arteriole (blue line within the lumen reflects the specific length of the arteriole that is used for analyses) and venule (wall of the venule is partly marked with blue color). (**b**) Algorithm of assessment of the outer arteriolar diameter borders (triangles) based on average of all of the blue lines (reflecting the measurements every 10 μm over the specific lengths of the arteriole in reflection images). (**c**) Algorithm of assessment of the inner arteriolar diameter borders (triangles) based on average of all of the green lines (reflecting the measurements every 10 μm over the specific lengths of the arteriole in perfusion images). (**d**) Results of arteriolar structural parameters (e.g., WLR)

Table 3.1 Development of the Heidelberg retina flowmetry image automated full field perfusion image analysis (AFFPIA) software

AFFPIA software	Retinal vessel morphology
SLDF version 3.3 [53]	NO
SLDF version 3.6/3.7 [14]	One pixel resolution: 10 μm × 10 μm
	OD with 10 μm accuracy/ID with 0.01 μm interpolated accuracy
SLDF version 4.0 [4]	One pixel resolution: 10 μm × 10 μm
	OD and ID with the adjusted accuracy (1 μm interpolated)

AFFPIA automated full field perfusion image analysis, *OD* outer (vessel) diameter, *ID* inner (lumen) diameter

has undergone further refinements thereby improving accuracy and reliability (Tables 3.1 and 3.2). Notably, in the current software version, the accuracy of lumen measurement decreased from 10 μm (physical resolution of the device) to subpixel accuracy (1 μm) due to the application of a mathematical curve fit model (parabolic function) to the measured data [4].

Thus, the coefficients of variation are now below the threshold set by the European and American Societies of Cardiology [5, 6].

Outer diameter (OD) and inner diameter (ID) are measured in reflection and perfusion images, respectively (Fig. 3.1). The software automatically compares the two images taken in the same retinal area. In more detail, the reflection image is created from the direct current of the Doppler signal and approximates the amount of reflected laser light of the nonmoving tissue. Based on the acute angle between

Table 3.2 Test-retest results of retinal arteriolar structural parameters (according [4, 40, 45, 51])

Retinal arteriolar parameter	CV (%, mean ± SD)	α-CC
OD		
Mean	5.8 ± 4.0	0.91
Systole	6.44 ± 3.7	0.83
Diastole	6.61 ± 4.4	0.83
ID		
Mean	6.8 ± 4.1	0.90
Systole	8.00 ± 3.6	0.84
Diastole	7.98 ± 3.0	0.88
WT	8.2 ± 4.1	0.92
WLR	10.0 ± 3.4	0.91
WCSA	12.5 ± 7.6	0.91
ICD	3.6 ± 3.0	0.93
CapA	8.53 ± 5.4	0.87
RVR	7.75 ± 2.1	0.90

CV coefficient of variation, *α-CC* α-Crombach's reliability coefficient, *OD* outer diameter, *ID* inner diameter, *WT* wall thickness, *WLR* wall-to-lumen ratio, *WCSA* wall cross-sectional area, *ICD* intercapillary distance, *CapA* capillary area, *RVR* retinal vascular resistance

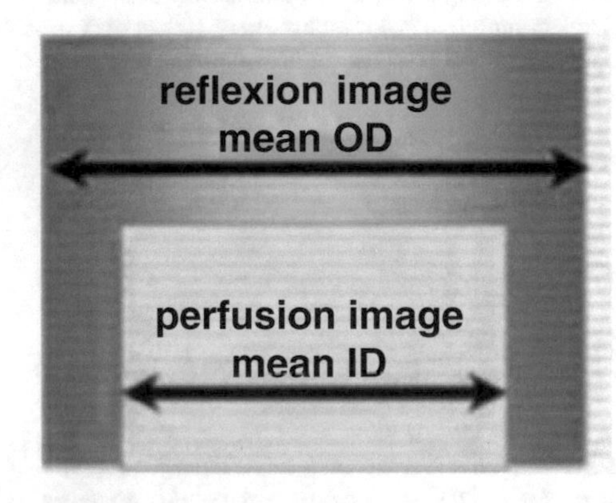

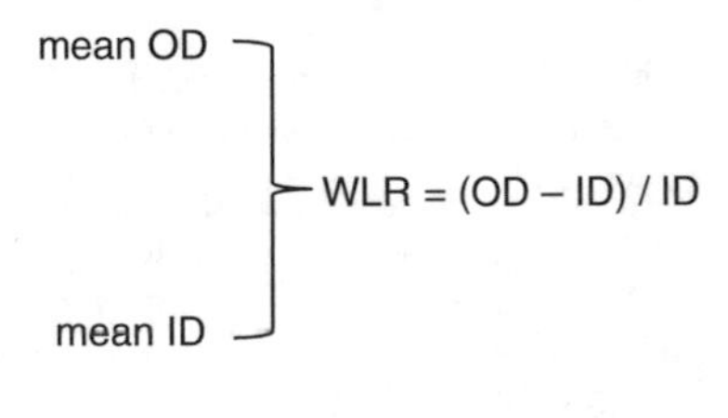

Fig. 3.3 Scheme of assessment of outer and inner diameter and calculation of WLR (adapted from [52])

light direction and vessel wall border, at the outer vessel wall border is the weakest reflection. The turning points with maximal slope (triangles) are considered for the definition of outer vessel wall border with respect to the largest difference of reflectivity between two points lying side by side. The perfusion image is generated by the Doppler effect caused by moving blood corpuscles. Because blood flow velocity is greatest in the center of the blood vessel, a Poisson velocity distribution of the blood stream within the vessel can be adapted. The crossing points of the parabolic curves (reflecting the velocity distribution of blood flow within the blood vessel) and straight lines (reflecting baseline reflectivity) define the inner vessel wall border (triangles) [7, 8] (Fig. 3.2). Based on such measurement of OD and ID, further retinal structural parameters can be calculated:

– Wall-to-lumen ratio (WLR): (OD – ID)/ID (Figure 3.3)

- Wall thickness (WT): (OD – ID)/2
- Wall cross-sectional area (WCSA): $(\pi/4) \times (OD^2 - ID^2)$

No mydriatic drugs, with possible interfering effects as well as comprising daily life for the patient, have to be applied for the assessment of retinal arteriolar morphology by SLDF. This is noteworthy to mention, since locally administered tropicamide profoundly affects, by about one-third decrement of retinal capillary flow, and hence pupil dilatation impairs any assessment of retinal microcirculation [9].

Assessment of retinal circulation using SLDF offers several advantages compared to other methods dealing with small (functional and) structural arteriolar alterations, and its scientific value has been proven over the last decades. However, current limitations have to be discussed. The HRF is no longer available on the market, and thus the technical service support of existing devices may be limited over a long time. Also clinical development may be thwarted.

Assessment of ID is based on Doppler flow images. Therefore, it has to be taken into account that lower corpuscle velocities at the border of the flow column as well as the plasma edge may impact on assessed diameter [7]. However, it was confirmed that already SLDF version 3.3 is a reliable method to study retinal perfusion [10].

It has to be considered that arteriolar diameters are not assessed in fully relaxed state, as given on wire myograph, but this should be considered as an advantage since ex vivo data may only partially and potentially incorrectly reflect the situation in vivo with its natural metabolic and nerval environment.

Nowadays, SLDF with other software (e.g., [2]) as well as other approaches are introduced focusing on retinal arteriolar morphology. For comparison see Table 3.3. However, it has to be kept in mind that methodologies and types of used software differ, and hence, raw data and experiences of various centers may not be simply interchangeable [11].

3.3 Wall-to-Lumen Ratio

In the meantime, several studies have used SLDF-based structural parameters in different conditions and diseases including investigating treatment effects. In this part of this chapter, we will focus on these findings with WLR, as key parameter of retinal arteriolar morphology. There are several reasons for that:

- Media-to-lumen ratio and WLR are the only structural parameters that are independent of vessel dimension, and thus free from possible sampling/assessment bias [12].
- A close relationship of WLR assessed by SLDF (noninvasive retinal arterioles in vivo) with media-to-lumen ratio measured with myograph (invasively taken subcutaneous small arteries and analyzed in vitro) was demonstrated even when hypertensive patients ($r = 0.80$, $p < 0.001$) and normotensive participants ($r = 0.58$, $p < 0.05$) were considered separately. Therefore, it can be

Table 3.3 Comparison of retinal structural parameters assessed with scanning laser Doppler flowmetry (SLDF), adaptive optics, and optical coherence tomography (OCT)

	SLDF (Ritt et al [13])		SLDF (Rizzoni et al [2])		Adaptive optics (Koch et al [54])		OCT (Muraoka et al [55])	
	Normotensive	Hypertensive	Normotensive	Hypertensive	Normotensive	Hypertensive	Normotensive	Hypertensive
	(*n* = 29)	(*n* = 21)	(*n* = 16)	(*n* = 24)	(*n* = 30)	(*n* = 19)	(*n* = 83)	(*n* = 103)
Age (years)	36.7 ± 5.9		59.3 ± 14		42.3 ± 15		68.5 ± 7.8	
BMI (kg/m²)	31.5 ± 2.3		25.6 ± 4.4		23.8 ± 4.5			
Systolic BP (mmHg)	129 ± 6.9		125 ± 17		118 ± 13			
Diastolic BP (mmHg)	78 ± 7.6		71 ± 12		74 ± 9.5			
WLR (–)	0.28 ± 0.1		0.26 ± 0.1		0.29 ± 0.1		0.41[a] (mean)	
Outer (vessel) diameter (μm)	109 ± 15		93.6 ± 19		107[b] (mean)		123 ± 9.3	
Inner (lumen) diameter (μm)	85.3 ± 11		74.4 ± 16		83.5 ± 11		87.3 ± 8.3	
Age (years)		39.1 ± 5.4		57.7 ± 15		48 ± 11		69.1 ± 8.1
BMI (kg/m²)		33.1 ± 4.4		27.4 ± 5.1		26.4 ± 4		
Systolic BP (mmHg)		145 ± 6.8		139 ± 17		154 ± 14		
Diastolic BP (mmHg)		88 ± 8.3		89 ± 10		100 ± 10		
WLR (–)		0.36 ± 0.1		0.37 ± 0.1		0.36 ± 10		0.41[a] (mean)
Outer (vessel) diameter (μm)		111 ± 9.6		81.7 ± 20		100[b] (mean)		125 ± 11
Inner (lumen) diameter (μm)		81.8 ± 7.8		59.6 ± 13		74 ± 13		88.5 ± 11

aCalculated on the published mean values ([outer diameter – inner diameter]/inner diameter)
bCalculated on the published mean values (lumen diameter + parietal thickness)

suggested that SLDF provides similar information regarding microvascular alterations as subcutaneous small arteries [2].

- Multiple findings based on media-to-lumen ratio of invasive taken subcutaneous biopsies were confirmed by noninvasively assessed observations with SLDF.

3.3.1 Arterial Hypertension

In primary arterial hypertension, we observed a significantly greater WLR in never-treated hypertensive patients compared to normotensive controls, and a close correlation of WLR with office blood pressure (BP) [13]. Expanding this line, we demonstrated that in patients with poor BP control WLR was higher compared to hypertensive patients with controlled BP [14]. Nowadays, it is acknowledged that central BP in the aorta (being the true perfusion pressure to key organs rather than peripheral pressure) is of greater prognostic relevance than the peripheral BP [15–17]. Taken this into account, we were again able to demonstrate by multiple regression analysis that WLR is independently related with central pulse pressure, but not with other classical cardiovascular risk factors [18].

In another study, WLR was significantly related to peripheral, central, and 24-h ambulatory BP, but multiple regression analysis revealed that only 24-h ambulatory BP was independently associated with increased WLR, indicating that 24-h pressure load may be a stronger determinant than single measurement either central or peripheral [19].

First evaluation in patients with secondary causes of hypertension revealed that, compared to hypertensive controls, WLR is greater in both patients suffering from pheochromocytoma ($n = 24$) [20] and primary aldosteronism ($n = 30$) [21], indicating that high catecholamine levels and excessive aldosterone have deleterious effects on retinal arterioles by exaggerating the hypertensive response to increased BP.

Interestingly, in the ongoing Polish Registry for Fibromuscular Dysplasia (ARCADIA-POL-Study), patients (n = 41) with fibromuscular dysplasia affecting carotid and/or intracranial arteries had lower WLR than matched patients with essential hypertension [22].

Prospective studies assessing antihypertensive treatment effects on retinal structural arteriolar morphology are scarce. A first tiny study reported, that both aliskiren ($n = 9$) and ramipril ($n = 7$) (hydrochlorothiazide could be added to achieve equivalent BP control) induced significant reduction of WLR after 1-year treatment, without difference between the groups [23]. In a second unblinded study [24] with previously published extension, [25] hypertensive patients (total n = 30) were treated with lercanidipine alone for 4 weeks, and thereafter randomized to enalapril or hydrochlorothiazide on top for 24 weeks. Already single therapy with lercanidipine resulted in a decrement of WLR, which was further reduced only with enalapril, but not with hydrochlorothiazide. Notably, both studies reported surprisingly high baseline values of WLR (>0.5) and achieved a total reduction of about 50% (below values seen in healthy, normotensive participants). In general, analyses relying on the assessment in vascular remodeling of subcutaneous small arteries are by far smaller [26], and in the former study also lower values have been given [23]. This inconsistency has not been explained so far.

In a double-blind, randomized, placebo-controlled study (n = 114), the renin inhibitor aliskiren given on top of a preexisting standardized open-label therapy with valsartan for 4 weeks showed no clear effect on vascular structural parameters. However, restricting analyses to patients with some evidence of vascular remodeling (based on median of WLR [0.3326]), there was a significant reduction of WLR after 8 weeks of treatment with on-top aliskiren compared to placebo [27].

3.3.2 Cerebrovascular Event

In a smaller study, we analyzed different approaches of determining vascular remodeling in patients with cerebrovascular disease. Using arteriole-to-venule (A/V) ratio, we were not able to discriminate between patients with cerebral event (transient ischemic attack or lacunar stroke) and both normotensive and hypertensive patients. In contrast, WLR was significantly higher in patients with cerebrovascular events, and thus allowed discrimination between patients with cerebrovascular event versus without [28]. Thus, in our hands, the assessment of WLR (and WCSA [parameter of hypertrophic vascular remodeling]) was more sensitive than digitized funduscopic images providing the A/V ratio.

Interestingly, a recent meta-analysis documented the predictive value of carotid stiffness for stroke, independent of age, BP and other cardiovascular risk factors [29]. In normo- and hypertensive patients, carotid stiffness assessed by echotracking was independently from possible confounders, significantly associated with WLR, providing a mechanistic explanation of the underlying relationship between increased (local) carotid stiffness, retinal vascular changes, and cerebrovascular damage [30].

3.3.3 Chronic Kidney Disease

Urinary albumin excretion is an acknowledged parameter not only for renal but also more in general of microvascular damage in the systemic circulation. Repeatedly, e.g., in the Prevention of Renal and Vascular End-Stage Disease (PREVEND) study, it was shown that increased urinary albumin excretion is a predictor of cardiovascular events [31]. We were able to demonstrate that urinary albumin-to-creatinine ratio (UACR) was related with WLR, independent of other cardiovascular risk factors [32].

In the German Chronic Kidney Disease (CKD) cohort, patients (n = 76) with estimated glomerular filtration rate 30–60 ml/min per 1.73 m^2 or proteinuria >500 mg/g creatinine revealed a significantly greater WLR compared to healthy control subjects. Moreover, in these patients with CKD a correlation of WLR with serum phosphate levels, but not with 24-h ambulatory BP, was documented, indicating a BP-independent mechanism of hyperparathyroidism on retinal arteriolar structure [33].

3.3.4 Diabetes

Analyzing retinal arteriolar morphology in patients with type 2 diabetes mellitus revealed that, after adjustments, WLR tended ($p = 0.08$) to be greater compared to healthy controls but similar to hypertensive patients. Since disease duration may impact on amount of alterations, analyzed patients with type 2 diabetes mellitus were further stratified according median (< vs. >60 months). Again there was a trend ($p = 0.08$) toward higher gender and age-adjusted WLR values in patients with longer than shorter disease duration [34]. Similarly, patients with type 1 diabetes mellitus and disease duration >10 years showed a higher WLR compared to patients with shorter duration [35]. In a placebo-controlled study in patients with type 2 diabetes mellitus, no effect of saxagliptin monotherapy for 6 weeks on WLR was observed [36]. In another open-label study, vildagliptin but not glimepride given on top of preexisting metformin therapy, resulted in a significant decrement of WLR after 24 weeks, whereas no significant diffcrcncc of WLR was observed after shorter treatment phase of 12 weeks [37]. This indicates that treatment duration is also of importance.

3.4 Pulsatile Structural Parameters

It is well established that an increased pulsatile pressure induces as well as aggravates (micro-)vascular damage, indicating a close relationship of structural alterations in large and small vessels [38]. Even more, although after adjustments of known cardiovascular risk factors, pulse pressure, and media-to-lumen ratio of subcutaneous arterioles were significantly and independently associated with the occurrence of cardiovascular events [39]. Based on the pathophysiological concept, one of the further advantages is that SLDF enables a dynamic assessment of retinal circulation.

By using SLDF, we were able to provide a reliable tool for the noninvasive assessment of pulsatile characteristics of retinal arteriolar structural parameters (Fig. 3.1, Table 3.2). By doing so, the applicability of pulsatile structural components in systole and diastole in clinical research was investigated in two hypertensive groups of different severity, namely, primary hypertension grade 1 and 2 (HTN1–2) and treatment-resistant hypertension (TRH). The measured parameters OD and ID as well as derived parameter WT did not differ between the groups neither in systole nor diastole. In contrast, pulsatile changes of OD and ID were exaggerated, and of WT diminished in TRH compared to HTN1–2, irrespective whether expressed as absolute term or percentage change. In accordance, the pulsatile changes were different between the groups. In HTN1–2 there was no change in OD between systole and diastole, but we observed a significant decrement of ID and increment of WT in diastole compared to systole. In contrast, in TRH both OD and ID were significantly higher in systole compared to diastole, which results in an unaltered WT between heart phases. These findings imply stiffer wall properties of retinal arterioles in latter patients [40].

3.5 Capillary Rarefaction

Nowadays, it is acknowledged that capillary rarefaction is of crucial importance in the pathogenesis of end-organ damage due to modulation of pressure and blood flow pattern. Moreover, it affects metabolism [41] and vascular resistance (i.e., BP) [42].

Capillary rarefaction is due to either structural capillary rarefaction, originating from anatomical absence, or functional capillary rarefaction caused by non-perfused capillaries. In retinal circulation it is postulated that loss of pericytes results in vulnerable capillaries and rarefaction [43] as well as decrement of spontaneously perfused capillaries (due to enhanced contractile activity) [44].

Recently, the SLDF technique of measuring retinal capillary flow (RCF) has been expanded toward assessment of retinal capillary rarefaction. Two parameters have been proposed: intercapillary distance (ICD) and capillary area (CapA) of retinal circulation. Both variables can reliable be assessed (Table 3.2). The assessment of both parameters (ICD and CapA) of capillary rarefaction is taken from perfusion images of pixels. By definition, ICD represents the distance between any pixel outside and the next pixel inside the vessel and is given in the unit µm (Fig. 3.4). The smallest dot of optic solution, where flow can be detected, is defined as one pixel.

Fig. 3.4 Measurement of intercapillary distance in perfusion image with SLDF and calculation of intercapillary distance using AFFPIA. Based on the vessel size, pixels were categorized into pixel inside vessels >20 µm (non-capillary pixel, e.g., arteriolar pixel), pixel inside vessel ≤20 µm (capillary pixel), and pixel outside a vessel (intercapillary pixel)

SLDF optic solution works in a confocal manner with pixel size 10 × 10 μm, and pixels are located in a depth of 300 μm.

Retinal CapA was defined as an area with predominance of vessels ≤20 μm, hence subtracting areas with vessels >20 μm from the total area. CapA is given in number of pixels.

In a first study, it was reported that ICD was greater and CapA was smaller in patients with type 2 diabetes mellitus compared to healthy, but comparable to extents in hypertensive patients [45]. Taken into account that diseased duration may impact on these early markers of retinal alterations, we have further investigated retinal capillary rarefaction in patients with short- and long-term HTN1–2 compared to healthy controls. There was a significant increment of ICD between healthy subjects and patients with short-term disease duration. Furthermore, ICD was significantly increased in long-term versus short-term duration of hypertension. In contrast, CapA was only significantly smaller in hypertensive patients with long-term disease duration compared to healthy controls [46].

In patients with CKD, even after adjustment of clinical characteristics, ICD was greater compared to healthy controls [33].

Drugs targeting the renin-angiotensin system have been shown to ameliorate capillary rarefaction [47]. Therefore, we have expanded our studies regarding retinal capillary rarefaction and analyzed the treatment effect of an angiotensin-receptor blocker (valsartan) in hypertensive patients. We found that, in accordance with existing literature, valsartan treatment in hypertensive patients decreased ICD and increased CapA compared to baseline values, indicating diminished (functional) retinal capillary rarefaction. In addition, treatment improved both ICD and CapA toward a level that was not different from healthy controls [48].

It has to be kept in mind that SLDF parameters of capillary density depend on perfusion of capillaries and therefore cannot discriminate between structural and functional retinal rarefaction. Nevertheless, the assessment of capillary rarefaction represents an advantage of SLDF compared to other optic systems, such as adaptive optics imaging, that cannot determine perfusion of retinal vessels. Overall, SLDF allows the assessment of several parameters of retinal circulation, even in very early phase of vascular remodeling when structural alterations are not yet assessable by other techniques due to its ability of measuring perfusion of arterioles (down to 20 μm).

3.6 Vascular Resistance

From a pathophysiological viewpoint, structural and functional alterations of small resistance vessels occur early in hypertension, and increased peripheral resistance in the systemic circulation is the hemodynamic hallmark of arterial hypertension [49]. We have applied this concept to the retinal circulation and investigated whether early remodeling of retinal circulation lead to an increased retinal vascular resistance (RVR). It can be assumed that among others (e.g., capillary rarefaction, see above) decreased ID may be caused by rearrangement of smooth muscle cells, but without growth response in the early stage of hypertension, as demonstrated in subcutaneous arterioles [50]. Using SLDF, we are able to assess established structural parameters (OD, ID, and WLR) but also to assess RVR, defined as mean arterial pressure divided by RCF, as possible early parameter of vascular remodeling in the retinal circulation. After reliability was proven (Table 3.2), we could show for the first time that RVR was significantly higher in hypertensive compared to normotensive patients, although WLR did not differ. Moreover, in both groups a correlation of RVR with WLR (normotensive: $r = 0.25$, $p = 0.09$; hypertensive: $r = 0.26$, $p = 0.004$) was found, suggesting that RVR may be a more sensitive marker of early vascular remodeling in retinal arterioles [51].

3.7 Conclusion/Summary

The application of SLDF in vascular disease (i.e., cardiovascular risk factors) allows to assess early functional and structural changes of vascular remodeling in the retinal circulation. For the assessment of morphology, WLR is the key parameter. WLR is increased in hypertension and more sensitive than A/V ratio and corresponds nicely to the media-to-lumen ratio obtained ex vivo by in vitro measurements. The uniqueness of SLDF relies in its ability to measure retinal capillary perfusion. Pulsatile analysis, parameters of capillary rarefaction, and retinal vascular perfusion emerge as early indicators of vascular remodeling in the retinal circulation.

References

1. Agabiti-Rosei E, Rizzoni D. Microvascular structure as a prognostically relevant endpoint. J Hypertens. 2017;35:914–21.
2. Rizzoni D, Porteri E, Duse S, De Ciuceis C, Rosei CA, La Boria E, Semeraro F, Costagliola C, Sebastiani A, Danzi P, Tiberio GA, Giulini SM, Docchio F, Sansoni G, Sarkar A, Rosei EA. Relationship between media-to-lumen ratio of subcutaneous small arteries and wall-to-lumen ratio of retinal arterioles evaluated noninvasively by scanning laser doppler flowmetry. J Hypertens. 2012;30:1169–75.
3. Keith NM, Wagener HP, Barker NW. Some different types of essential hypertension: their course and prognosis. Am J Med Sci. 1939;197:332–43.
4. Harazny JM, Raff U, Welzenbach J, Ott C, Ritt M, Lehmann M, Michelson G, Schmieder RE. New software analyses increase the reliability of measurements of retinal arterioles morphology by scanning laser Doppler flowmetry in humans. J Hypertens. 2011;29:777–82.
5. Apple FS, Wu AH, Jaffe AS. European society of cardiology and american college of cardiology guidelines for redefinition of myocardial infarction: how to use existing assays clinically and for clinical trials. Am Heart J. 2002;144:981–6.
6. Alpert JS, Thygesen K, Antman E, Bassand JP. Myocardial infarction redefined--a consensus document of the joint european society of cardiology/american college of cardiology committee for the redefinition of myocardial infarction. J Am Coll Cardiol. 2000;36:959–69.
7. Michelson G, Warntges S, Baleanu D, Welzenbach J, Ohno-Jinno A, Pogorelov P, Harazny J. Morphometric age-related evaluation of small retinal vessels by scanning laser doppler flowmetry: determination of a vessel wall index. Retina. 2007;27:490–8.
8. Ritt M, Schmieder RE. Wall-to-lumen ratio of retinal arterioles as a tool to assess vascular changes. Hypertension. 2009;54:384–7.

9. Harazny JM, Schmieder RE, Welzenbach J, Michelson G. Local application of tropicamide 0.5% reduces retinal capillary blood flow. Blood Press. 2013;22:371–6.
10. Kreis AJ, Nguyen T, Rogers S, Wang JJ, Harazny J, Michelson G, Farouque HM, Wong TY. Reliability of different image analysis methods for scanning laser doppler flowmetry. Curr Eye Res. 2008;33:493–9.
11. Harazny JM, Schmieder RE. Interpretation of noninvasive retinal microvascular studies: the individual source of the automatic full field imaging analysis program has to be taken into account. J Hypertens. 2018;36:2277.
12. Schiffrin EL, Hayoz D. How to assess vascular remodelling in small and medium-sized muscular arteries in humans. J Hypertens. 1997;15:571–84.
13. Ritt M, Harazny JM, Ott C, Schlaich MP, Schneider MP, Michelson G, Schmieder RE. Analysis of retinal arteriolar structure in never-treated patients with essential hypertension. J Hypertens. 2008;26:1427–34.
14. Harazny JM, Ritt M, Baleanu D, Ott C, Heckmann J, Schlaich MP, Michelson G, Schmieder RE. Increased wall:lumen ratio of retinal arterioles in male patients with a history of a cerebrovascular event. Hypertension. 2007;50:623–9.
15. Roman MJ, Devereux RB, Kizer JR, Lee ET, Galloway JM, Ali T, Umans JG, Howard BV. Central pressure more strongly relates to vascular disease and outcome than does brachial pressure: the strong heart study. Hypertension. 2007;50:197–203.
16. Wang KL, Cheng HM, Chuang SY, Spurgeon HA, Ting CT, Lakatta EG, Yin FC, Chou P, Chen CH. Central or peripheral systolic or pulse pressure: which best relates to target organs and future mortality? J Hypertens. 2009;27:461–7.
17. Huang CM, Wang KL, Cheng HM, Chuang SY, Sung SH, Yu WC, Ting CT, Lakatta EG, Yin FC, Chou P, Chen CH. Central versus ambulatory blood pressure in the prediction of all-cause and cardiovascular mortalities. J Hypertens. 2011;29:454–9.
18. Ott C, Raff U, Harazny JM, Michelson G, Schmieder RE. Central pulse pressure is an independent determinant of vascular remodeling in the retinal circulation. Hypertension. 2013;61:1340–5.
19. Salvetti M, Agabiti Rosei C, Paini A, Aggiusti C, Cancarini A, Duse S, Semeraro F, Rizzoni D, Agabiti Rosei E, Muiesan ML. Relationship of wall-to-lumen ratio of retinal arterioles with clinic and 24-hour blood pressure. Hypertension. 2014;63:1110–5.
20. Prejbisz A, Harazny J, Szymanek K, et al. Retinal arteriolar structure in patients with pheochromocytoma. J Hypertens. 2015;33:e102.
21. Gosk-Przybylek M, Harazny J, Binczyk E, et al. Retinal arteriolar structure in patients with primary aldosteronism. J Hypertens. 2015;33:e103.
22. Warchol-Celinska E, Gosk-Przybylek M, Harazny J, et al. Evaluation of retinal microperfusion and arteriolar structure in patients with fibromuscular dysplasia—the polish registry for Fibromuscular dysplasia (ARCADIA-POL STUDY). J Hypertens. 2017;35:e267.
23. De Ciuceis C, Savoia C, Arrabito E, Porteri E, Mazza M, Rossini C, Duse S, Semeraro F, Agabiti Rosei C, Alonzo A, Sada L, La Boria E, Sarkar A, Petroboni B, Mercantini P, Volpe M, Rizzoni D, Agabiti RE. Effects of a long-term treatment with aliskiren or ramipril on structural alterations of subcutaneous small-resistance arteries of diabetic hypertensive patients. Hypertension. 2014;64:717–24.
24. De Ciuceis C, Salvetti M, Rossini C, Muiesan ML, Paini A, Duse S, La Boria E, Semeraro F, Cancarini A, Rosei CA, Sarkar A, Ruggeri G, Caimi L, Ricotta D, Rizzoni D, Rosei EA. Effect of antihypertensive treatment on microvascular structure, central blood pressure and oxidative stress in patients with mild essential hypertension. J Hypertens. 2014;32:565–74.
25. De Ciuceis C, Salvetti M, Paini A, Rossini C, Muiesan ML, Duse S, Caletti S, Coschignano MA, Semeraro F, Trapletti V, Bertacchini F, Brami V, Petelca A, Agabiti Rosei E, Rizzoni D, Agabiti RC. Comparison of lercanidipine plus hydrochlorothiazide vs. Lercanidipine plus enalapril on micro and macrocirculation in patients with mild essential hypertension. Intern Emerg Med. 2017;12:963–74.
26. Agabiti-Rosei E, Heagerty AM, Rizzoni D. Effects of antihypertensive treatment on small artery remodelling. J Hypertens. 2009;27:1107–14.

27. Jumar A, Ott C, Kistner I, Friedrich S, Schmidt S, Harazny JM, Schmieder RE. Effect of aliskiren on vascular remodelling in small retinal circulation. J Hypertens. 2015;33:2491–9.
28. Baleanu D, Ritt M, Harazny J, Heckmann J, Schmieder RE, Michelson G. Wall-to-lumen ratio of retinal arterioles and arteriole-to-venule ratio of retinal vessels in patients with cerebrovascular damage. Invest Ophthalmol Vis Sci. 2009;50:4351–9.
29. van Sloten TT, Sedaghat S, Laurent S, London GM, Pannier B, Ikram MA, Kavousi M, Mattace-Raso F, Franco OH, Boutouyrie P, Stehouwer CDA. Carotid stiffness is associated with incident stroke: a systematic review and individual participant data meta-analysis. J Am Coll Cardiol. 2015;66:2116–25.
30. Paini A, Muiesan ML, Agabiti-Rosei C, Aggiusti C, De Ciuceis C, Bertacchini F, Duse S, Semeraro F, Rizzoni D, Agabiti-Rosei E, Salvetti M. Carotid stiffness is significantly correlated with wall-to-lumen ratio of retinal arterioles. J Hypertens. 2018;36:580–6.
31. Hillege HL, Fidler V, Diercks GF, van Gilst WH, de Zeeuw D, van Veldhuisen DJ, Gans RO, Janssen WM, Grobbee DE, de Jong PE, Prevention of R, Vascular End Stage Disease Study G. Urinary albumin excretion predicts cardiovascular and noncardiovascular mortality in general population. Circulation. 2002;106:1777–82.
32. Ritt M, Harazny JM, Ott C, Schneider MP, Schlaich MP, Michelson G, Schmieder RE. Wall-to-lumen ratio of retinal arterioles is related with urinary albumin excretion and altered vascular reactivity to infusion of the nitric oxide synthase inhibitor n-monomethyl-l-arginine. J Hypertens. 2009;27:2201–8.
33. Bosch A, Scheppach JB, Harazny JM, Raff U, Eckardt KU, Schmieder RE, Schneider MP. Retinal capillary and arteriolar changes in patients with chronic kidney disease. Microvasc Res. 2018;118:121–7.
34. Jumar A, Ott C, Kistner I, Friedrich S, Michelson G, Harazny JM, Schmieder RE. Early signs of end-organ damage in retinal arterioles in patients with type 2 diabetes compared to hypertensive patients. Microcirculation. 2016;23:447–55.
35. Stefanski A, Harazny J, Wolf J et al. Impact of type 1 diabetes and its duration on wall-to-lumen ratio of retinal arterioles. submitted.
36. Ott C, Raff U, Schmidt S, Kistner I, Friedrich S, Bramlage P, Harazny JM, Schmieder RE. Effects of saxagliptin on early microvascular changes in patients with type 2 diabetes. Cardiovasc Diabetol. 2014;13:19.
37. Berndt-Zipfel C, Michelson G, Dworak M, Mitry M, Loffler A, Pfutzner A, Forst T. Vildagliptin in addition to metformin improves retinal blood flow and erythrocyte deformability in patients with type 2 diabetes mellitus—results from an exploratory study. Cardiovasc Diabetol. 2013;12:59.
38. O'Rourke MF, Safar ME. Relationship between aortic stiffening and microvascular disease in brain and kidney: cause and logic of therapy. Hypertension. 2005;46:200–4.
39. Rizzoni D, Porteri E, Boari GE, De Ciuceis C, Sleiman I, Muiesan ML, Castellano M, Miclini M, Agabiti-Rosei E. Prognostic significance of small-artery structure in hypertension. Circulation. 2003;108:2230–5.
40. Harazny JM, Ott C, Raff U, Welzenbach J, Kwella N, Michelson G, Schmieder RE. First experience in analysing pulsatile retinal capillary flow and arteriolar structural parameters measured noninvasively in hypertensive patients. J Hypertens. 2014;32:2246–52. discussion 2252
41. Clark MG, Barrett EJ, Wallis MG, Vincent MA, Rattigan S. The microvasculature in insulin resistance and type 2 diabetes. Semin Vasc Med. 2002;2:21–31.
42. Serne EH, Gans RO, ter Maaten JC, Tangelder GJ, Donker AJ, Stehouwer CD. Impaired skin capillary recruitment in essential hypertension is caused by both functional and structural capillary rarefaction. Hypertension. 2001;38:238–42.
43. Schrimpf C, Teebken OE, Wilhelmi M, Duffield JS. The role of pericyte detachment in vascular rarefaction. J Vasc Res. 2014;51:247–58.
44. Debbabi H, Uzan L, Mourad JJ, Safar M, Levy BI, Tibirica E. Increased skin capillary density in treated essential hypertensive patients. Am J Hypertens. 2006;19:477–83.
45. Jumar A, Harazny JM, Ott C, Friedrich S, Kistner I, Striepe K, Schmieder RE. Retinal capillary rarefaction in patients with type 2 diabetes mellitus. PLoS One. 2016;11:e0162608.

46. Bosch AJ, Harazny JM, Kistner I, Friedrich S, Wojtkiewicz J, Schmieder RE. Retinal capillary rarefaction in patients with untreated mild-moderate hypertension. BMC Cardiovasc Disord. 2017;17:300.
47. Battegay EJ, de Miguel LS, Petrimpol M, Humar R. Effects of anti-hypertensive drugs on vessel rarefaction. Curr Opin Pharmacol. 2007;7:151–7.
48. Jumar A, Harazny JM, Ott C, Kistner I, Friedrich S, Schmieder RE. Improvement in retinal capillary rarefaction after valsartan treatment in hypertensive patients. J Clin Hypertens. 2016;18:1112–8.
49. Folkow B. Regulation of the peripheral circulation. Br Heart J. 1971;33(Suppl):27–31.
50. Schiffrin EL. Remodeling of resistance arteries in essential hypertension and effects of antihypertensive treatment. Am J Hypertens. 2004;17:1192–200.
51. Kannenkeril D, Harazny JM, Bosch A, Ott C, Michelson G, Schmieder RE, Friedrich S. Retinal vascular resistance in arterial hypertension. Blood Press. 2018;27:82–7.
52. Ott C, Schmieder RE. Retinal circulation in arterial disease. In: Berbari A, Mancia G, editors. Arterial disorders: Springer; 2015.
53. Michelson G, Welzenbach J, Pal I, Harazny J. Automatic full field analysis of perfusion images gained by scanning laser doppler flowmetry. Br J Ophthalmol. 1998;82:1294–300.
54. Koch E, Rosenbaum D, Brolly A, Sahel JA, Chaumet-Riffaud P, Girerd X, Rossant F, Paques M. Morphometric analysis of small arteries in the human retina using adaptive optics imaging: relationship with blood pressure and focal vascular changes. J Hypertens. 2014;32:890–8.
55. Muraoka Y, Tsujikawa A, Kumagai K, Akiba M, Ogino K, Murakami T, Akagi-Kurashige Y, Miyamoto K, Yoshimura N. Age- and hypertension-dependent changes in retinal vessel diameter and wall thickness: an optical coherence tomography study. Am J Ophthalmol. 2013;156:706–14.

Assessment of Retinal Arteriolar Morphology by Adaptive Optics Ophthalmoscopy

4

Antonio Gallo, Xavier Girerd, M. Pâques, D. Rosenbaum, and Damiano Rizzoni

4.1 Introduction

The use of retinal digital image analysis has become increasingly common over the past decade thanks to the development of novel imaging techniques that are more reliable than previous micrometric.

In 2010 a novel and extremely promising approach was made commercially available: the direct measurement of wall-to-lumen ratio (WLR) of retinal arterioles using an adaptive optics ophthalmoscopy (AOO) imaging system [1, 2]. This is a considerably improved version of a traditional fundus camera based on an approach originally applied to correct for aberrations in astronomic optical systems [3], allowing to investigate vessels from 20 μm to over 150 μm of diameter [4].

A. Gallo (✉) · D. Rosenbaum
Cardiovascular Prevention Unit, Groupe Hospitalier Pitié-Salpêtrière, APHP, Paris, France

Sorbonne Université, UPMC Université Paris 06, INSERM 1146, CNRS 7371, Laboratoire d'imagerie Biomédicale, Paris, France
e-mail: antonio.gallo@aphp.fr

X. Girerd
Cardiovascular Prevention Unit, Groupe Hospitalier Pitié-Salpêtrière, APHP, Paris, France
e-mail: xavier.girerd@aphp.fr

M. Pâques
Unité INSERM 968 Institut de la vision – Centre d'Investigation Clinique 503 Centre Hospitalier National des Quinze-Vingts, Assistance Publique-Hôpitaux de Paris, Paris, France

D. Rizzoni
Clinica Medica, Department of Clinical and Experimental Sciences, University of Brescia, Brescia, Italy

Division of Medicine, Istituto Clinico Città di Brescia, Brescia, Italy
e-mail: damiano.rizzoni@unibs.it

E. Agabiti-Rosei et al. (eds.), *Microcirculation in Cardiovascular Diseases*, Updates in Hypertension and Cardiovascular Protection,
https://doi.org/10.1007/978-3-030-47801-8_4

Adaptive optics ophthalmoscopy allows a differentiation of the arteriolar wall from other perivascular structures providing more information about the vessel diameter than just the transition in contrast at the border of the blood column [5].

4.2 Measured Variables for Retinal Microcirculation Analysis

The near-histologic resolution of AOO images allows the measurement of many arteriolar and venular parameters, as detailed in Table 4.1. Wall thickness (WT) is directly measured as the difference between the outer diameter and the inner diameter. A venular wall is not visible, which makes the analysis of venular wall not possible. Both internal (ID) and outer diameters (OD) are automatically measured by an algorithm of localization that is based on the gradient of intensity of the reflected light from the lumen (central axial reflection) and the walls (of smaller intensity). If the ID may be available for both arterioles and venules, this is not the case for the OD, available only for arterioles due to the lack of a visible venular wall along the vessel. Measured variables include: WLR, the main validated remodeling index, which consists of the ratio between the directly measured arteriolar WT and the arteriolar ID; wall cross-sectional area (WCSA), derived from both arteriolar ID and OD, which is an indirect measurement of the vascular mass; arterio-venous ratio (AVR), which is expressed as the ratio between both arteriolar and venular IDs, possibly an indirect measurement of arterio-venous interaction. Being wall and lumen directly measured, a coefficient of variation is also possible to calculate, although this measurement is still not automatically performed by the machine. It is derived by three consecutive measurements realized along the vascular segment captured on the same image and is an index of vascular wall or diameter variability. A detailed procedure of vascular image acquisition with AOO is detailed in the Appendix 4.1.

4.2.1 Validation of AOO Measurements

Validation studies in humans by direct comparison of AOO within single and multiple operators have been performed. A coefficient of variation <5% was found for the intra-observer studies and <10% in inter-observer study [11, 12] .

Most importantly, a direct comparison of the assessment of the media-to-lumen ratio (MLR) of subcutaneous small resistance arteries evaluated by wire micromyography and the WLR evaluated by either AOO or SLDF was performed in 41 subjects (12 normotensive lean controls, 12 essential hypertensive lean patients, 9 normotensive obese patients, and 8 hypertensive obese patients) [12]. The evaluation of the WLR of retinal arterioles by AOO showed a clear advantage over SLDF, due to a better reproducibility and a closer correlation with the MLR of subcutaneous small resistance arteries ($r = 0.84$, $p < 0.001$ AOO vs $r = 0.52$, $p < 0.001$ SLDF; $p < 0.01$ for AOO vs SLDF) (Fig. 4.2).

Different softwares are available for the measurement of retinal microvascular parameters, among these are the AOV (Institut Supérieur d'Electronique, Paris, France) and AoDetect ® (Imagine Eyes, Orsay, France). They were both developed using custom software running under Matlab (Mathworks, Natick, Massachusetts, USA), using filters that allow smoothing the blood vessel while preserving the contrast along their edges. The enhancement of the axial reflection and the detection of the darkest region represent the first-step segmentation of the retinal vessel, followed by an algorithm of extraction of the borders of the vessel. While the first system performs measures that are averaged on a 250 μm vessel segment, the second focuses on a 38 μm vessel length (given an eye axial length of 24 mm, which is not adjustable). Table 4.2 provides mean values of retinal morphological parameters evaluated in a primary prevention cohort. Reference values on a larger population are not yet available.

4.3 Adaptive Optics Ophthalmoscopy Principles

When light propagates through an optically imperfect media, the shape of light wave is affected by distortions, which introduce blur and reduce image quality. This happens when observing the stars using ground-based telescopes: the light coming from the stars is distorted by aberrations introduced by the Earth turbulent atmosphere, which degrades image quality. Astrophysicists invented AOO to eliminate distortions from light waves so that sharp images can be captured even from the ground.

Like Earth's atmosphere, the eye is not optically perfect. When capturing retinal images, the light backscattered by the retina is then distorted by fluids in movement, irregularities on the cornea surface, and other optical defects. Because of those aberrations, the transverse resolution of retinal images is limited to 15–20 μm, even with a camera that would theoretically enable to resolve smaller details.

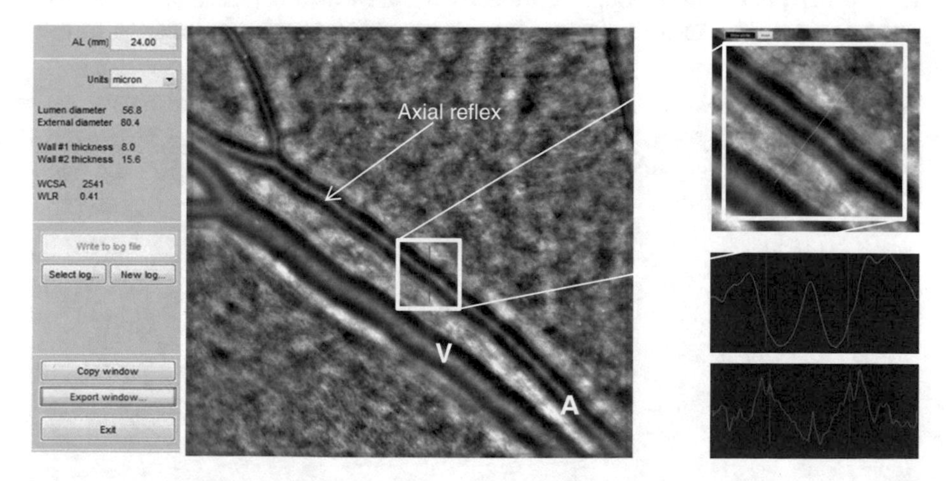

Fig. 4.1 On the left, measures of the selected segment of a retinal arteriole. In the center, AOO imaging and segmentation of superotemporal retinal arteriole and vein. On the right, Image magnification of the selected vessel for segmentation. The yellow lines represent the intensity of the reflection, while the blue lines represent the gradient of the intensity. When the measurement is not rightly performed, the operator can refer to the graphic description of gradient intensity in order to semiautomatically set the measurement point on the peak gradient. A = arteriole; V = vein

In the 1990s, scientific groups have adapted the AOO technology used in astronomy for retinal imaging, to enable in vivo visualization of structural details which previously remained invisible [6]. Several laboratory prototypes successfully resolved structures of a few microns in size, such as capillaries [7]. However, the complexity and large footprint of such systems limited the use of AOO retinal imaging in the hospital.

In 2010, Imagine Eyes (Orsay, France) launched the rtx1 – rtx1-e in its current version – the first commercially available AO retinal imaging camera which can be used in any clinical environment. It combines an integrated AOO system with a near-infrared flood-illumination fundus camera. As for most astronomical AOO devices, it is based on three key components that operate together in a closed-loop feedback configuration:

- A wave front sensor repetitively measures shape distortions in light waves.
- A deformable mirror constantly reshapes its reflecting surface in order to compensate for the distortions.
- Computer algorithms process the information from the sensor and drive the deformable mirror.

The AOO camera illuminates the fundus with a noncoherent light delivered by a superluminescent diode at 840 nm, covering a 4° × 4° area (i.e., approximately 1.2 mm × 1.2 mm in emmetropic eyes).

This light is backscattered by the retina and then collected by the optical system after exiting, distorted, from the eye. The information collected by the wave front sensor using the light beam is processed by the computer, which calculates the necessary correction and drives the deformable mirror to adapt its shape. As AOO system operates in a closed-loop, when the wave front sensor is placed after the wave front corrector, the measured wave front is sent back to the controller to further

Table 4.1 Directly measured and calculated variables of retinal microcirculation by AOO

	Direct measure	Calculated measure (formula)	Arteriole	Venule
Internal diameter (ID)	x		x	x
Outer diameter (OD)	x		x	
Wall thickness (WT)	x		x	
Wall-to-lumen ratio (WLR)		WT (μm)/ID (μm)	x	
Wall cross-sectional area (WCSA)		$\pi * ((OD/2)^2 - (ID/2)^2)$	x	
Arterio-venous ratio (AVR)		ID(arteriole)/ID(venule)	x	x
Lumen variability index (ID VI)		SD/mean ID	x	x
Wall variability index (WT VI)		SD/ mean WT	x	

reduce the aberration theoretically up to the diffraction limit [7]. This AOO correction loop works at a 50 Hz rate, resulting in real-time compensation of the distortions in the imaging beam.

In 2 s, a series of 40 images is acquired by a charge-coupled device (CCD) camera. These images are registered using a cross-correlation method and averaged to produce a final image with improved signal-to-noise ratio. The raw images that show artifacts due to eye blinking and saccades are automatically eliminated before averaging (Fig. 4.1).

The AOO correction system and image processing described above are fully automated so that the imaging process is comparable to a standard commercial fundus camera.

Adaptive optics system has been also coupled with other two main existing ophthalmic modalities: confocal scanning laser ophthalmoscopy (SLO) and ophthalmic optical coherence tomography (OCT). The first technique is achieved by placing a spatial pinhole at a retinal conjugate plane to include only direct backscattered light and eliminate multiply scattered light (out of focus) simultaneously [8]. It allows an analyzable picture of the retinal vasculature in vivo and is able to detect pathological features such as non-perfusion, infarctions, edema, neovascularization, microaneurysms, and leakage at blood-retinal barrier ruptures. The second relies on a contrast

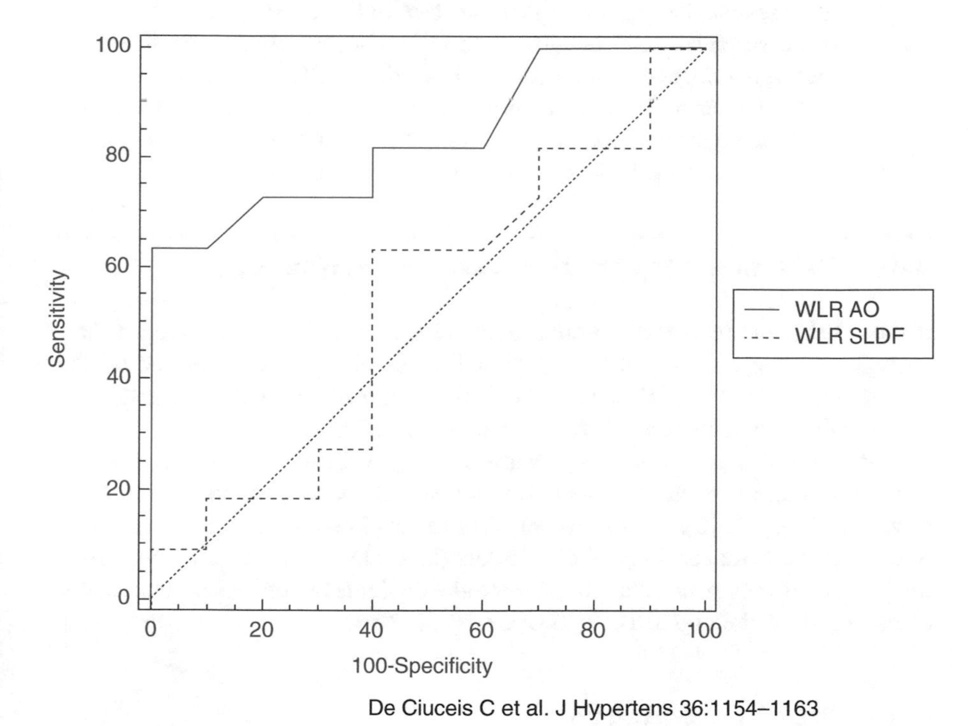

Fig. 4.2 Receiver-operating characteristic curves of retinal arteriolar WLR evaluated by AOO and SLDF for discriminating patients and controls according to subcutaneous small resistance arterioles' media-to-lumen ratio reference values

Table 4.2 Directly measured and calculated variables of retinal microcirculation by AOO

	Mean	SD
Arteriolar internal diameter (μm)	91.6	12.9
Arteriolar outer diameter (μm)	116.4	14.5
Arteriolar wall thickness (μm)	24.5	4.1
Arteriolar wall-to-lumen ratio	0.272	0.05
Arteriolar wall cross-sectional area (μm^2)	4060.5	972.3
Arterio-venous ratio (AVR)	0.738	0.14
Arteriolar lumen variability index (%)	3.3	2.8
Arteriolar wall variability index (%)	9.4	5.9
Venular internal diameter (μm)	126.5	17.7
Venular lumen variability (%)	3.0	1.8

Data extracted from a population of 356 subjects, male/female ratio 1.2 and median age 57 (46–65) years old. Hypertension 67%, dyslipidemia 45%, and diabetes 23%. Data are expressed as mean ± standard deviation (SD). Measurements realized with AoDetect® software. Personal data

generated by light scattering and allows for imaging with micrometer scale resolution over a depth of several millimeters (or even centimeters in low-scattering media) [9]. Due to relatively fast image acquisition speeds, thanks to simultaneous parallel detection of photons scattered from the entire imaging depth, OCT does not only provide cross-sectional images of tissue but can yield full 3D information of the sample within seconds. AOO coupled with OCT allows a tighter depth of focus which allows a superior separation between vascular beds. This represents an advantage of AO-OCT over AO-SLO and simple OCT angiography [10]. Both AO-SLO and AO-OCT, although very promising, are still relatively new to the clinical practice and less commonly approachable than the flood-illumination fundus camera.

4.4 Determinants of Retinal Vascular Morphology

If the WLR is the marker of vascular remodeling, it must be acknowledged that it averages the contributions of both wall and lumen changes. Through the detailed measurement of WT and ID, the study of retinal remodeling can be further detailed: which is the role of the lumen? Which that of the wall?

The study of a large primary prevention cohort with AO has started providing answers to these questions allowing the in vivo validation of previously observed in vitro phenomena [2]. A more recent work has explored in detail the contribution of each of the main cardiometabolic factors (high blood pressure and hyperglycemia) on top of ageing, disentangling their individual role on retinal microcirculatory changes (https://doi.org/10.1007/s00392-020-01680-3).

4.4.1 Arterial Hypertension

In arterial hypertension an inward remodeling takes place, characterized by a decreased in the vessel caliber without any changes in vascular mass [13]. It is considered as a proper feature of essential hypertension as in secondary hypertension a hypertrophic remodeling is found, due to smooth muscle cells rearrangement.

The study of retinal microcirculation confirmed a strong predictive role of blood pressure on WLR values (Table 4.3). In arterial hypertension WLR is increased mainly as a consequence of a decrease in ID, without any changes in WCSA thus confirming the presence of a eutrophic remodeling. This phenomenon is an *in vivo* manifestation of an increased myogenic tone, typical of hypertension and resulting from the increase in peripheral resistances and the systemic adrenergic overdrive. In young subjects with an increased blood pressure only a significant lumen narrowing is observed, confirming that the interaction between age and blood pressure seems to play a role only on lumen diameter.

The possibility to visualize an arteriolar remodeling of retinal arterioles might help in the diagnosis of arterial hypertension. In a cohort of 1500 subjects, a WLR higher than 0.31 suggests the presence of arterial hypertension, as well as a lumen diameter lower than 78 μm seems to evoke a masked hypertension [14].

A comparison among 276 subjects with different patterns of hypertension (77 controlled, 52 masked, 45 white coat, and 102 sustained hypertensive subjects) showed that WLR was increased in all hypertensive groups as compared to normotensive subjects ($p < 0.0001$) [14]. The increase in WLR was explained by an increase in WT in sustained hypertension and by a significant arteriolar narrowing in masked hypertension. Also, an increased WCSA was found in sustained hypertension compared to masked hypertension ($p < 0.01$). This suggests that a double-mechanism drives retinal arteriolar remodeling, being WLR the expression of both structural variations in vascular wall and of myogenic tone overdrive.

The study of the effects of antihypertensive treatment has provided information about the reversibility of the retinal remodeling process. A short-term reduction in blood pressure obtained by antihypertensive treatment determined a WLR decrease

Table 4.3 Role of aging, high blood pressure, and hyperglycemia on retinal microcirculatory changes

Aging	Hyperglycemia	High blood pressure	Age, hyperglycemia, and high blood pressure
Euluminal growth	Outward growth	Inward eutrophic remodeling	Outward hypertrophic remodeling and growth
↑ WLR	= WLR	↑ WLR	↑ WLR
↑ WCSA	↑ WCSA	= WCSA	↑ WCSA
↑ WT	= WT	↑ WT	↑ WT
= ID	↑ ID	↓ ID	↑ ID

Empty circles represent the reference arteriole, full circles represent the arteriole on the influence of each factor. *WLR* wall-to-lumen ratio, *WCSA* wall cross-sectional area, *WT* wall thickness, *ID* internal diameter. Adapted from DOI: https://doi.org/10.1007/s00392-020-01680-3

due to lumen dilatation rather than to wall thickness changes in 45 uncontrolled hypertensive subjects (WLR % reduction −6% vs 1.1% in the stable blood pressure group. $p < 0.001$) after a mean follow-up of 50 days [2]. In treated and controlled hypertensive subjects on monotherapy a reduction in WLR is observed, to be ascribed to combined wall decrease and lumen dilatation independently of antihypertensive pharmacological classes. This suggests that both short-term and long-term changes contribute to retinal remodeling reversal, the first being ascribable to myogenic tone regulation, the latter mostly to structural changes. An adequate control of blood pressure may provide protection form microvascular alterations.

Arterial hypertension is characterized by a sympathetic overdrive. This led to the development of novel implantable devices that target carotid baroreflex. The retina is devoid of adrenergic innervation, which enables an autoregulation that is only dependent on local vascular and metabolic stimuli [15]. Apart from pharmacological treatment, the baroreflex activation therapy in resistant hypertensive subjects with a previous failure of renal denervation was associated with a reverse hypotrophic remodeling, as confirmed by a significant reduction in WCSA, after 6-month follow-up. Interestingly, also in this case, switching off the device in newly implanted patients did not lead to any changes in retinal microvascular remodeling, despite the acute increase in blood pressure [16].

4.4.2 Aging

During physiologic aging, wall thickening is the main driver of WLR increase, while an unchanged arteriolar diameter is observed in elderly subjects, resulting in microvascular hypertrophy rather than remodeling (Table 4.3). This also results in an increase in vascular mass, as confirmed by the increase in WCSA [2]. As recently observed stated before, blood pressure influences lumen diameter in young and adult subjects: with ageing however, the effect of blood pressure on lumen diameter is not found anymore.

4.4.3 Diabetes

Diabetes and female gender are associated with an increased ID and WT: in both these conditions WLR is not altered due to the homogenous increase in both WT and ID: this results in an increased WCSA, corresponding to retinal microvascular outward hypertrophy [2] (https://doi.org/10.1007/s00392-020-01680-3) (Fig. 4.3). Acromegaly, an endocrinopathy associated with an increased risk of hypertension and diabetes, has made the object of extensive studies at the microcirculatory level [27], showing mostly a hypertrophic remodeling in small subcutaneous arteries [28]. AOO was used to study in vivo microvascular changes in this condition [29]: a eutrophic inward remodeling was found to be associated with active disease. Interestingly, achieving a good control of disease was associated with a similar arteriolar morphology of non-acromegaly age and sex-matched subjects. Indeed, IGF-1 levels were independent predictors of arteriolar WLR and ID (Fig. 4.3).

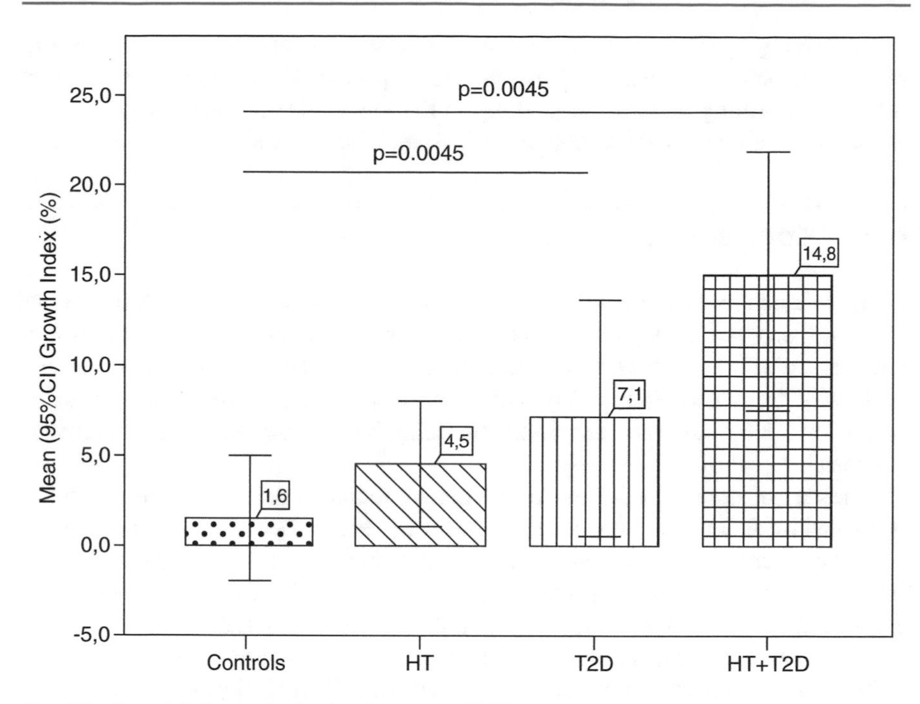

Fig. 4.3 Growth index in the four study groups. WCSA growth index was calculated according to the following formula [26]: GI = ($WCSA_{RP}$ -$WCSA_{NT}$)/$WCSA_{NT}$, where $WCSA_{RP}$ refers to individual WCSA mean levels and $WCSA_{NT}$ indicated the WCSA reference value derived from a subgroup of healthy subjects (73 subjects from our cohort selected according to the absence of office and home high BP, absence of T2D and naivety from any antihypertensive/antidiabetic therapy). Personal data

Further studies are ongoing to confirm these data on larger cohorts and reference values are still to be established.

4.5 Qualitative Studies

AOO allows the differentiation of two patterns of focal vascular changes: focal arteriolar narrowing (FAN) and arterio-venous nicking (AVN). These are known to be correlated with age, blood pressure, and inflammation biomarkers [17], and it has been reported that they are a sign of a dynamic process rather than an index of an irreversible degenerative process. After analysis with AOO, neither FANs nor AVNs seemed to involve parietal growth as their primary cause. In AVNs, AOO revealed a combination of loss of retinal transparency and presence of focal venous narrowing upstream and downstream of the arterio-venous crossing. Moreover, AOO images of AVNs in which the arterio-venous interface could be observed showed that venous nicking could occur even in the absence of arterio-venous contact [18].

AOO imaging showed also a very good precision in detecting extensive vascular involvement in vasculitis, characterized by parietal thickening, with different possible sheathing patterns depending on the vasculitis etiology. The remission of vascular inflammation corresponded to a decrease in the sheathing pattern [19].

4.6 Conclusions

There is undoubtedly a need for noninvasive approaches in the evaluation of microvascular morphology that may provide us with a better risk stratification of patients, as well as with further important information about the effects of antihypertensive drugs, especially in terms of possible prognostic impact of the regression of microvascular structural alterations that may be obtained with antihypertensive treatment.

The demonstration provided that WLR of retinal arterioles, especially when evaluated by AOO, is closely correlated with MLR of subcutaneous small arteries, the most potent predictor of cardiovascular and cerebrovascular events also in multivariate analyses [20, 21], strongly supports the clinical interest in a noninvasive evaluation of microvascular structure, and the possibility to transfer such measurements to the bedside [22], considering indices of microvascular morphology as intermediate end point in the evaluation of antihypertensive patients [23] and ready for prognostic translation [24]. A direct demonstration of the prognostic relevance of the easement of WLR in retinal arterioles with AOO is still lacking [21]. Data available with AOO are limited, at present, to the observation of correlations of WLR with age [2, 25] and blood pressure [1] or of parallelisms between changes in WLR during treatment and extent of blood pressure reduction [2].

In conclusion, thanks to the high-quality information on retinal microcirculation provided by AOO, the interest in a noninvasive evaluation of microvascular structure is progressively increasing, and it could potentially represent in the near future an easy tool for early diagnosis, cardiovascular risk stratification, treatment decision, and management in arterial hypertension, provided that a demonstration of a prognostic value of noninvasive measures of microvascular structure is made available. We have however too little information for unequivocally stating that the proposed approach provides exactly the same information obtained with the micromyographic approach, although the possible advantages related to the evaluation of WLR of retinal arterioles with AOO are clear, in terms of noninvasiveness and consequent possibility to obtain in the future prognostic data about regression of microvascular alterations by antihypertensive treatment.

Appendix 4.1 How to Measure Retinal Microvascular Parameters with AoDetect

Up-to-date measurements of retinal arteriolar parameters have been performed on the superotemporal retinal arteriole of the right eye.

Move the blue square along the 400 pixels segment of the arteriole (A) to be analyzed (310 μm, given an eye axial length of 24 mm).

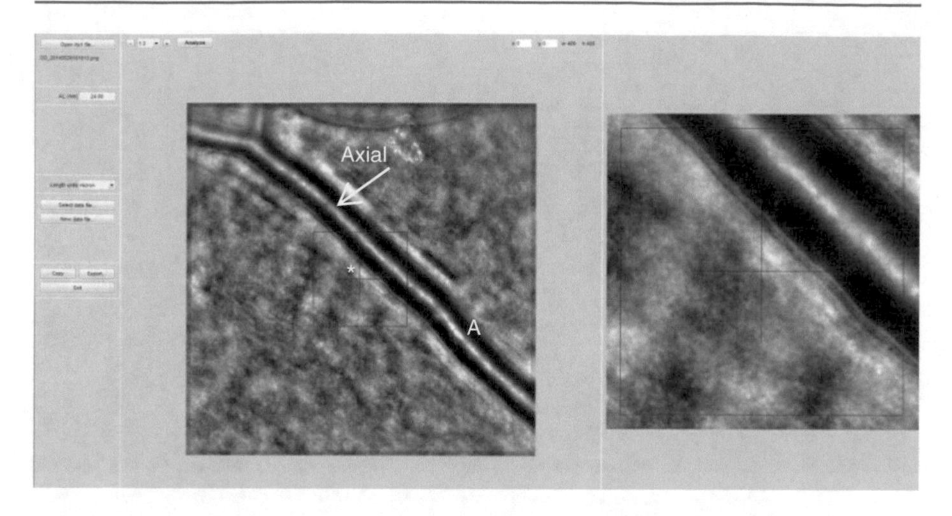

Use the enhanced zoom image on the right side of the screen to optimize the individuation of the arterial segment and the best positioning of the pointer.

Set the pointer (*) in the middle of the arterial lumen (along the axial reflex, white arrow) in order to allow the best algorithm of calculation of the wall thickness and diameter.

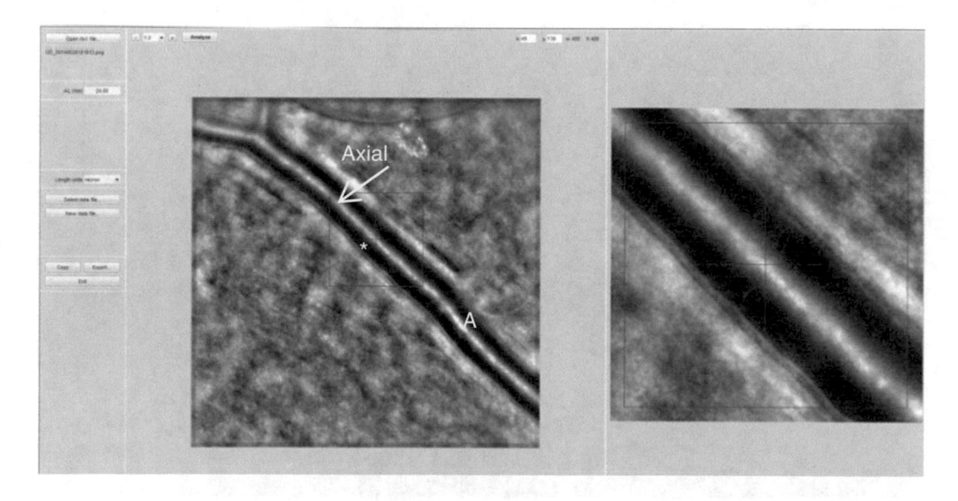

Click on the right mouse button to freeze the image.

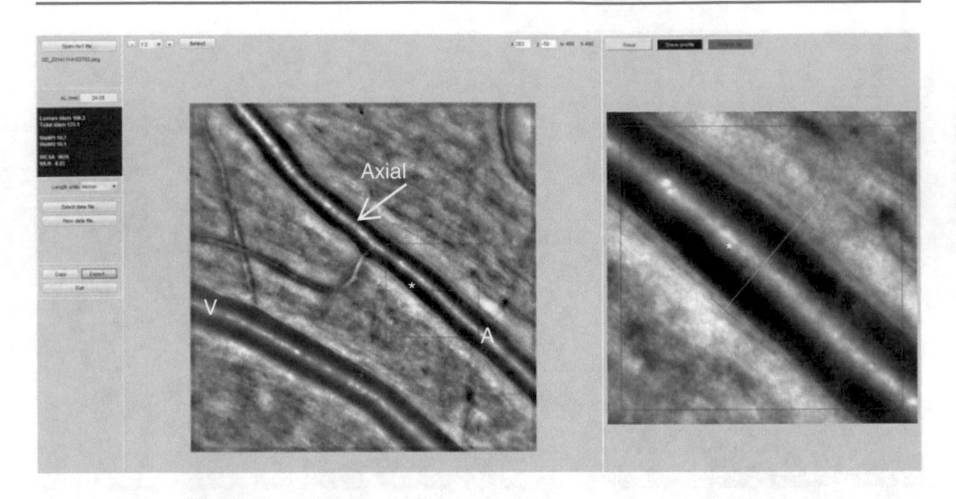

The blue square will turn red, a yellow line on the transversal axe of the vessel at the height of the central pointer will appear and two light-blue and dark-blue longitudinal lines will delimit, respectively, the internal and external walls along a 50 pixels length (corresponding to 38.7 μm given an eye axial length of 24 mm). These lines are automatically located on the highest gradient peak. To confirm it, click on Show profile:

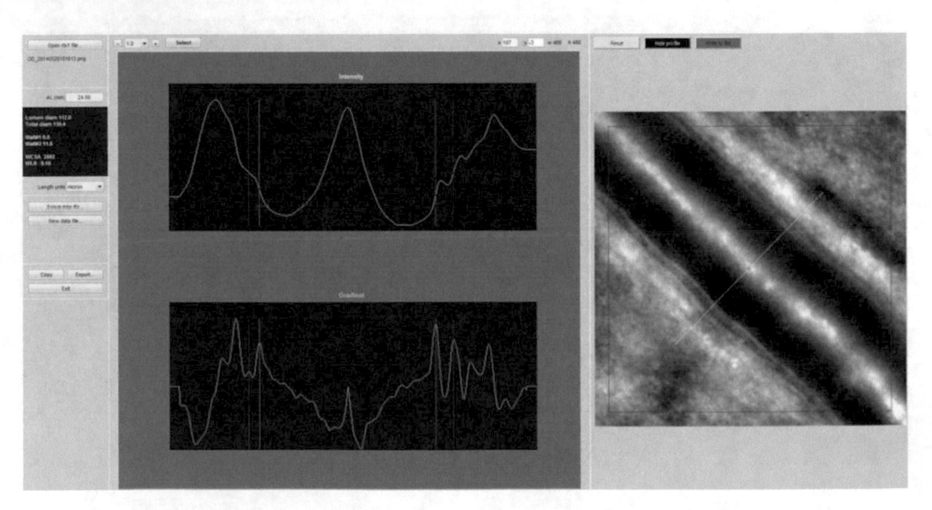

A manual adjustment of the measure can be made on the froze image by viewing the intensity and gradient wave profile. The pointer is moved toward another gradient peak that fits with the delimitation of the wall observed by the operator:

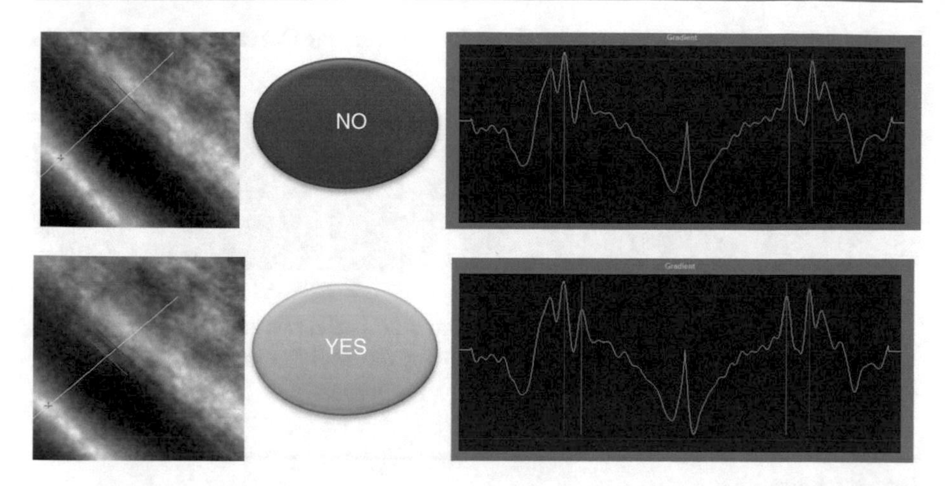

Axial orientation on the yellow line can be equally made. Manual adjustments must be kept at their minimum and only used when clear miscalculations have been automatically made by the software.

Once the measurement has been confirmed, final results of internal and external diameter, both walls' thickness, wall-to-lumen ratio, and wall cross-sectional area will be displayed in yellow on the left side of the screen (red arrow).

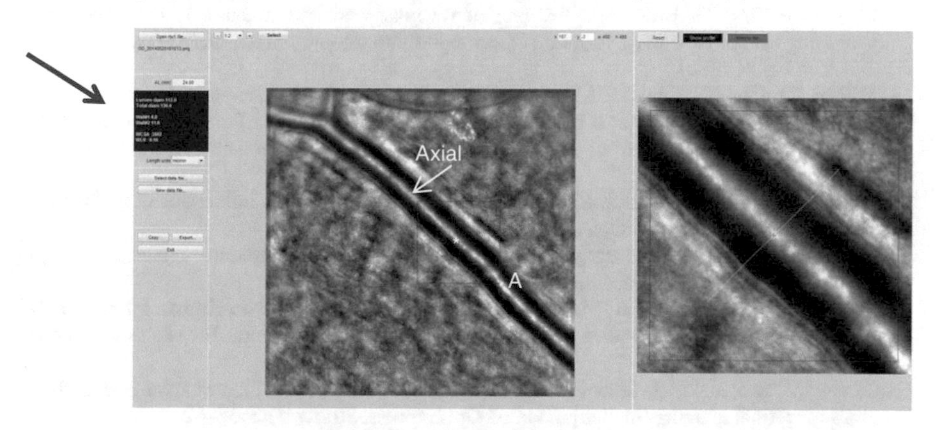

This procedure should be repeated on three consecutive segments granting up to 50% overlap between each consecutive image, as shown in the following pictures.

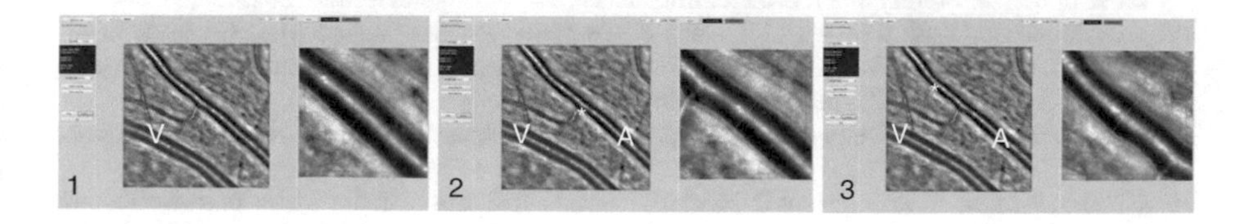

The same measurements can be done for the venular (V) segment, taking into account only the internal diameter.

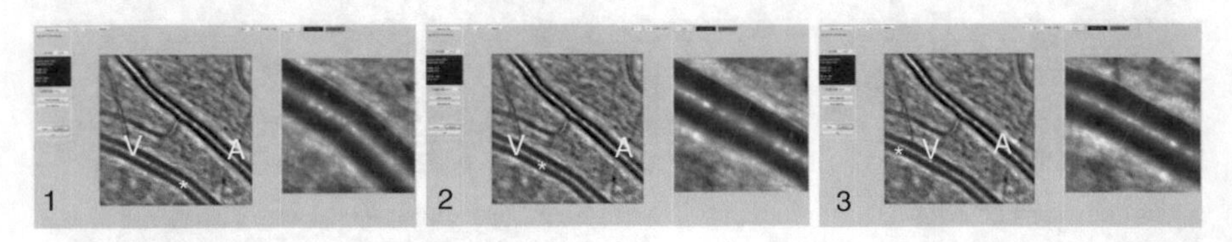

The three measurements are averaged, and standard deviation can be calculated in order to obtain the coefficient of variation for the wall thickness and the internal diameter.

References

1. Koch E, Rosenbaum D, Brolly A, Sahel J-A, Chaumet-Riffaud P, Girerd X, et al. Morphometric analysis of small arteries in the human retina using adaptive optics imaging: relationship with blood pressure and focal vascular changes. J Hypertens. 2014;32(4):890–8.
2. Rosenbaum D, Mattina A, Koch E, Rossant F, Gallo A, Kachenoura N, et al. Effects of age, blood pressure and antihypertensive treatments on retinal arterioles remodeling assessed by adaptive optics. J Hypertens. 2016;34(6):1115–22.
3. Chui TYP, Gast TJ, Burns SA. Imaging of vascular wall fine structure in the human retina using adaptive optics scanning laser ophthalmoscopy. Invest Ophthalmol Vis Sci. 2013;54(10):7115–24.
4. Hillard JG, Gast TJ, Chui TYP, Sapir D, Burns SA. Retinal arterioles in hypo-, normo-, and hypertensive subjects measured using adaptive optics. Transl Vis Sci Technol. 2016;5(4):16.
5. Bek T. Diameter changes of retinal vessels in diabetic retinopathy. Curr Diab Rep. 2017;17(10):82.
6. Dreher AW, Bille JF, Weinreb RN. Active optical depth resolution improvement of the laser tomographic scanner. Appl Opt. 1989;28(4):804–8.
7. Lombardo M, Serrao S, Devaney N, Parravano M, Lombardo G. Adaptive optics technology for high-resolution retinal imaging. Sensors. 2012;13(1):334–66.
8. Chui TYP, Mo S, Krawitz B, Menon NR, Choudhury N, Gan A, et al. Human retinal microvascular imaging using adaptive optics scanning light ophthalmoscopy. Int J Retina Vitreous. 2016;2:11.
9. Pircher M, Zawadzki RJ. Review of adaptive optics OCT (AO-OCT): principles and applications for retinal imaging [invited]. Biomed Opt Express. 2017;8(5):2536–62.
10. Salas M, Augustin M, Ginner L, Kumar A, Baumann B, Leitgeb R, et al. Visualization of micro-capillaries using optical coherence tomography angiography with and without adaptive optics. Biomed Opt Express. 2017;8(1):207–22.
11. Rosenbaum D, Koch E, Girerd X, Rossant F, Pâques M. Imaging of retinal arteries with adaptative optics, feasibility and reproducibility. Ann Cardiol Angeiol (Paris). 2013;62(3):184–8.
12. De Ciuceis C, Agabiti Rosei C, Caletti S, Trapletti V, Coschignano MA, GAM T, et al. Comparison between invasive and noninvasive techniques of evaluation of microvascular structural alterations. J Hypertens. 2018;36(5):1154–63.
13. Schiffrin EL. Vascular Remodeling in hypertension: mechanisms and treatment. Hypertension. 2012;59(2):367–74.
14. Gallo A, Mattina A, Rosenbaum D, Koch E, Paques M, Girerd X. Retinal arteriolar remodeling evaluated with adaptive optics camera: relationship with blood pressure levels. Ann Cardiol Angeiol (Paris). 2016;65(3):203–7.

15. Laties AM. Central retinal artery innervation. Absence of adrenergic innervation to the intraocular branches. Arch Ophthalmol. 1967;77(3):405–9.
16. Gallo A, Rosenbaum D, Kanagasabapathy C, Girerd X. Effects of carotid baroreceptor stimulation on retinal arteriole remodeling evaluated with adaptive optics camera in resistant hypertensive patients. Ann Cardiol Angeiol (Paris). 2017;66(3):165–70.
17. Klein R, Sharrett AR, Klein BE, Chambless LE, Cooper LS, Hubbard LD, et al. Are retinal arteriolar abnormalities related to atherosclerosis?: the atherosclerosis risk in communities study. Arterioscler Thromb Vasc Biol. 2000;20(6):1644–50.
18. Paques M, Brolly A, Benesty J, Lermé N, Koch E, Rossant F, et al. Venous nicking without Arteriovenous contact: the role of the arteriolar microenvironment in Arteriovenous Nickings. JAMA Ophthalmol. 2015;133(8):947–50.
19. Mahendradas P, Vala R, Kawali A, Akkali MC, Shetty R. Adaptive optics imaging in retinal vasculitis. Ocul Immunol Inflamm. 2018;26(5):760–6.
20. Rizzoni D, Porteri E, Boari GEM, De Ciuceis C, Sleiman I, Muiesan ML, et al. Prognostic significance of small-artery structure in hypertension. Circulation. 2003;108(18):2230–5.
21. Agabiti-Rosei E, Rizzoni D. Microvascular structure as a prognostically relevant endpoint. J Hypertens. 2017;35(5):914–21.
22. Schiffrin EL, Touyz RM. From bedside to bench to bedside: role of renin-angiotensin-aldosterone system in remodeling of resistance arteries in hypertension. Am J Physiol Heart Circ Physiol. 2004;287(2):H435–46.
23. Mulvany MJ. Small artery structure: time to take note? Am J Hypertens. 2007;20(8):853–4.
24. Heagerty AM. Changes in small artery structure in hypertension: ready for prognostic translation? J Hypertens. 2017;35(5):945–6.
25. Meixner E, Michelson G. Measurement of retinal wall-to-lumen ratio by adaptive optics retinal camera: a clinical research. Graefes Arch Clin Exp Ophthalmol. 2015;253(11):1985–95.
26. Heagerty AM, Aalkjaer C, Bund SJ, Korsgaard N, Mulvany MJ. Small artery structure in hypertension. Dual processes of remodeling and growth. Hypertension. 1993;21(4):391–97.
27. Pietro Maffei, Francesca Dassie, Alexandra Wennberg, Matteo Parolin, Roberto Vettor. The Endothelium in Acromegaly. Frontiers in Endocrinology 10, 2019.
28. Rizzoni D. Acromegalic Patients Show the Presence of Hypertrophic Remodeling of Subcutaneous Small Resistance Arteries. Hypertension. 2004;43(3):561–65.
29. Antonio Gallo, Emmanuelle Chaigneau, Christel Jublanc, David Rosenbaum, Alessandro Mattina, Michel Paques, Florence Rossant, Xavier Girerd, Monique Leban, Eric Bruckert, IGF-1 is an independent predictor of retinal arterioles remodeling in subjects with uncontrolled acromegaly. European Journal of Endocrinology. 2020;182(3):375–83.

The Cerebral Microcirculation

5

Anne-Eva van der Wijk, Ed VanBavel,
and Erik N. T. P. Bakker

5.1 Introduction

The high energy demand of the brain and its limited capacity for anaerobic metabolism requires a well-regulated perfusion. To fulfill this requirement, a dense vascular network, with strong autoregulation and other additional locally acting mechanisms, exists. For a more extensive review on this topic, the reader is referred to a previous excellent paper by Cipolla [1]. The circulation of the brain is unique in several aspects. First, the brain is encased in the skull, which limits volume changes. Increases in volume raise intracranial pressure, which causes a reduction in perfusion pressure and potentially limits cerebral blood flow. Second, the brain endothelium forms the blood-brain barrier, which greatly restricts the access of blood-borne substances into the neuronal tissue. This feature also dramatically changes the water exchange between the blood and the brain. Third, the brain parenchyma lacks a true lymphatic system. However, the role of the lymphatics may be taken over by the cerebrospinal fluid, which fills the ventricles, the subarachnoid space, and the paravascular spaces around meningeal vessels and penetrating arterioles.

5.2 Vascular Architecture

In mammals, cerebral blood flow is supplied via the carotid and vertebral arteries, which join to form the circle of Willis at the base of the brain. From there, the leptomeningeal vessels form a network of anastomosing branches, embedded in the subarachnoid space, which cover the brain surface. The degree of collateralization

A.-E. van der Wijk · E. VanBavel · E. N. T. P. Bakker (✉)
Amsterdam UMC, University of Amsterdam, Biomedical Engineering and Physics,
Amsterdam Cardiovascular Sciences, Amsterdam, The Netherlands
e-mail: a.e.vanderwijk@amc.uva.nl; e.vanbavel@amc.uva.nl; n.t.bakker@amc.uva.nl

E. Agabiti-Rosei et al. (eds.), *Microcirculation in Cardiovascular Diseases*,
Updates in Hypertension and Cardiovascular Protection,
https://doi.org/10.1007/978-3-030-47801-8_5

at the brain surface varies markedly among strains of mice [2] and is genetically determined [3]. The number of collaterals decreases with age, which makes the brain more vulnerable for ischemic insults in later life [4]. Characteristically, penetrating arterioles branch off perpendicularly from the leptomeningeal network and dive into the tissue to perfuse a column of tissue [5]. Crucially, occlusion neither of a leptomeningeal artery nor of a subsurface microvessel leads to ischemia, but occlusion of a penetrating arteriole (typically 10–30 μm in diameter) creates a distinct infarct in rodents [6, 7]. Apparently, this design results in a particular vulnerability at the level of these arterioles. The microvasculature that is fed by these penetrating arterioles consists of a network of capillaries, arterioles, and venules that range from 5 to 10 μm.

In peripheral vascular networks, resistance is mainly located in the smaller arteries and arterioles [8]. In the brain however, resistance appears to be more evenly distributed along the vascular tree. Early work in dogs indicated that 20–40% of resistance is located at the major feeding arteries that form the circle of Willis [9]. This relatively high proximal resistance allows regulation of blood flow and pressure before entering the brain tissue, providing additional protection against systemic influences. On the other hand, model work, based on realistic vascular trees obtained in mice, indicates that another major part of resistance is located at the capillary level [10]. Perhaps, the latter can be understood as an adaptation that allows very localized regulation of blood flow, rather than the more uniform distribution of flow that would suffice in other organs.

5.3 Cerebral Artery Composition

Cerebral arteries from humans and rodents are grossly similar to systemic arteries, with an intimal layer of endothelial cells, covered by a glycocalyx [11], and an internal elastic lamina. However, the medial smooth muscle cell layer is relatively thin, which is probably related to the lower pressure in the cerebral circulation, while the external elastic layer is missing. The adventitia is also relatively thin and accommodates nerve fibers [12]. Regarding the role of elastin, it is interesting to note that the cerebral arteries are devoid of longitudinally orientated fibers. In line with this, ex vivo experiments on isolated, cannulated cerebral arteries subjected to elastase treatment showed that elastin does not bear a longitudinal load, in contrast to systemic arteries [13]. This is probably a result of the fact that within the skull, arteries are not subjected to changes in length. The internal elastic lamina contains fenestrations, which allow myoendothelial junctions, and possibly facilitates diffusion across the vessel wall.

The extracellular matrix of cerebral arteries is a composite of numerous proteins. Apart from collagens and elastin, which are major load-bearing proteins, proteomic analysis revealed 94 different proteins, of which many fulfill functions that are incompletely understood [14]. The basement membrane supports the vascular cells and consists of a network formed by several proteins, including laminins, nidogen, collagens, and proteoglycans. It is distinct from the interstitial matrix, which mainly

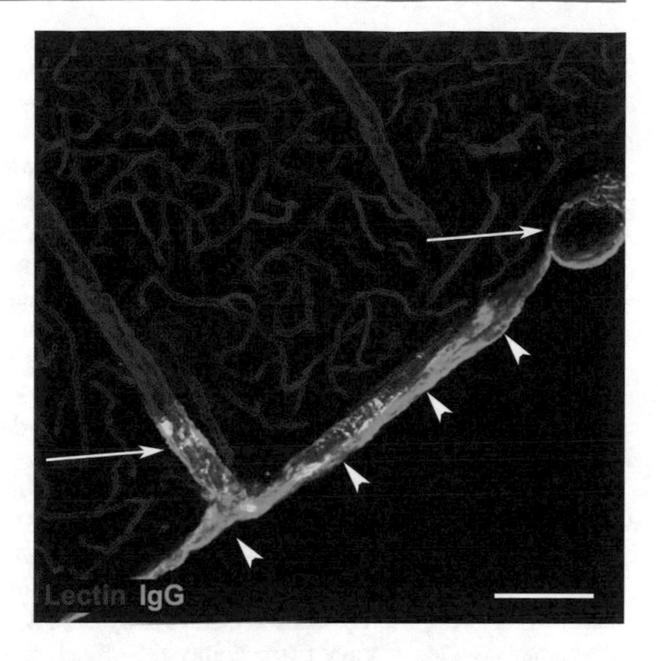

Fig. 5.1 Confocal image of a section of rat brain, with IgG staining (green) and lectin staining to identify the vasculature (red). Immunoglobulins are present in the cerebrospinal fluid that fills the subarachnoid space (SAS, arrowheads in figure). Paravascular spaces (indicated by arrows) are continuous with the SAS and extend along penetrating arteries into the brain. Scale bar = 100 μm

consists of glycoproteins, proteoglycans, and hyaluronan [15]. The basement membranes of cerebral arteries have been suggested to act as a pathway for interstitial fluid flow and waste removal from the brain [16]. Theoretical studies, however, suggest that resistance in this compartment is too high to allow significant flow [17].

A unique feature of cerebral arteries is the presence of a paravascular space, located between the adventitia and glia limitans, historically known as the Virchow-Robin space. We consider this a space that is continuous with the subarachnoid space, extending into the brain along penetrating arteries [18], becoming smaller around more distal vessels, and disappearing proximal to the capillaries (Fig. 5.1). There are conflicting views on whether or not the paravascular spaces sustain a bulk flow into the brain interstitium across astrocyte end feet [19–21]. There is, however, a broadly supported view that these paravascular spaces facilitate exchange between the cerebrospinal fluid and interstitial fluid, including waste removal. Perhaps, mixing in the paravascular space that results from arterial pulsations facilitates solute exchange between interstitial fluid and cerebrospinal fluid [21].

5.4 Regulation of Arterial Tone

Autoregulation of blood flow is particularly strong in the brain, in comparison to other vascular beds. It depends on changes in arterial tone induced by pressure, with myogenic reactivity accounting for constriction at higher pressures and dilation at lower pressure levels. Isolated, cannulated penetrating arteries from the rat show significant (≈30%) myogenic tone at 60 mmHg [22], which can be modulated by changes in pressure and various agonists. In humans, isolated pial arteries often show rhythmic oscillations in contraction [23], which is referred to as vasomotion. Decreases in flow may stimulate the release of vasodilatory metabolic substances from nearby tissue [1, 24] and decrease oxygen and increase carbon dioxide levels, both of which are potent vasodilator stimuli [1].

Interestingly, flow has been shown to induce constriction in isolated cerebral arteries of man and rats [25]. This is opposite to the vasodilatory effect of flow that is reported by most investigators in systemic arteries. Koller and Toth [24] argued that the combination of myogenic- and flow-induced constriction provides a stronger autoregulatory response, which is needed in the brain, as increases in cerebral blood volume could raise intracranial pressure due to the limitations of the skull.

5.5 The Neurovascular Unit and the Blood-Brain Barrier

A distinct feature of the CNS circulation is the presence of several blood-tissue barriers: the blood-CSF barrier, the blood-brain barrier (BBB), and the blood-retinal barrier (BRB). Whereas the blood-CSF barrier is formed by epithelium of the choroid plexus, the latter two are highly analogous endothelial barriers and share many common features [26]. The BBB regulates entry of molecules into the brain parenchyma and controls vascular permeability. Of critical importance for functioning of the BBB is the interplay between several cell types that compose the neurovascular unit (NVU), a concept which emerged almost two decades ago and acknowledges the intimate relationship between brain cells and the cerebral vasculature (for a recent review, see Iadecola [27]). Key players of the NVU are pericytes, astrocytes, and capillary endothelial cells, which form the main physical barrier between blood and tissue (Fig. 5.2). Cerebral endothelial cells are of a continuous barrier-type endothelium, which means that they lack fenestrations and have a high number of tight and adherens junctions to limit paracellular transport and a paucity of pinocytotic vesicles [28] to keep nonspecific transcytosis across the endothelium to a minimum [29]. In order to meet the metabolic needs of the brain and to remove toxic waste products, cerebral endothelial cells express a large number of selective solute carriers and receptors that mediate entry and removal of specific substances, such as glucose and amino acids. The crucial role of pericytes in BBB function was shown by two studies in which absolute pericyte coverage of vessels in the mouse brain directly correlated to BBB permeability due to increased endothelial transcytosis of macromolecules [30, 31]. In addition to regulating BBB permeability, pericytes stabilize capillaries [32] and regulate angiogenesis during CNS development and vascular remodeling [33]. Moreover, they may be involved in the regulation of capillary diameter [34, 35] and consequently cerebral blood flow [34], although this is still controversial [35, 36]. Endothelial cells and pericytes share their basal lamina, and the pericyte-to-endothelial ratio is high in the brain (1:3) compared to peripheral microvascular beds (1:10), and even higher in the retina (1:1) [37], underscoring the importance of pericytes in the CNS. In addition, astrocytes envelope cerebral blood vessels with their astrocytic end feet and thus are in close apposition to pericytes and endothelial cells. Astrocytic end feet are highly polarized structures and

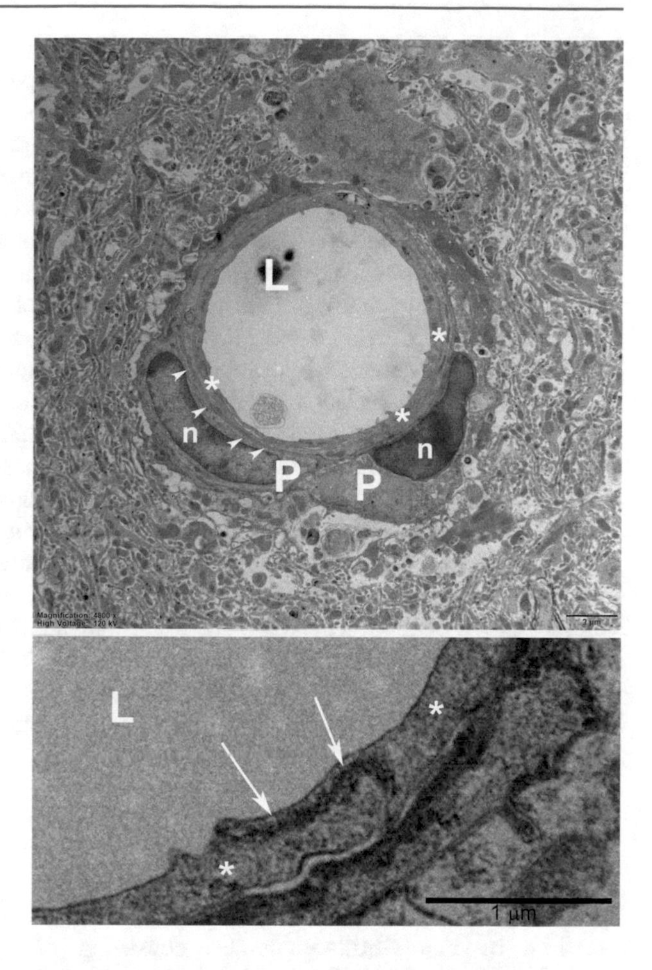

Fig. 5.2 Ultrastructure of the neurovascular unit of the blood-brain barrier. Upper panel: electron microscopy image of a rat brain capillary, ensheathed by pericytes. Endothelial cells and pericytes share a basal lamina. Scale bar = 2 µm. Lower panel: endothelial cells of the blood-brain barrier have a high number of tight junctions, which seal the intracellular space in between the cells. Scale bar = 1 µm. L = lumen, * = endothelial cell, P = pericyte, n = nucleus, arrowheads = basal lamina, arrows = tight junction

regulate water and ion homeostasis at the NVU through expression of the water channel aquaporin-4 and potassium channel Kir4.1 [38]. Moreover, astrocytes may enhance BBB function through paracrine signaling, e.g., they are a major source of Norrin [39], a protein that activates canonical Wnt signaling which is indispensable in CNS vascularization and barrier maintenance [40].

5.6 Pathology: Hypertension

Hypertension has a strong impact on cerebrovascular diseases, which are discussed in the sections below. In humans, particularly midlife hypertension negatively affects cognitive function and Alzheimer pathology in later life and increases the risk for stroke [41]. Cerebral resistance arteries of hypertensive subjects show remodeling, with an increase in media/lumen ratio [42]. This can be seen as an adaptation to normalize wall stress. In addition, a decrease in capillary density was found in the same study [42]. In a longitudinal study, patients with untreated hypertension and poorly controlled hypertension showed a decline in cerebral blood flow, which could be normalized by angiotensin receptor blockers, but not other antihypertensive medication [43].

The cerebral vasculature has been studied in several animal models of hypertension, including the spontaneously hypertensive rat (SHR), DOCA-salt rats, and angiotensin II-treated mice, the pathology of which partly concurs with findings in humans (for review, see [44]). The stroke-prone spontaneously hypertensive rat (SHR-SP) is a subline of SHR that shows very high blood pressure levels and mixed cerebral vascular pathology in later life, including stroke. These animals develop cerebral artery remodeling during life, of which inward remodeling accounts for most of the encroachment on the lumen and medial hypertrophy contributes to a lesser extent [45]. Contrary to widespread belief, arterial distensibility was not altered in this model [45]. Endothelium-dependent dilation of pial arterioles was reduced in SHR-SP [46]. The SHR displays a milder phenotype as compared to the SHR-SP, with BBB dysfunction reported in some studies [47], although earlier work indicated that SHR were actually protected against pressure-induced BBB disruption [48]. Capillary volume was found to be increased in SHR by some investigators [47]. In our hands however, we did not detect changes in either capillary volume or BBB disruption in adult SHR when compared to the normotensive Wistar Kyoto strain [49]. In contrast to structural changes, SHR do appear to develop impaired neurovascular coupling. Thus, a progressive impairment of functional hyperemia with aging was found using multimodal imaging [50].

5.7 Alzheimer's Disease

Alzheimer's disease (AD) is the leading cause of dementia and cognitive impairment and has traditionally been considered a neurodegenerative disease. AD is characterized by neurofibrillary tangles consisting of hyperphosphorylated and misfolded tau protein, the formation of which is largely paralleled by neuronal and synapse loss, and by amyloid plaques resulting from extracellular amyloid-β (Aβ) deposits in the brain parenchyma [51]. However, an increasing body of research links dementia and AD to vascular pathology. For instance, the large majority of AD patients also present with some degree of cerebral amyloid angiopathy (CAA), which is characterized by Aβ deposits in the tunica media of leptomeningeal arteries and cortical capillaries, small arterioles, and medium-sized arteries [52]. Moreover, cerebrovascular lesions such as microinfarcts [53] (further discussed in *Silent brain infarcts*), macroinfarcts, and small-vessel disease are common in patients with AD [54]. Yet to date, there is no consensus on the contribution of vascular lesions on the risk and burden of dementia.

Decreased regional cerebral blood flow in AD patients has been found to be associated with faster cognitive decline [55, 56], and perfusion deficiencies are already present in the very early preclinical phases of AD [57]. Furthermore, vascular-related risk factors such as ischemic stroke, atherosclerosis, hypertension, diabetes, and cardiac diseases have been reported to contribute to AD pathology [58], and all cause cerebral hypoperfusion [57]. Together, these findings have put forward the alternative "vascular hypothesis" of AD pathogenesis, which argues that cerebral hypoperfusion resulting from vascular disease and aging leads to neuronal death and cognitive dysfunction in AD, as opposed to the "amyloid cascade hypothesis" where overproduction and deposition of Aβ fragments are considered the primary cause of AD [59]. In the vascular hypothesis, it is suggested that persistent cerebral hypoperfusion leads to diminished oxygen and nutrient delivery causing microglial and astrocytic activation and eventually neuronal damage [60]. This mild but constant ischemic-hypoxic state may also stimulate Aβ production in endothelial cells and neighboring neurons, leading to the generation of CAA [61]. Moreover, ischemic conditions trigger the upregulation of hypoxia-induced factor (HIF)-1α, which was found to be elevated in the microcirculation of AD patients [62] and may in turn lead to secretion of angiogenic molecules, such as vascular endothelial growth factor (VEGF). Whereas VEGF is an important neurotrophic factor, its upregulation may also induce (pathological) neovascularization and vascular permeability, leading to BBB disruption [63]. BBB disruption can induce inflammation, and this inflammatory response may initiate a feedback loop by increasing secretion of angiogenic molecules, reducing amyloid clearance, and increasing Aβ accumulation and pathological angiogenesis which generates fragile and leaky vessels [64]. BBB disruption, demonstrated by a pathological CSF/serum albumin ratio, was evident in patients with different types of dementia (i.e., early- and late-onset AD, vascular dementia, dementia with Lewy bodies, mixed AD, and vascular dementia) [65]. A statistically significant increase in CSF/serum albumin ratio in AD patients compared to healthy controls was also found in a recent meta-analysis, but with a small effect size, precluding its use as a diagnostic marker for AD [66]. Results from postmortem brain studies on BBB disruption are conflicting, with some studies showing perivascular staining of plasma proteins such as fibrin and IgG in the brain parenchyma of AD patients [67] and others showing no differences in serum protein extravasation between AD patients and control individuals [68]. These discrepancies could be attributable to differences in methodology but more likely to the heterogeneity of concomitant vascular pathology in AD patients.

5.8 Ischemic Stroke and the No-Reflow Phenomenon

Stroke is the second leading cause of death worldwide and the third most common cause of long-term or even permanent disability in survivors [69]. In over 85% of all cases, stroke is caused by thrombosis or embolism within a blood vessel that supplies blood to the brain, leading to an interruption of blood flow, and hence this type is called acute ischemic stroke. Ischemic stroke therapeutics aim to improve the patient outcome months and years after the event, which can be achieved by recanalization of the affected vessel to initiate reperfusion and limit the tissue at risk. The therapeutic time window to treat acute ischemic stroke is very limited. Until recently, the only therapy with proven efficacy was intravenous thrombolysis using tissue plasminogen activator, which should be administered within 3 h after stroke onset

[70]. Nowadays, endovascular thrombectomy, mechanically removing the clot with a stent or aspiration device to recanalize the vessel, is increasingly becoming common practice [71]. However, a large part of patients receiving acute reperfusion therapies still experience significant long-term disability. One reason for this adverse outcome after acute therapy may be the no-reflow phenomenon.

The no-reflow phenomenon is a well-known concept in the ischemic heart and occurs when restoration of flow by opening the occluded artery fails to ensure reperfusion of the microvasculature of that organ. The rise of thrombectomy as a treatment for ischemic stroke now shows that there is a similar phenomenon occurring in the brain. Although the first evidence for no-reflow in the cerebral microcirculation after cerebral ischemia comes from experimental animal data [72], there is increasing clinical evidence that suggests that restoration of the microcirculatory blood flow is a more accurate predictor of outcome in acute ischemic stroke patients than recanalization [73, 74].

Several mechanisms causing impaired microcirculatory reperfusion have been suggested, based on data from animal experiments. Whereas previous studies demonstrated impaired perfusion using histological techniques or intravital imaging of the cortical surface [72, 75], the phenomenon has recently been shown to occur throughout almost the entire depth of mouse cortex in vivo, using optical coherence tomography [76]. In a number of studies, obstruction of microvessels has been reported, resulting from leukocyte plugging [77] or erythrocyte trapping [78]. In addition, narrowing of the vascular lumen has been suggested as an alternative mechanism underlying this process, either through endothelial cell swelling or swelling of the astrocytic end feet [79], although these events are not consistently observed. One group showed that there was persistent contraction of pericytes leading to capillary constriction, even after reopening of the occluded vessel [75]. Considering the small capillary diameter (~5 μm), additional narrowing due to capillary constriction may result in entrapment of erythrocytes (with a diameter of ~6.5 μm) and obstruct the microcirculation [75]. However, the contribution of pericytes to capillary constriction and the no-reflow phenomenon is still under debate [36, 80], which may relate to the controversy regarding mural cell nomenclature [36, 81]. Moreover, it should be noted that most of these studies (except for [75]) were conducted in the late twentieth century. Since then, the no-flow phenomenon has received little attention in (pre)clinical studies, likely as a result of the shortcomings of current imaging modalities to reliably image the microcirculation in vivo in human patients. As efforts are being put into its detection in a clinical setting [82], we can hopefully increase our knowledge regarding the occurrence of microvascular no-reflow after recanalization therapy in stroke.

5.9 Silent Brain Infarcts and Embolus Clearance

In the cerebral microcirculation, small embolic events may cause occlusions and lead to the development of silent brain infarcts (SBI), which are defined as ischemic events detected by brain imaging, without any overt clinical symptoms. Although these microsized lesions do not cause acute clinical symptoms, they are associated with cognitive decline, dementia [83], and increased risk of stroke and overall mortality [84]. Conversely, ischemic stroke is also a risk factor for the development of SBI, as is age, hypertension, and atherosclerosis [85].

A large portion of microemboli can be cleared from the vasculature, and therefore not all microemboli lead to infarction and tissue damage. The principal clearing mechanisms of occlusions in brain vasculature include fibrinolysis and subsequent washout by hemodynamic forces [86]. Moreover, the continuity of the capillary network in the brain provides a protective mechanism, since perfusion of microvessels adjacent to occluded arterioles or capillaries may be sufficient to nourish the surrounding tissue. However, this is dependent on the location of the occluded vessel (as discussed in vascular architecture), and, crucially, not all microemboli are susceptible to fibrinolysis (e.g., cholesterol crystals dislodging from atherosclerotic plaques). An alternative mechanism of embolus clearance has been described, coined angiophagy [87] which entails embolus extravasation from the vessel lumen. This was first described in mice by Lam et al. (2010), using high-resolution fixed tissue and transcranial imaging with two-photon microscopy in living animals [87]. In that study, mice were exposed to microemboli of different sources (microspheres, fibrin clots, and cholesterol emboli) through infusion into the internal carotid artery, which then lodge in the cerebral microvasculature. Remarkably, Lam et al. reported that between 2- and 6-day post-embolization, the large majority of emboli had extravasated from the vessel lumen into the brain parenchyma, regardless of the source of the occluding material. In a follow-up study, angiophagy was also observed in other organs, including the heart and the retina, suggesting that it is an ubiquitous mechanism of recanalization [88].

The mechanism through which angiophagy takes place has not yet been fully elucidated but involves endothelial cell remodeling. Within hours after embolization, the endothelium projects lamellipodia toward the embolus and eventually envelopes the obstructing material [87, 88]. Following this envelopment by newly formed endothelial membranes, the original endothelium may undergo remodeling to retract, in order to allow for translocation of the embolus into the parenchyma. Once in the parenchyma, extravasated emboli appeared to be degraded by pericytes and microglia [88].

Although it is a highly interesting and potentially effective protective mechanism to dispose of microemboli in the brain, angiophagy has received little attention to date. One study reported extravasation of 3–5 μm microspheres from occluded brain capillaries, yet only in 2% of the cases [89]. In contrast, we found that ~80% of 15 μm microspheres had extravasated from the lumen of cerebral arterioles 7-day post-infusion and all microspheres 28-day post-infusion in a rat model of SBI (Fig. 5.3, manuscript under review) [90]. In fact, it may very well be that the occurrence of angiophagy is one of the reasons why SBI generally do not cause clinical symptoms. Nevertheless, when the capacity of the brain to dispose of microsized emboli is exceeded by the occurrence of embolisms, this may lead to an accumulation of SBI. Indeed, the rate of embolus extravasation was significantly diminished in aged mice [87], and considering that age is a major risk factor for SBI, this may explain the association between SBI and manifest cognitive decline and dementia.

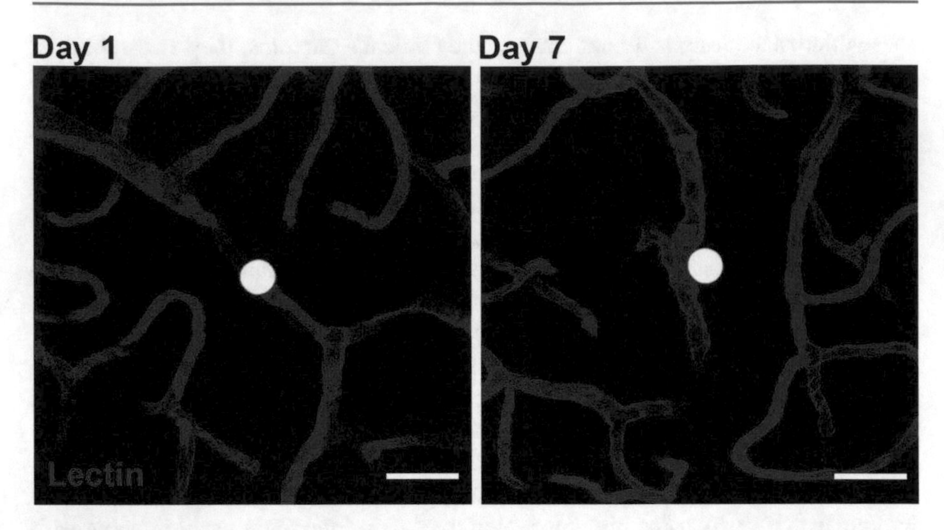

Fig. 5.3 Extravasation of microspheres from cerebral vessels in the rat. Fluorescent microspheres of 15 μm (white) were injected into the internal carotid artery at the day of surgery. At day 1 postsurgery, lectin (red) was injected i.v. to stain the vasculature, and microspheres were found lodged inside the vessel lumen. In contrast, at day 7 postsurgery, the majority of the microspheres had extravasated from the vessel lumen, allowing for recanalization of the vessels. Scale bar = 30 μm

5.10 Conclusions and Future Directions

The neuronal tissue is vulnerable for ischemia due to its limited energy storage capacity and high metabolic demand. Hypoperfusion, either globally or locally, is therefore a continuous threat. Aging and vascular pathologies, such as hypertension, microemboli, stroke, and subsequent no-reflow phenomena, are conditions where blood flow regulation and protective mechanisms may fall short. In this chapter, we identified several processes with considerable potential therapeutic gain that warrant further study. These include angiophagy, paravascular spaces and their role in waste removal, and microcirculatory obstructions following recanalization after stroke.

References

1. Cipolla MJ. Integrated systems physiology: from molecule to function. The cerebral circulation. San Rafael, CA: Morgan & Claypool Life Sciences. Copyright (c) 2010 by Morgan & Claypool Life Sciences; 2009.
2. Chalothorn D, Clayton JA, Zhang H, Pomp D, Faber JE. Collateral density, remodeling, and VEGF-A expression differ widely between mouse strains. Physiol Genomics. 2007;30(2):179–91.
3. Wang S, Zhang H, Dai X, Sealock R, Faber JE. Genetic architecture underlying variation in extent and remodeling of the collateral circulation. Circ Res. 2010;107(4):558–68.

4. Faber JE, Zhang H, Lassance-Soares RM, Prabhakar P, Najafi AH, Burnett MS, et al. Aging causes collateral rarefaction and increased severity of ischemic injury in multiple tissues. Arterioscler Thromb Vasc Biol. 2011;31(8):1748–56.
5. Blinder P, Shih AY, Rafie C, Kleinfeld D. Topological basis for the robust distribution of blood to rodent neocortex. Proc Natl Acad Sci U S A. 2010;107(28):12670–5.
6. Shih AY, Ruhlmann C, Blinder P, Devor A, Drew PJ, Friedman B, et al. Robust and fragile aspects of cortical blood flow in relation to the underlying angioarchitecture. Microcirculation. 2015;22(3):204–18.
7. Shih AY, Blinder P, Tsai PS, Friedman B, Stanley G, Lyden PD, et al. The smallest stroke: occlusion of one penetrating vessel leads to infarction and a cognitive deficit. Nat Neurosci. 2013;16(1):55–63.
8. Christensen KL, Mulvany MJ. Location of resistance arteries. J Vasc Res. 2001;38(1):1–12.
9. Heistad DD, Marcus ML, Abboud FM. Role of large arteries in regulation of cerebral blood flow in dogs. J Clin Invest. 1978;62(4):761–8.
10. Gould IG, Tsai P, Kleinfeld D, Linninger A. The capillary bed offers the largest hemodynamic resistance to the cortical blood supply. J Cereb Blood Flow Metab. 2017;37(1):52–68.
11. Haeren RH, van de Ven SE, van Zandvoort MA, Vink H, van Overbeeke JJ, Hoogland G, et al. Assessment and imaging of the cerebrovascular Glycocalyx. Curr Neurovasc Res. 2016;13(3):249–60.
12. Lee RM. Morphology of cerebral arteries. Pharmacol Ther. 1995;66(1):149–73.
13. Clifford PS, Ella SR, Stupica AJ, Nourian Z, Li M, Martinez-Lemus LA, et al. Spatial distribution and mechanical function of elastin in resistance arteries: a role in bearing longitudinal stress. Arterioscler Thromb Vasc Biol. 2011;31(12):2889–96.
14. Badhwar A, Stanimirovic DB, Hamel E, Haqqani AS. The proteome of mouse cerebral arteries. J Cereb Blood Flow Metab. 2014;34(6):1033–46.
15. Thomsen MS, Routhe LJ, Moos T. The vascular basement membrane in the healthy and pathological brain. J Cereb Blood Flow Metab. 2017;37(10):3300–17.
16. Carare RO, Bernardes-Silva M, Newman TA, Page AM, Nicoll JA, Perry VH, et al. Solutes, but not cells, drain from the brain parenchyma along basement membranes of capillaries and arteries: significance for cerebral amyloid angiopathy and neuroimmunology. Neuropathol Appl Neurobiol. 2008;34(2):131–44.
17. Faghih MM, Sharp MK. Is bulk flow plausible in perivascular, paravascular and paravenous channels? Fluids Barriers CNS. 2018;15(1):17.
18. Bedussi B, van der Wel NN, de Vos J, van Veen H, Siebes M, VanBavel E, et al. Paravascular channels, cisterns, and the subarachnoid space in the rat brain: a single compartment with preferential pathways. J Cereb Blood Flow Metab. 2017;37(4):1374–85.
19. Iliff JJ, Wang M, Liao Y, Plogg BA, Peng W, Gundersen GA, et al. A paravascular pathway facilitates CSF flow through the brain parenchyma and the clearance of interstitial solutes, including amyloid beta. Sci Transl Med. 2012;4(147):147ra11.
20. Smith AJ, Verkman AS. The "glymphatic" mechanism for solute clearance in Alzheimer's disease: game changer or unproven speculation? FASEB J. 2018;32(2):543–51.
21. Bakker EN, Bacskai BJ, Arbel-Ornath M, Aldea R, Bedussi B, Morris AW, et al. Lymphatic clearance of the brain: perivascular, Paravascular and significance for neurodegenerative diseases. Cell Mol Neurobiol. 2016;36(2):181–94.
22. Dacey RG Jr, Duling BR. A study of rat intracerebral arterioles: methods, morphology, and reactivity. Am J Phys. 1982;243(4):H598–606.
23. Gokina NI, Bevan RD, Walters CL, Bevan JA. Electrical activity underlying rhythmic contraction in human pial arteries. Circ Res. 1996;78(1):148–53.
24. Koller A, Toth P. Contribution of flow-dependent vasomotor mechanisms to the autoregulation of cerebral blood flow. J Vasc Res. 2012;49(5):375–89.
25. Toth P, Rozsa B, Springo Z, Doczi T, Koller A. Isolated human and rat cerebral arteries constrict to increases in flow: role of 20-HETE and TP receptors. J Cereb Blood Flow Metab. 2011;31(10):2096–105.

26. van der Wijk AE, Vogels IMC, van Veen HA, van Noorden CJF, Schlingemann RO, Klaassen I. Spatial and temporal recruitment of the neurovascular unit during development of the mouse blood-retinal barrier. Tissue Cell. 2018;52:42–50.
27. Iadecola C. The neurovascular unit coming of age: a journey through neurovascular coupling in health and disease. Neuron. 2017;96(1):17–42.
28. Reese TS, Karnovsky MJ. Fine structural localization of a blood-brain barrier to exogenous peroxidase. J Cell Biol. 1967;34(1):207–17.
29. De Bock M, Van Haver V, Vandenbroucke RE, Decrock E, Wang N, Leybaert L. Into rather unexplored terrain-transcellular transport across the blood-brain barrier. Glia. 2016;64(7):1097–123.
30. Daneman R, Zhou L, Kebede AA, Barres BA. Pericytes are required for blood-brain barrier integrity during embryogenesis. Nature. 2010;468(7323):562–6.
31. Armulik A, Genove G, Mae M, Nisancioglu MH, Wallgard E, Niaudet C, et al. Pericytes regulate the blood-brain barrier. Nature. 2010;468(7323):557–61.
32. Lindahl P, Johansson BR, Leveen P, Betsholtz C. Pericyte loss and microaneurysm formation in PDGF-B-deficient mice. Science (New York, NY). 1997;277(5323):242–5.
33. Park DY, Lee J, Kim J, Kim K, Hong S, Han S, et al. Plastic roles of pericytes in the blood-retinal barrier. Nat Commun. 2017;8:15296.
34. Peppiatt CM, Howarth C, Mobbs P, Attwell D. Bidirectional control of CNS capillary diameter by pericytes. Nature. 2006;443(7112):700–4.
35. Fernandez-Klett F, Offenhauser N, Dirnagl U, Priller J, Lindauer U. Pericytes in capillaries are contractile in vivo, but arterioles mediate functional hyperemia in the mouse brain. Proc Natl Acad Sci U S A. 2010;107(51):22290–5.
36. Hill RA, Tong L, Yuan P, Murikinati S, Gupta S, Grutzendler J. Regional blood flow in the normal and ischemic brain is controlled by arteriolar smooth muscle cell contractility and not by capillary pericytes. Neuron. 2015;87(1):95–110.
37. Stewart PA, Tuor UI. Blood-eye barriers in the rat: correlation of ultrastructure with function. J Comp Neurol. 1994;340(4):566–76.
38. Abbott NJ, Ronnback L, Hansson E. Astrocyte-endothelial interactions at the blood-brain barrier. Nat Rev Neurosci. 2006;7(1):41–53.
39. Ye X, Smallwood P, Nathans J. Expression of the Norrie disease gene (Ndp) in developing and adult mouse eye, ear, and brain. Gene Expr Patterns. 2011;11(1–2):151–5.
40. Zhou Y, Wang Y, Tischfield M, Williams J, Smallwood PM, Rattner A, et al. Canonical WNT signaling components in vascular development and barrier formation. J Clin Invest. 2014;124(9):3825–46.
41. Iadecola C, Yaffe K, Biller J, Bratzke LC, Faraci FM, Gorelick PB, et al. Impact of hypertension on cognitive function: a scientific statement from the American Heart Association. Hypertension. 2016;68(6):e67–94.
42. Rizzoni D, De Ciuceis C, Porteri E, Paiardi S, Boari GE, Mortini P, et al. Altered structure of small cerebral arteries in patients with essential hypertension. J Hypertens. 2009;27(4):838–45.
43. Muller M, van der Graaf Y, Visseren FL, Mali WP, Geerlings MI. Hypertension and longitudinal changes in cerebral blood flow: the SMART-MR study. Ann Neurol. 2012;71(6):825–33.
44. Pires PW, Dams Ramos CM, Matin N, Dorrance AM. The effects of hypertension on the cerebral circulation. Am J Physiol Heart Circ Physiol. 2013;304(12):H1598–614.
45. Baumbach GL, Heistad DD. Remodeling of cerebral arterioles in chronic hypertension. Hypertension. 1989;13(6 Pt 2):968–72.
46. Yang ST, Mayhan WG, Faraci FM, Heistad DD. Endothelium-dependent responses of cerebral blood vessels during chronic hypertension. Hypertension. 1991;17(5):612–8.
47. Kaiser D, Weise G, Moller K, Scheibe J, Posel C, Baasch S, et al. Spontaneous white matter damage, cognitive decline and neuroinflammation in middle-aged hypertensive rats: an animal model of early-stage cerebral small vessel disease. Acta Neuropathol Commun. 2014;2:169.
48. Mueller SM, Heistad DD. Effect of chronic hypertension on the blood-brain barrier. Hypertension. 1980;2(6):809–12.

49. Naessens DMP, de Vos J, VanBavel E, Bakker E. Blood-brain and blood-cerebrospinal fluid barrier permeability in spontaneously hypertensive rats. Fluids Barriers CNS. 2018;15(1):26.
50. Calcinaghi N, Wyss MT, Jolivet R, Singh A, Keller AL, Winnik S, et al. Multimodal imaging in rats reveals impaired neurovascular coupling in sustained hypertension. Stroke. 2013;44(7):1957–64.
51. Serrano-Pozo A, Frosch MP, Masliah E, Hyman BT. Neuropathological alterations in Alzheimer disease. Cold Spring Harb Perspect Med. 2011;1(1):a006189.
52. Brenowitz WD, Nelson PT, Besser LM, Heller KB, Kukull WA. Cerebral amyloid angiopathy and its co-occurrence with Alzheimer's disease and other cerebrovascular neuropathologic changes. Neurobiol Aging. 2015;36(10):2702–8.
53. Brundel M, de Bresser J, van Dillen JJ, Kappelle LJ, Biessels GJ. Cerebral microinfarcts: a systematic review of neuropathological studies. J Cereb Blood Flow Metab. 2012;32(3):425–36.
54. Azarpazhooh MR, Avan A, Cipriano LE, Munoz DG, Sposato LA, Hachinski V. Concomitant vascular and neurodegenerative pathologies double the risk of dementia. Alzheimers Dement. 2018;14(2):148–56.
55. Hanyu H, Sato T, Hirao K, Kanetaka H, Iwamoto T, Koizumi K. The progression of cognitive deterioration and regional cerebral blood flow patterns in Alzheimer's disease: a longitudinal SPECT study. J Neurol Sci. 2010;290(1–2):96–101.
56. Benedictus MR, Leeuwis AE, Binnewijzend MA, Kuijer JP, Scheltens P, Barkhof F, et al. Lower cerebral blood flow is associated with faster cognitive decline in Alzheimer's disease. Eur Radiol. 2017;27(3):1169–75.
57. Austin BP, Nair VA, Meier TB, Xu G, Rowley HA, Carlsson CM, et al. Effects of hypoperfusion in Alzheimer's disease. J Alzheimers Dis. 2011;26(Suppl 3):123–33.
58. Viswanathan A, Rocca WA, Tzourio C. Vascular risk factors and dementia: how to move forward? Neurology. 2009;72(4):368–74.
59. Kelleher RJ, Soiza RL. Evidence of endothelial dysfunction in the development of Alzheimer's disease: is Alzheimer's a vascular disorder? Am J Cardiovasc Dis. 2013;3(4):197–226.
60. de la Torre J. The vascular hypothesis of Alzheimer's disease: a key to preclinical prediction of dementia using neuroimaging. J Alzheimers Dis. 2018;63(1):35–52.
61. Salminen A, Kauppinen A, Kaarniranta K. Hypoxia/ischemia activate processing of amyloid precursor protein: impact of vascular dysfunction in the pathogenesis of Alzheimer's disease. J Neurochem. 2017;140(4):536–49.
62. Grammas P, Samany PG, Thirumangalakudi L. Thrombin and inflammatory proteins are elevated in Alzheimer's disease microvessels: implications for disease pathogenesis. J Alzheimers Dis. 2006;9(1):51–8.
63. Olsson AK, Dimberg A, Kreuger J, Claesson-Welsh L. VEGF receptor signalling—in control of vascular function. Nat Rev Mol Cell Biol. 2006;7(5):359–71.
64. Chakraborty A, de Wit NM, van der Flier WM, de Vries HE. The blood brain barrier in Alzheimer's disease. Vasc Pharmacol. 2017;89:12–8.
65. Skillback T, Delsing L, Synnergren J, Mattsson N, Janelidze S, Nagga K, et al. CSF/serum albumin ratio in dementias: a cross-sectional study on 1861 patients. Neurobiol Aging. 2017;59:1–9.
66. Olsson B, Lautner R, Andreasson U, Ohrfelt A, Portelius E, Bjerke M, et al. CSF and blood biomarkers for the diagnosis of Alzheimer's disease: a systematic review and meta-analysis. Lancet Neurol. 2016;15(7):673–84.
67. Halliday MR, Rege SV, Ma Q, Zhao Z, Miller CA, Winkler EA, et al. Accelerated pericyte degeneration and blood-brain barrier breakdown in apolipoprotein E4 carriers with Alzheimer's disease. J Cereb Blood Flow Metab. 2016;36(1):216–27.
68. Tomimoto H, Akiguchi I, Suenaga T, Nishimura M, Wakita H, Nakamura S, et al. Alterations of the blood-brain barrier and glial cells in white-matter lesions in cerebrovascular and Alzheimer's disease patients. Stroke. 1996;27(11):2069–74.
69. Feigin VL, Norrving B, Mensah GA. Global burden of stroke. Circ Res. 2017;120(3):439–48.

70. Wardlaw JM, Murray V, Berge E, del Zoppo G, Sandercock P, Lindley RL, et al. Recombinant tissue plasminogen activator for acute ischaemic stroke: an updated systematic review and meta-analysis. Lancet (London, England). 2012;379(9834):2364–72.
71. Berkhemer OA, Fransen PS, Beumer D, van den Berg LA, Lingsma HF, Yoo AJ, et al. A randomized trial of intraarterial treatment for acute ischemic stroke. N Engl J Med. 2015;372(1):11–20.
72. Ames A 3rd, Wright RL, Kowada M, Thurston JM, Majno G. Cerebral ischemia. II. The no-reflow phenomenon. Am J Pathol. 1968;52(2):437–53.
73. De Silva DA, Fink JN, Christensen S, Ebinger M, Bladin C, Levi CR, et al. Assessing reperfusion and recanalization as markers of clinical outcomes after intravenous thrombolysis in the echoplanar imaging thrombolytic evaluation trial (EPITHET). Stroke. 2009;40(8):2872–4.
74. Soares BP, Tong E, Hom J, Cheng SC, Bredno J, Boussel L, et al. Reperfusion is a more accurate predictor of follow-up infarct volume than recanalization: a proof of concept using CT in acute ischemic stroke patients. Stroke. 2010;41(1):e34–40.
75. Yemisci M, Gursoy-Ozdemir Y, Vural A, Can A, Topalkara K, Dalkara T. Pericyte contraction induced by oxidative-nitrative stress impairs capillary reflow despite successful opening of an occluded cerebral artery. Nat Med. 2009;15(9):1031–7.
76. Lee J, Gursoy-Ozdemir Y, Fu B, Boas DA, Dalkara T. Optical coherence tomography imaging of capillary reperfusion after ischemic stroke. Appl Opt. 2016;55(33):9526–31.
77. del Zoppo GJ, Schmid-Schonbein GW, Mori E, Copeland BR, Chang CM. Polymorphonuclear leukocytes occlude capillaries following middle cerebral artery occlusion and reperfusion in baboons. Stroke. 1991;22(10):1276–83.
78. Liu S, Connor J, Peterson S, Shuttleworth CW, Liu KJ. Direct visualization of trapped erythrocytes in rat brain after focal ischemia and reperfusion. J Cereb Blood Flow Metab. 2002;22(10):1222–30.
79. Garcia JH, Liu KF, Yoshida Y, Chen S, Lian J. Brain microvessels: factors altering their patency after the occlusion of a middle cerebral artery (Wistar rat). Am J Pathol. 1994;145(3):728–40.
80. Vates GE, Takano T, Zlokovic B, Nedergaard M. Pericyte constriction after stroke: the jury is still out. Nat Med. 2010;16(9):959. author reply 60
81. Damisah EC, Hill RA, Tong L, Murray KN, Grutzendler J. A fluoro-Nissl dye identifies pericytes as distinct vascular mural cells during in vivo brain imaging. Nat Neurosci. 2017;20(7):1023–32.
82. Ng FC, Coulton B, Chambers B, Thijs V. Persistently elevated microvascular resistance Postrecanalization. Stroke. 2018;49(10):2512–5.
83. Goldberg I, Auriel E, Russell D, Korczyn AD. Microembolism, silent brain infarcts and dementia. J Neurol Sci. 2012;322(1–2):250–3.
84. Vermeer SE, Longstreth WT Jr, Koudstaal PJ. Silent brain infarcts: a systematic review. Lancet Neurol. 2007;6(7):611–9.
85. Wang Z, van Veluw SJ, Wong A, Liu W, Shi L, Yang J, et al. Risk factors and cognitive relevance of cortical cerebral microinfarcts in patients with ischemic stroke or transient ischemic attack. Stroke. 2016;47(10):2450–5.
86. Fisher MJ. Brain regulation of thrombosis and hemostasis: from theory to practice. Stroke. 2013;44(11):3275–85.
87. Lam CK, Yoo T, Hiner B, Liu Z, Grutzendler J. Embolus extravasation is an alternative mechanism for cerebral microvascular recanalization. Nature. 2010;465(7297):478–82.
88. Grutzendler J, Murikinati S, Hiner B, Ji L, Lam CK, Yoo T, et al. Angiophagy prevents early embolus washout but recanalizes microvessels through embolus extravasation. Sci Transl Med. 2014;6(226):226ra31.
89. Reeson P, Choi K, Brown CE. VEGF signaling regulates the fate of obstructed capillaries in mouse cortex. elife. 2018;7:e33670.
90. van der Wijk A-E, Lachkar N, de Vos J, Grootemaat AE, van der Wel NN, Hordijk PL, Bakker ENTP, vanBavel E. Extravasation of microspheres in a rat model of silent brain infarcts. Stroke. 2019;50(6):1590–4.

6 Role of Inflammation in Microvascular Damage

Carmine Savoia

6.1 Introduction

Essential hypertension is characterized by increased peripheral vascular resistance to blood flow, due in large part to functional and structural alterations of small resistance arteries, which may play an important role in the development of hypertension [1] and may contribute to complications induced by hypertension [2].

Essential hypertension is characterized by increased peripheral vascular resistance to blood flow, which occurs mostly as a result of energy dissipation in small resistance arteries, particularly in younger individuals. Enhanced constriction of resistance arteries may increase peripheral resistance in hypertension by reducing lumen diameter [3].

In hypertension, resistance arteries undergo vascular remodeling (reduced lumen with increased media width) that may be structural, mechanical, or functional. Small artery remodeling may be the first manifestation of target organ damage in hypertension, since all stage I hypertensive patients present small artery remodeling, whereas almost half of them have endothelial dysfunction and left ventricular hypertrophy [4]. Small artery remodeling has demonstrated prognostic significance [5] since hypertensive patients with the highest media-to-lumen ratio had increased incidence of cardiovascular events.

Chronic vasoconstriction of the arteries may contribute to the remodeling of extracellular matrix which further enhances the vasoconstricted state that might become permanent, and therefore the vessels may not return to their vasodilated state [6]. This is mainly due to the influence of myogenic tone, the renin-angiotensin-aldosterone system (RAAS), endothelins, catecholamines, and growth factors which lead to reduced vascular lumen with increased media thickness. Furthermore,

C. Savoia (✉)
Clinical and Molecular Medicine Department, Cardiology Unit, Sant'Andrea Hospital, Sapienza University of Rome, Rome, Italy

E. Agabiti-Rosei et al. (eds.), *Microcirculation in Cardiovascular Diseases*, Updates in Hypertension and Cardiovascular Protection,
https://doi.org/10.1007/978-3-030-47801-8_6

endothelial dysfunction may participate to the increased vascular tone in hypertension [7], with reduced vasodilation associated with a proinflammatory and prothrombotic state.

Low-grade inflammation in the vascular wall is an important contributor to remodeling of resistance arteries and endothelial dysfunction as well as to the pathophysiology of hypertension [8] and hypertension-related vascular damage and the development of cardiovascular disease (CVD) [9]. Extracellular matrix deposition and inflammation are critically involved in vascular remodeling. Inflammation contributes to vascular remodeling by promoting cell growth and proliferation of vascular smooth muscle cells (VSMCs). This is supported by the increased expression in the vascular wall of adhesion molecules and ligands, leukocyte extravasation, increased oxidative stress, cytokine production, activation of immune cells, and proinflammatory signaling pathways. Greater expression of adhesion molecules (VCAM-1, ICAM-1) on the endothelial cell membrane, accumulation of monocyte/macrophages, dendritic cells, natural killer cells, and B and T lymphocytes participate to the inflammatory response in the vascular wall [10, 11].

In this chapter, we will focus on the mechanisms of inflammation described in vascular remodeling of resistance arteries.

6.2 Endothelial Dysfunction and Vascular Remodeling

Endothelial dysfunction is one of the early determinants in the development of hypertension and in the progression to atherosclerosis, and it is independently associated with increased cardiovascular risk [12]. Dysfunction of the endothelium is often associated with elevation of blood pressure as well as with cardiovascular risk factors related to hypertension, including smoking, obesity, dyslipidemia, insulin resistance, metabolic syndrome, and diabetes mellitus. Endothelial dysfunction may affect different functions of the endothelium and is characterized by an inflammatory phenotype of endothelial cells with altered morphology, increased proliferation, production of inflammatory mediators (including C-reactive protein, monocyte chemotactic protein 1 and plasminogen activator inhibitor 1, upregulated adhesion molecules), and enhanced thrombogenicity [7, 13, 14]. Thus, endothelial dysfunction may be involved in the initiation of vascular inflammation, in the development of vascular remodeling, and in the progression to atherosclerosis by increasing lipoprotein oxidation, smooth muscle cell proliferation, extracellular matrix deposition, cell adhesion, and thrombus formation [15–17], and it is independently associated with increased cardiovascular risk [7, 9, 12]. In particular, endothelial dysfunction promotes vascular inflammation by inducing the production of vasoconstrictor agents, adhesion molecules, and growth factors including angiotensin II and endothelin-1 [15, 16]. In turn, angiotensin II, one of the final products of the RAAS, is actively involved in the pathophysiology of the vascular alterations described in hypertension [18]. It may be responsible for triggering endothelial dysfunction and vascular inflammation by inducing oxidative stress, resulting in upregulation of inflammatory mediators and cell growth. Furthermore, endothelial dysfunction is catheterized by reduced endothelium-dependent vasodilation [19] and may participate to the increased vascular tone in hypertension [8].

Chronically elevated blood pressure and stretch-induced complex signal transduction cascades lead to vascular remodeling that contributes not only to elevation of blood pressure but also to hypertensive complications [20]. Increased oxidative stress, inflammation, and extracellular matrix deposition are critically involved in the pathophysiology of vascular remodeling. With chronic vasoconstriction due to endothelial dysfunction as well as increased myogenic tone, RAAS and endothelin activation, catecholamines, and growth factor production, vessels may become embedded in the remodeled extracellular matrix and may not return to their vasodilated state [6]. Thus, in hypertension, resistance arteries undergo vascular remodeling (reduced lumen with increased media thickness) that may be structural, mechanical, or functional. Two types of remodeling have been described, inward eutrophic or inward hypertrophic remodeling which is characterized by the enlargement of the media cross-sectional area (true hypertrophy) [3, 21–23]. Eutrophic remodeling is usually found in essential hypertension in humans and spontaneously hypertensive rats (SHR). Even mildly activation of the renin-angiotensin system largely contributes to this type of vascular remodeling.

Hypertrophic remodeling has been described in secondary hypertension (in renovascular hypertension, primary aldosteronism, or pheochromocytoma) [24]. Moreover also in hypertension associated with diabetes mellitus [25, 26] and in acromegaly [27], hypertrophic remodeling has been described. Hypertrophic remodeling is associated to activation of the endothelin system. Indeed in salt-dependent hypertension in rodents [28–30], diabetes mellitus, and malignant hypertension, endothelin system is activated [31]. Hyperplasia of VSMC is found in small arteries of hypertensive rats, whereas in the aorta, VSMC hypertrophy has been reported [32, 33].

The mechanisms leading to inward eutrophic remodeling are poorly understood. In eutrophic remodeling in essential hypertension, cell volume and number in small arteries are similar to those of normotensive subjects [34]. Several experimental studies have shown deposition of collagen and fibronectin with increased collagen-to-elastin ratio in small vessels from hypertensive humans and rodents [35, 36] which may be induced by endothelins [37], angiotensin II, and aldosterone [38] acting in an endocrine or paracrine fashion. As a result, the vessels undergo to rearrangement of cellular and fibrillar components attributed to changes in interaction of these structures through integrins [39] and to inward growth associated with peripheral apoptosis or from vasoconstriction embedded in increased extracellular matrix [6].

6.3 Inflammation and Vascular Remodeling

Low-grade inflammation of both vascular wall and perivascular fat participates in arterial remodeling [40], and it is an important contributor to the pathophysiology of hypertension [8], atherosclerosis, and the development of CVD [9, 14]. This process is characterized by inflammatory cell infiltration, upregulation of inflammatory mediators (vascular cellular adhesion molecule 1, intercellular adhesion molecule 1, nuclear factor (NF)-KB, monocyte chemoattractant factor 1, and plasminogen activator inhibitor 1), and enhanced oxidative stress (increased production of reactive oxygen species, ROS). The main source of ROS in the vascular wall is the NADPH (nicotinamide adenine dinucleotide phosphate) oxidase which is expressed in endothelial cells, VSMC, fibroblasts, and monocytes/macrophages [41, 42].

Blood pressure itself or RAAS activation [11, 43] may induce oxidative stress and the inflammatory process, which participates to vascular remodeling and may contribute to accelerated vascular damage in aging and CVD. Inflammation contributes to vascular remodeling promoting cell growth and proliferation of VSMC, as well as increasing collagen deposition and rearrangement of extracellular matrix in the vascular wall.

Experimental studies have shown increased oxidative stress and production of proinflammatory cytokines, adhesion molecules, and ligands as well as proinflammatory signaling pathways in the vascular wall of resistance arteries in different experimental models of hypertension. In hypertensive conditions, it has been shown a greater expression of adhesion molecules (VCAM-1, ICAM-1) on the endothelial cell membrane, accumulation of monocyte/macrophages, dendritic cells, natural killer cells, and B and T lymphocytes participate to the inflammatory response in the vascular wall [44].

Innate immunity may play also a role in the mechanisms that contribute to the low-grade inflammatory response in hypertension. In particular, T lymphocytes may participate in hypertension and peripheral inflammation possibly by the increased production of oxidative stress [45]. The central and pressor effects of angiotensin II are critical for T-cell activation and the development of vascular inflammation [46]. Interestingly, in mice deficient in vascular macrophages as well as in T and B lymphocytes, angiotensin II and deoxycorticosterone acetate-(DOCA-) salt were not able to induce hypertension or vascular remodeling [47, 48]. The lack of response to angiotensin II in T- and B-deficient mice was corrected by effector T cell but not by B lymphocyte adoptive transfer.

Furthermore, also Treg adoptive transfer may lower blood pressure and protect from vascular remodeling in mice infused with either angiotensin II or aldosterone, as well as ameliorated cardiac damage and improved eutrophic remodeling in angiotensin II-infused mice, independently of blood-pressure-lowering effects [49, 50]. It has been showed that adoptive immunity may be enhanced as a result of a genetic predisposition with loci on chromosome 2 (which carries many proinflammatory genes) [51]. The presence of normotensive chromosome 2 was associated with upregulation of Treg markers (CD8+ and CD4+, CD25+) in a consomic strain of rats, which contains the genetic background of hypertensive Dahl salt-sensitive rats and chromosome 2 from Brown-Norway normotensive rats. Hence, different subsets of T lymphocytes may be involved in the mechanisms leading to the inflammatory response described in cardiac and metabolic diseases when an imbalance exists between the proinflammatory Th1, Th2, and Th17 and the anti-inflammatory Treg subsets [49, 50].

6.4 Role of Renin-Angiotensin-Aldosterone System

Angiotensin II plays a key role in the pathophysiology of hypertension [11, 13] and CVD. Several mechanisms may be activated by angiotensin II such as endothelial dysfunction, vasoconstriction, cell growth, oxidative stress, inflammation, and vascular remodeling. Increased angiotensin II-induced ROS generation is involved in the process of vascular remodeling. This occurs by VSMC proliferation and hypertrophy and collagen deposition and by modulating cytokine release and proinflammatory transcription factors in small resistance arteries of patients with essential hypertension and hypertensive rats. Angiotensin II may enhance basal superoxide production in the vascular wall by the activation of reduced NADPH oxidase and expression of its subunits via cSrc, PKC (protein kinase C), PLA2 (phospholipase A2), and PLD (phospholipase D) pathways [13, 52–55].

ROS act as signaling molecules, modulating vascular tone and structural changes in the vasculature [53], and participate in the development and progression of atherosclerosis [56] by activating multiple signaling molecules. Among those are activated mitogen-activated protein kinases (MAP kinases); nonreceptor tyrosine kinases such as Src, JAK-2 (Janus kinase 2), STAT (signal transducer and activator of transcription), p21Ras, Pyk-2 (proline-rich tyrosine kinase 2), and Akt; receptor tyrosine kinases (EGFR (epidermal growth factor receptor), IGFR (insulin-like growth factor receptor 1), and PDGFR (platelet-derived growth factor receptor); and protein tyrosine phosphatases and redox-sensitive transcription factors (NF-κB, activator protein 1 (AP)-1, and hypoxia-inducible factor 1 (HIF-1)) [42, 56].

Ang II-induced ROS production increases ICAM-1 expression, macrophage infiltration, monocyte chemotactic protein (MCP)-1 production, and vascular hypertrophy, independently of blood pressure elevation [57]. Macrophages which infiltrate the adventitia or the media of blood vessels may generate oxidative stress by NADPH oxidase [57, 58]. Ang II-induced growth and profibrotic effects are partially modulated by endogenous production of mitogenic factors (including TGF-β, PDGF (platelet-derived growth factor), EGF (epidermal growth factor), IGF-1 (insulin-like growth factor 1), and endothelin-1) [59, 60]. In particular TGF-β is produced by macrophages, lymphocytes, fibroblasts, and VSMCs and is overexpressed in many cardiovascular and renal disorders associated with activation of the RAAS. TGF-β increases extracellular matrix biosynthesis, downregulates matrix degradative enzymes, and influences integrin receptors [61, 62]. p38MAP kinase and connective tissue growth factor are among the major downstream profibrogenic mediators of TGF-β. Inhibition of the RAAS with ACE inhibitors or ARBs is directly correlated with suppression of TGF-β production and amelioration of fibrosis [61, 63].

Increased ROS production reduces also the vascular bioavailability of NO [10, 64] which is associated with impaired endothelium-dependent vascular relaxation and increased vascular contractile responses. Moreover, this is also associated to structural changes in the wall of small arteries and to the increased peripheral vascular resistance [65].

Interestingly, inflammation may activate RAAS, which in turn may further contribute to vascular remodeling in hypertension.

Angiotensin II may modulate also innate and adaptive immune responses in part through the autonomic nervous system. Angiotensin II administration into the lateral cerebral ventricles was shown to increase mRNA expression of proinflammatory splenic cytokines (IL-1b and IL-6) [66]. These responses were abrogated by splenic sympathetic denervation.

Similar effects to those elicited by angiotensin II in the vasculature may result from mineralocorticoid receptor activation by aldosterone. Furthermore an interaction between the two systems has been described.

Ang II induces aldosterone synthesis which in turn increases tissue ACE activity [67] and upregulates angiotensin receptors [68]. The activation of mineralocorticoid receptors may contribute to cardiovascular dysfunction, inflammation, fibrosis, and vascular damage. Several experimental studies have confirmed that aldosterone may induce ROS formation and endothelial dysfunction, and therefore it can cause injury of the vasculature of the brain, heart, and kidneys [69]. Endothelial dysfunction and inflammation induced by aldosterone may involve the activation of COX-2 (cyclooxygenase-2) in normotensive and hypertensive rats [70]. Mineralocorticoid antagonism attenuates the aldosterone-induced damage by reducing the direct proinflammatory and profibrotic effects of aldosterone that may be mediated via activation of the endothelin system [71–74]. Mineralocorticoid receptor blockade may also improve endothelial function and reduce oxidative stress in angiotensin II-infused rats [75]. This suggests that aldosterone may induce in part actions usually attributed to direct effects of angiotensin II in the vasculature.

An interaction between the actions induced by angiotensin II and aldosterone has been described. In vitro and in vivo studies showed that nuclear translocation of NF-KB requires AT1aR, AT1bR, and mineralocorticoid receptors in response to aldosterone or angiotensin II, whereas aldosterone-induced c-fos gene transcription requires AT1aR. Moreover, deficiency in AT1aR prevented aldosterone-induced mesenteric resistance artery hypertrophic remodeling and endothelial dysfunction in mice [76]. Moreover, angiotensin II-induced inflammation via NF-κ B and AP-1 activation involves in part endothelin receptors [77].

ROS are potent stimulators of endothelin-1 synthesis by endothelial cells and VSMC [78]. Endothelin-1 in turn activates NADPH oxidase as well as other sources of ROS, such as xanthine oxidase and mitochondria, in VSMCs and blood vessels [79–81]. Endothelin-1-induced oxidative stress promotes inflammatory responses which contribute to the vascular remodeling and endothelial dysfunction in animal models of hypertension that present an endothelin-mediated component [82].

Severe elevation of blood pressure seems to trigger the expression of endothelin-1 in the endothelium, and, therefore, patients with stage 2 primary hypertension exhibited increased expression of prepro-endothelin-1 mRNA in the endothelium of small arteries from biopsies of gluteal subcutaneous tissue [83].

In mice overexpressing human prepro-endothelin-1 in the endothelium, small arteries exhibited hypertrophic remodeling, increased stiffness, and endothelial dysfunction independently of blood pressure modification [81]. Increased activity of NADPH oxidase with enhanced NADPH oxidase-derived ROS and upregulation of inflammatory mediators [77, 84] were also found. Exposed to a high-salt diet, those mice presented a greater increase in blood pressure compared with wild-type mice [85]. High-salt diet further deteriorated endothelial function and oxidative stress. Endothelin A receptor antagonism partially improved the effects of high-salt diet, whereas endothelin B receptor antagonism worsened them. Moreover, it has been showed that endothelin A receptor antagonism decreases oxidative stress, normalizes hypertrophic remodeling, decreases collagen and fibronectin deposition, and reduces adhesion molecules in the vasculature of aldosterone-infused rats [71]. Also in humans, exogenous endothelin-1 has been shown to increase arterial stiffness [86, 87].

In salt-sensitive hypertensive models (deoxycorticosterone acetate-salt rats), enhanced endothelin-1 expression in the endothelium [88] is associated with hypertrophic remodeling and endothelial dysfunction that could be corrected by endothelin antagonists [28]. Thus hypertrophic remodeling is a feature of endothelin involvement in the vascular alterations described in hypertension [30]. Aldosterone infusion also increased endothelin-1 expression and induced hypertrophic vascular remodeling with extracellular matrix deposition of collagen types I and III and fibronectin [37, 89].

6.5 Conclusions

Functional and structural alterations of resistance arteries are the earliest vascular alterations that may occur in hypertension. Low-grade inflammation participates in the development and pathogenesis of the alterations of microvasculature in hypertensive subjects. RAAS activation as well as endothelin system activation plays a key role in the pathogenesis of oxidative stress increase and inflammation in the vascular wall, leading to endothelial dysfunction and vascular remodeling. In particular angiotensin II activates redox-sensitive pathways promoting cell growth and inflammation. Selective antihypertensive agents might partially correct the vascular remodeling and impaired endothelial function of small resistance arteries. In particular, it has been shown that antihypertensive drugs that antagonize RAAS, including ACE inhibitors, ARBs, and mineralocorticoid receptor antagonists, embody valid therapeutic tools to improve vascular function and reduce vascular remodeling and cardiovascular risk, as assessed in several clinical studies in hypertensive patients.

References

1. Schiffrin EL. Reactivity of small blood vessels in hypertension: relation with structural changes: state of the art lecture. Hypertension. 1992;19(2):II1–9.
2. Schiffrin EL. Resistance arteries as endpoints in hypertension. Blood Press. 1997;6(2):S24–30.
3. Schiffrin EL. Remodeling of resistance arteries in essential hypertension and effects of antihypertensive treatment. Am J Hypertens. 2004;17(12):1192–200.

4. Park JB, Schiffrin EL. Small artery remodeling is the most prevalent (earliest?) form of target organ damage in mild essential hypertension. J Hypertens. 2001;19:921–30.
5. Rizzoni D, Porteri E, Boari GE, et al. Prognostic significance of small-artery structure in hypertension. Circulation. 2003;108:2230–5.
6. Bakker ENTP, van der Meulen ET, van den Berg BM, et al. Inward remodeling follows chronic vasoconstriction in isolated resistance arteries. J Vasc Res. 2002;39:12–20.
7. Savoia C, Sada L, Zezza L, et al. Vascular inflammation and endothelial dysfunction in experimental hypertension. Int J Hypertens. 2011;2011:281240.
8. Savoia C, Battistoni A, Calvez V, et al. Microvascular Alterations in Hypertension and Vascular Aging. Curr Hypertens Rev. 2017;13(1):16–23.
9. Savoia C, Schiffrin EL. Inflammation in hypertension. Curr Opin Nephrol Hypertens. 2006;15(2):152–8.
10. Touyz RM, Schiffrin EL. Signal transduction mechanisms mediating the physiological and pathophysiological actions of angiotensin II in vascular smooth muscle cells. Pharmacol Rev. 2000;52:639–72.
11. Savoia C, Burger D, Nishigaki N, et al. Angiotensin II and the vascular phenotype in hypertension. Expert Rev Mol Med. 2011;13:e11.
12. Lerman A, Zeiher AM. Endothelial function: cardiac events. Circulation. 2005;111(3):363–8.
13. Savoia C, Schiffrin EL. Inhibition of the renin angiotensin system: implications for the endothelium. Curr Diab Rep. 2006;6(4):274–8.
14. Savoia C, Schiffrin EL. Reduction of C-reactive protein and the use of anti-hypertensives. Vasc Health Risk Manag. 2007;3(6):975–83.
15. Durier S, Fassot C, Laurent S, et al. Physiological genomics of human arteries: quantitative relationship between gene expression and arterial stiffness. Circulation. 2003;108(15):1845–51.
16. Yu Y, Fukuda N, Yao EH, et al. Effects of an ARB on endothelial progenitor cell function and cardiovascular oxidation in hypertension. Am J Hypertens. 2008;21:72–7.
17. Ross R. Atherosclerosis-an inflammatory disease. New Engl J Med. 1999;340:115–26.
18. Kranzhofer R, Schmidt J, Pfeiffer CAH, et al. Angiotensin induces inflammatory activation of human vascular smooth muscle cells. Arterioscler Thromb Vasc Biol. 1999;19(7):1623–9.
19. Deng LY, Li JS, Schiffrin EL. Endothelium-dependent relaxation of small arteries from essential hypertensive patients: mechanisms and comparison with normotensive subjects and with responses of vessels from spontaneously hypertensive rats. Clin Sci. 1995;88:611–22.
20. Schiffrin EL, Touyz RM. From bedside to bench to bedside: role of renin-angiotensin-aldosterone system in remodeling of resistance arteries in hypertension. Am J Phys. 2004;287(2):H435–46.
21. Heagerty AM, Aalkjaer C, Bund SJ, et al. Small artery structure in hypertension: dual processes of remodeling and growth. Hypertension. 1993;21:391–7.
22. Schiffrin EL, Deng LY, Larochelle P. Morphology of resistance arteries and comparison of effects of vasoconstrictors in mild essential hypertensive patients. Clin Invest Med. 1993;16:177–86.
23. Mulvany MJ, Baumbach GL, Aalkjaer C, et al. Vascular remodeling. Hypertension. 1996;28:505–6.
24. Rizzoni D, Porteri E, Castellano M, et al. Vascular hypertrophy and remodeling in secondary hypertension. Hypertension. 1996;28:785–90.
25. Rizzoni D, Porteri E, Guelfi D, et al. Structural alterations in subcutaneous small arteries of normotensive and hypertensive patients with non-insulin-dependent diabetes mellitus. Circulation. 2001;103:1238–44.
26. Endemann D, Pu Q, De Ciuceis C, et al. Persistent remodeling of resistance arteries in type 2 diabetic patients on antihypertensive treatment. Hypertension. 2004;43:399–404.
27. Rizzoni D, Porteri E, Giustina A, et al. Acromegalic patients show the presence of hypertrophic remodeling of subcutaneous small resistance arteries. Hypertension. 2004;43:561–5.
28. Li JS, Lariviere R, Schiffrin EL. Effect of a nonselective endothelin antagonist on vascular remodeling in deoxycorticosterone acetate-salt hypertensive rats: evidence for a role of endothelin in vascular hypertrophy. Hypertension. 1994;24:183–8.

29. Schiffrin EL, Lariviere R, Li JS, Sventek P. Enhanced expression of the endothelin-1 gene in blood vessels of DOCA-salt hypertensive rats: correlation with vascular structure. J Vasc Res. 1996;33:235–48.
30. D'Uscio LV, Barton M, Shaw S, et al. Structure and function of small arteries in salt-induced hypertension: effects of chronic endothelin-subtype-A-receptor blockade. Hypertension. 1997;30:905–11.
31. Schiffrin EL, Touyz RM. From bedside to bench to bedside: role of renin-angiotensin-aldosterone system in remodeling of resistance arteries in hypertension. Am J Physiol Heart Circ Physiol. 2004;287:H435–46.
32. Lee RMKW, Garfield RE, Forrest JB, Daniel EE. Morphometric study of structural changes in the mesenteric blood vessels of spontaneously hypertensive rats. Blood Vessels. 1983;20:57–71.
33. Mulvany MJ, Baandrup U, Gundersen HJ. Evidence for hyperplasia in mesenteric resistance vessels of spontaneously hypertensive rats using a three-dimensional dissector. Circ Res. 1985;57:794–800.
34. Korsgaard N, Aalkjær C, Heagerty AM, et al. Histology of subcutaneous small arteries from patients with essential hypertension. Hypertension. 1993;22:523–6.
35. Intengan HD, Deng LY, Li JS, Schiffrin EL. Mechanics and composition of human subcutaneous resistance arteries in essential hypertension. Hypertension. 1999;33:569–74.
36. Intengan HD, Thibault G, Li JS, Schiffrin EL. Resistance artery mechanics, structure, and extracellular components in spontaneously hypertensive rats: effects of angiotensin receptor antagonism and converting enzyme inhibition. Circulation. 1999;100:2267–75.
37. Pu Q, Neves MF, Virdis A, et al. Endothelin antagonism on aldosterone-induced oxidative stress and vascular remodeling. Hypertension. 2003;42:49–55.
38. Neves MF, Virdis A, Schiffrin EL. Resistance artery mechanics and composition in angiotensin II-infused rats: effects of aldosterone antagonism. J Hypertens. 2003;21:189–98.
39. Brassard P, Amiri F, Thibault G, Schiffrin EL. Role of angiotensin type-1 and angiotensin type 2 receptors in the expression of vascular integrins in angiotensin II-infused rats. Hypertension. 2006;47:122–7.
40. Intengan HD, Schiffrin EL. Vascular remodeling in hypertension: roles of apoptosis, inflammation, and fibrosis. Hypertension. 2001;38:581–7.
41. Griendling KK, Sorescu D, Lassegue B, Ushio-Fukai M. Modulation of protein kinase activity and gene expression by reactive oxygen species and their role in vascular physiology and pathophysiology. Arterioscler Thromb Vasc Biol. 2000;20(10):2175–83.
42. Touyz RM, Chen X, Tabet F, et al. Expression of a functionally active gp91phox-containing neutrophil-type NAD(P)H oxidase in smooth muscle cells from human resistance arteries: regulation by angiotensin II. Circ Res. 2002;90:1205–13.
43. Blake GJ, Ridker PM. Novel clinical markers of vascular wall inflammation. Circ Res. 2001;89(9):763–71.
44. Wung BS, Cheng JJ, Chao YJ, et al. Cyclical strain increases monocyte chemotactic protein-1 secretion in human endothelial cells. Am J Phys. 1996;270(4):H1462–8.
45. Lob HE, Marvar PJ, Guzik TJ, et al. Induction of hypertension and peripheral inflammation by reduction of extracellular superoxide dismutase in the central nervous system. Hypertension. 2010;55(2):277–83.
46. Marvar PJ, Thabet SR, Guzik TJ, et al. Central and peripheral mechanisms of T-lymphocyte activation and vascular inflammation produced by angiotensin II-induced hypertension. Circ Res. 2010;107(2):263–70.
47. Guzik TJ, Hoch NE, Brown KA, et al. Role of the T cell in the genesis of angiotensin II-induced hypertension and vascular dysfunction. J Exp Med. 2007;204(10):2449–60.
48. De Ciuceis C, Amiri F, Brassard P, et al. Reduced vascular remodeling, endothelial dysfunction, and oxidative stress in resistance arteries of angiotensin II-infused macrophage colony-stimulating factor-deficient mice: evidence for a role in inflammation in angiotensin-induced vascular injury. Arterioscler Thromb Vasc Biol. 2005;25(10):2106–13.
49. Harrison DG, Vinh A, Lob H, Meena S. Madhur Role of the adaptive immune system in hypertension. Curr Opin Pharmacol. 2010;10(2):203–7.

50. Idris-Khodja N, Mian MOR, Paradis P, Schiffrin EL. Dual opposing roles of adaptive immunity in hypertension. Eur Heart J. 2014;35:1238–44.
51. Viel EC, Lemarie CA, Benkirane K, et al. Immune regulation and vascular inflammation in genetic hypertension. Am J Phys. 2010;298(3):H938–44.
52. Lassegue B, Clepmpus RE. NAD(P)H oxidase: specific features, expression and regulation. Am J Physiol. 2003;285:R277–97.
53. Touyz RM, Schiffrin EL. Reactive oxygen species in vascular biology: implications in hypertension. Histochem Cell Biol. 2004;122(4):339–52.
54. Touyz RM, Yao G, Schiffrin EL. C-Src induces phosphorylation and translocation of p47phox: role in superoxide generation by angiotensin II in human vascular smooth muscle cells. Arterioscler Thromb Vasc Biol. 2003;23(6):981–7.
55. Fukui T, Ishizaka N, Rajagopalan S, et al. p22phox mRNA expression and NADPH oxidase activity are increased in aortas from hypertensive rats. Circ Res. 1997;80(1):45–51.
56. Kalinina N, Agrotis A, Tararak E, et al. Cytochrome b558- dependent NAD(P)H oxidase-phox units in smooth muscle and macrophages of atherosclerotic lesions. Arterioscler Thromb Vasc Biol. 2002;22(12):2037–43.
57. Touyz RM, Tabet F, Schiffrin EL. Redox-dependent signalling by angiotensin II and vascular remodelling in hypertension. Clin Exp Pharmacol Physiol. 2003;30(11):860–6.
58. Schiffrin EL, Touyz RM. Inflammation and vascular hypertrophy induced by angiotensin II: role of NADPH oxidase-derived reactive oxygen species independently of blood pressure elevation? Arterioscler Thromb Vasc Biol. 2003;23(5):707–9.
59. Sarkar S, Vellaichamy E, Young D, Sen S. Influence of cytokines and growth factors in ANG II-mediated collagen upregulation by fibroblasts in rats: role of myocytes. Am J Phys. 2004;287(1):H107–17.
60. Satoh C, Fukuda N, Hu WY, et al. Role of endogenous angiotensin II in the increased expression of growth factors in vascular smooth muscle cells from spontaneously hypertensive rats. J Cardiovasc Pharmacol. 2001;37(1):108–18.
61. Azhar M, Schultz JEJ, Grupp I, et al. Transforming growth factor beta in cardiovascular development and function. Cytokine Growth Factor Rev. 2003;14(5):391–407.
62. Bobik A. Hypertension, transforming growth factor-β, angiotensin II and kidney disease. J Hypertens. 2004;22(7):1265–7.
63. Shin GT, Kim SJ, Ma KA, Kim SH, Kim D. ACE inhibitors attenuate expression of renal transforming growth factor-β1 in humans. Am J Kid Dis. 2000;36(5):894–902.
64. Rajagopalan S, Kurz S, Munzel T, et al. Angiotensin II-mediated hypertension in the rat increases vascular superoxide production via membrane NADH/NADPH oxidase activation: contribution to alterations of vasomotor tone. J Clin Invest. 1996;97(8):1916–23.
65. Jung O, Marklund SL, Geiger H, et al. Extracellular superoxide dismutase is a major determinant of nitric oxide bioavailability: in vivo and ex vivo evidence from ecSOD-deficient mice. Circ Res. 2003;93(7):622–9.
66. Ganta CK, Lu N, Helwig BG, et al. Central angiotensin II-enhanced splenic cytokine gene expression is mediated by the sympathetic nervous system. Am J Physiol Heart Circ Physiol. 2005;289:H1683–91.
67. Harada E, Yoshimura M, Yasue H, et al. Aldosterone induces angiotensin-converting-enzyme gene expression in cultured neonatal rat cardiocytes. Circulation. 2001;104(2):137–9.
68. Schiffrin EL, Gutkowska J, Genest J. Effect of angiotensin II and deoxycorticosterone infusion on vascular angiotensin II receptors in rats. Am J Phys. 1984;15(4):H614.
69. Joffe HV, Adler GK. Effect of aldosterone and mineralocorticoid receptor blockade on vascular inflammation. Heart Fail Rev. 2005;10(1):31–7.
70. Blanco-Rivero J, Cachofeiro V, Lahera V, et al. Participation of prostacyclin in endothelial dysfunction induced by aldosterone in normotensive and hypertensive rats. Hypertension. 2005;46(1):107–12.
71. Pu Q, Neves MF, Virdis A, Touyz RM, Schiffrin EL. Endothelin antagonism on aldosterone-induced oxidative stress and vascular remodeling. Hypertension. 2003;42:49–55.

72. Rocha R, Chander PN, Khanna K, Zuckerman A, Stier CT. Mineralocorticoid blockade reduces vascular injury in stroke-prone hypertensive rats. Hypertension. 1998;31(1):451–8.
73. Fiebeler A, Schmidt F, Muller DN, et al. Mineralocorticoid receptor affects AP-1 and nuclear factor-κB activation in angiotensin II-induced cardiac injury. Hypertension. 2001;37(2):787–93.
74. Ammarguellat FZ, Gannon PO, Amiri F, Schiffrin EL. Fibrosis, matrix metalloproteinases, and inflammation in the heart of DOCA-salt hypertensive rats: role of ETa receptors. Hypertension. 2002;39(2):679–84.
75. Virdis A, Neves MF, Amiri F, Viel E, et al. Spironolactone improves angiotensin-induced vascular changes and oxidative stress. Hypertension. 2002;40(4):504–10.
76. Briet M, Barhoumi T, Sierra C, et al. Aldosterone-induced small artery remodeling requires functional angiotensin type 1a receptors. Hypertension. 2011;58:e102.
77. Muller DN, Mervaala EMA, Schmidt F, et al. Effect of bosentan on nf-κb, inflammation, and tissue factor in angiotensin II-induced end-organ damage. Hypertension. 2000;36(2):282–90.
78. Kahler J, Ewert A, Weckmuller J, et al. Oxidative stress increases endothelin-1 synthesis in human coronary artery smooth muscle cells. J Cardiovasc Pharmacol. 2001;38(1):49–57.
79. Li L, Fink GD, Watts SW, et al. Endothelin-1 increases vascular superoxide via endothelinA-NADPH oxidase pathway in low-renin hypertension. Circulation. 2003;107(7):1053–8.
80. Iglarz M, Touyz RM, Amiri F, et al. Effect of peroxisome proliferator-activated receptor-α and -γ activators on vascular remodeling in endothelin-dependent hypertension. Arterioscler Thromb Vasc Biol. 2003;23(1):45–51.
81. Amiri F, Virdis A, Neves MF, et al. Endothelium-restricted overexpression of human endothelin-1 causes vascular remodeling and endothelial dysfunction. Circulation. 2004;110:2233–40.
82. Touyz RM, Yao G, Viel E, Amiri F, Schiffrin EL. Angiotensin II and endothelin-1 regulate MAP kinases through different redox-dependent mechanisms in human vascular smooth muscle cells. J Hypertens. 2004;22(6):1141–9.
83. Schiffrin EL, Deng LY, Sventek P, Day R. Enhanced expression of endothelin-1 gene in resistance arteries in severe human essential hypertension. J Hypertens. 1997;15:57–63.
84. Amiri F, Paradis P, Reudelhuber TL, Schiffrin EL. Vascular inflammation in absence of blood pressure elevation in transgenic murine model overexpressing endothelin-1 in endothelial cells. J Hypertens. 2008;26:1102–9.
85. Amiri F, Ko EA, Javeshghani D, et al. Deleterious combined effects of aging, salt-loading and endothelial cell restricted endothelin-1 overexpression on blood pressure and vascular function. J Hypertens. 2010;28:1243–51.
86. Vuurmans TJL, Boer P, Koomans HA. Effects of endothelin-1 and endothelin-1 receptor blockade on cardiac output, aortic pressure, and pulse wave velocity in humans. Hypertension. 2003;41(6):1253–8.
87. McEniery CM, Qasem A, Schmitt M, et al. Endothelin-1 regulates arterial pulse wave velocity in vivo. J Am Coll Cardiol. 2003;42(11):1975–81.
88. Lariviere R, Thibault G, Schiffrin EL. Increased endothelin-1 content in blood vessels of deoxycorticosterone acetate-salt hypertensive but not in spontaneously hypertensive rats. Hypertension. 1993;21:294–300.
89. Stow LR, Gumz ML, Lynch IJ, et al. Aldosterone modulates steroid receptor binding to the endothelin-1 gene (edn1). J Biol Chem. 2009;284:30087–96.

Immune Mechanisms in Vascular Remodeling in Hypertension

7

Ernesto L. Schiffrin

Small artery injury, where resistance to blood flow occurs and leads to rising peripheral resistance, is characteristic of arterial hypertension. In hypertension, large arteries become stiffer and develop outward hypertrophic remodeling as we age [1–3]. Remodeling of small arteries has classically been associated with increased media thickness. However, today we recognize two different types of small artery remodeling: inward eutrophic or inward hypertrophic remodeling [4]. The first is found mainly in essential or primary hypertension in humans and in spontaneously hypertensive rats (SHR). In it, the lumen is smaller, and the media to lumen ratio is similar to that of normotensive individuals [5–7]. Remodeling is eutrophic in situations where the renin-angiotensin-aldosterone system (RAAS) is slightly or inappropriately activated in relation to blood pressure (BP) elevation, which is found in primary hypertension and in SHR. Hypertrophic remodeling of small arteries occurs mainly in secondary forms of hypertension such as in renovascular hypertension, primary aldosteronism, or pheochromocytoma [8] and in hypertension found in diabetic patients [9, 10]. It has also been described in acromegaly [11], in salt-dependent, in mineralocorticoid hypertension, and in malignant hypertension in rodents [12–15]. All these conditions in which small artery hypertrophic remodeling is found are conditions in which the endothelin system is activated and may play a role in these changes [16–19]. Hyperplasia of vascular smooth muscle cells (VSMC) occurs in small arteries of hypertensive rats, whereas hypertrophy characterizes aortic VSMC [20, 21]. Volume and number of VSMC from small arteries in essential hypertension, however, remain similar to those in normotensive subjects [22]. The mechanisms for

E. L. Schiffrin (✉)
Department of Medicine, Sir Mortimer B. Davis-Jewish General Hospital, Montreal, QC, Canada

Hypertension and Vascular Research Unit, Lady Davis Institute for Medical Research, McGill University, Montreal, QC, Canada
e-mail: ernesto.schiffrin@mcgill.ca

E. Agabiti-Rosei et al. (eds.), *Microcirculation in Cardiovascular Diseases*, Updates in Hypertension and Cardiovascular Protection,
https://doi.org/10.1007/978-3-030-47801-8_7

inward growth are unclear but may be associated with peripheral apoptosis, or they could result from vasoconstriction embedded in an expanded extracellular matrix [23], as there is deposition of collagen and fibronectin with an increased collagen to elastin ratio in small arteries from hypertensive humans and rodents [22, 24]. These may be induced by endothelin [25], angiotensin II, and aldosterone [26]. Tissue transglutaminases via interactions of extracellular matrix fibrillar components with attachment sites on VSMC [27, 28] or matrix metaloproteinases (MMP) such as MMP2 [29, 30] may play an important role. We have previously suggested that rearrangement of cells and fibrillar components due to altered interaction of these through integrins may result in collagen fibers recruited at higher distending vessel pressures leading transiently to decreased stiffness [31]. However, with increased pulsatility impacting on small arteries, these eventually become stiffer as do larger conduit arteries [1, 32], compromising tissue perfusion [33].

Our studies demonstrated that small artery remodeling may be the first manifestation of target organ damage in hypertension [31]. All stage I hypertensive subjects present small artery remodeling, whereas endothelial dysfunction is found in 60% and left ventricular hypertrophy in 45%. Presence of small artery remodeling in fact has prognostic significance, since those hypertensive patients with the highest media to lumen ratio had the highest incidence of cardiovascular events [34].

Immune and inflammatory cell infiltration particularly of perivascular adipose tissue (PVAT) [35] accompanied by upregulation of inflammatory mediators such as vascular cellular adhesion molecule (VCAM)-1, intercellular adhesion molecule (ICAM)-1, nuclear factor (NF) kappaB, monocyte chemoattractant factor (MCP)-1, and plasminogen activator inhibitor (PAI)-1 in response to enhanced oxidative stress is all recognized participants in the remodeling process [32].

Dysfunction of endothelium is often but not always found in vessels of hypertensive subjects [31]. This change is probably a secondary phenomenon not involved in initiation of BP elevation but rather one that may contribute to progression of atherosclerosis and cardiovascular events in association with risk factors clustering with hypertension like dyslipidemia, obesity, metabolic syndrome, diabetes, and smoking. Endothelial dysfunction may be evaluated measuring different parameters such as reduced endothelium-dependent vasodilation [36] or expression of endothelin-1 or inflammatory or thrombogenic agents produced by the endothelium [37]. A dysfunctional endothelium is characterized by an inflammatory phenotype with increased endothelial cell proliferation, altered morphology, production of C-reactive protein, and other inflammatory and thrombogenic mediators including MCP-1 and plasminogen activator inhibitor (PAI)-1, upregulated adhesion molecules, enhanced thrombogenicity, and adhesiveness of circulating cells and anoikis [37].

7.1 Immunity, the Microbiome, and Cardiovascular Injury

Prevalence of cardiovascular disease [38] and hypertension [39] is enhanced in inflammatory and immune diseases such as rheumatoid arthritis, ankylosing spondylitis, and psoriasis. Indeed, low-grade inflammation participates as a mechanism of cardiovascular injury and BP elevation [40]. It is unclear how inflammation and immune mechanisms are activated in hypertension. Genetic susceptibility, obesity, and excess salt and other environmental triggers may act on the brain to stimulate the sympathetic nervous system inducing direct actions or acting indirectly via small rises in BP. The BP rise could generate damage-associated molecular patterns (DAMPs) recognized by pattern recognition receptors (PRRs) that activate innate immunity and also neo-antigens that could activate adaptive immunity. The enhanced sympathetic activity may by itself influence the immune system. There is increasing evidence of interactions of the sympathetic system, the intestinal microbiome, and the very abundant immune cells in the wall of the intestine that are a source of systemic T regulatory and Th17 lymphocytes, which may then migrate to secondary lymphoid organs such as lymph nodes or the spleen. Treg/T effector (Th1 or Th17) ratios may fluctuate with ratios of different gut bacteria, such as bacteroides (protective)/firmicutes (inflammatory), contributing to hypertensive or normotensive phenotypes [41]. Salt has recently been identified as a powerful modifier of gut microbiota. Salt intake results in decreased *Lactobacillus murinus* in the gut, which leads to enhanced activation of Th17 cells originating in the intestinal wall and contributes to increased BP [42]. As well, infectious agents acting through pathogen-associated molecular patterns (PAMPs) recognized by toll-like receptors (TLRs) could also contribute to activation of immune responses. Activation of dendritic cells leads to generation if isoketals that then stimulate other immune cells [43]. Activation of innate immunity is one of the first steps in the involvement of the immune system in hypertension and vascular injury. We demonstrated in osteopetrotic mice that have a mutation in the *monocyte/macrophage colony-stimulating factor* (*csf1*) gene that Ang II infusion [44] or DOCA-salt [45] fails to induce vascular injury and endothelial dysfunction or elevate BP in homozygous $csf1^{Op/Op}$ mice, whereas in heterozygous $csf1^{Op/+}$ effects were intermediate, which confirms that innate immunity is involved in BP elevation and vascular remodeling. Wenzel et al. [46] elegantly extended these results, demonstrating that in mice with inducible expression of diphtheria toxin receptor (LysM-DTR mice), depletion of lysozyme M-positive myelomonocytic cells by low-dose diphtheria toxin infusion prevented Ang II-induced vascular dysfunction, endothelial dysfunction, and BP elevation.

7.2 Adaptive Immunity in Hypertension

Adaptive immune mechanisms may also be involved in hypertension and vascular injury. Blunted Ang II-induced BP rise, aortic and small artery remodeling, and vascular oxidative stress were found in $Rag1^{-/-}$ mice, deficient in T and B lymphocytes [47]. Adoptive transfer of effector T cells but not B lymphocytes from control mice corrected these abnormal responses of $Rag1^{-/-}$ mice. The potential role of Th17 lymphocytes has been highlighted by the study that showed that Ang II infused into $IL17^{-/-}$ mice induced an elevation of BP that was less sustained, with less T cell infiltration in PVAT and superoxide production in aortic rings [48]. As well, Th17 lymphocytes are involved in the stiffening of the aorta that occurs in hypertensive mice [49]. Th17 may also play an important role in the response to salt. Earlier, the role of the gut microbiome was mentioned [42]. Salt however may act at other levels. Indeed, Norlander et al. [50] showed that deficiency of T cell serum- and glucocorticoid-regulated kinase 1 (SGK1) reduced the BP elevation and vascular and renal inflammation, as well as endothelial dysfunction and renal injury induced by Ang II and deoxycorticosterone acetate-salt (DOCA-salt) hypertension in mice. The authors demonstrated that the Na+-K+-2Cl- cotransporter 1 (NKCC1) is upregulated in Th17 cells by Ang II and plays a role in the salt-induced increase in SGK1, which leads to activation of Th17 lymphocytes and BP elevation and vascular and renal injury.

In Ang II-infused mice, CD8+ cells but not CD4+ cells in the kidney had changes in TCR transcript length in *Vβ3*, *8.1*, and *17* families as demonstrated by TCR spectratyping, whereas clonality was not observed in other tissues [51]. Moreover, Ang II-induced hypertension in $CD4^{-/-}$ and $MHCII^{-/-}$ mice similar to that in wild-type mice, whereas $CD8^{-/-}$ mice and $OT1xRag\text{-}1^{-/-}$ mice, which have only one TCR, exhibited a blunted hypertensive response to angiotensin II. Adoptive transfer of pan-T cells and CD8+ T cells but not CD4+/CD25− cells restored the hypertensive response to Ang II in $Rag\text{-}1^{-/-}$ mice. $CD8^{-/-}$ mice but not $CD4^{-/-}$ mice were protected from Ang II-induced vascular remodeling and endothelial dysfunction in the kidney.

7.3 T Regulatory Lymphocytes

The adaptive immune system has been shown to be enhanced in salt-sensitive hypertension, contributing to BP elevation and vascular damage. We studied consomic rats (SSBN2) bearing chromosome 2 from normotensive Brown Norway (BN) rats on a Dahl salt-sensitive genetic background. Chromosome 2 has loci for many pro-inflammatory genes such as IL-2, IL-6 receptor, fibroblast growth factor 2, VCAM-1, and angiotensin AT_{1b} receptor [52]. The consomic strain exhibited less vascular inflammation and remodeling [53], particularly in response to high salt intake [54], associated with upregulation of $CD4^{+}CD25^{+}$ and $CD8^{+}CD25^{+}$ lymphocytes (Treg) and their activity, expression of the transcription factor responsible for Treg maturation (Foxp3), and enhanced production by Treg of the anti-inflammatory mediators IL-10 and TGF-β. The vasculature of hypertensive Dahl salt-sensitive rats presented upregulated inflammatory cytokines (IL-1β, IL-2, IL-6, TNF-α, and IFN-γ), and low levels of Foxp3b, and did not produce the anti-inflammatory TGF-β and IL-10. We also demonstrated that adoptive transfer of Treg from untreated mice into Ang II-infused mice lowered systolic BP measured by telemetry, reduced small artery stiffness, decreased generation of superoxide and immune cell infiltration in vascular and PVAT, and decreased endothelial dysfunction [55]. When we injected $Rag1^{-/-}$ mice with T cells from wild-type or Scurfy mice (that lack Treg) or wild-type Treg alone or with Scurfy T cells, infusion of Ang II increased systolic BP in all groups but diastolic BP only in wild-type and Scurfy mice [56]. Ang II induced endothelial dysfunction, microvascular remodeling and stiffness, and oxidative stress in PVAT of small mesenteric arteries from wild-type T cell-injected *Rag1*$^{-/-}$ mice, but these effects were enhanced much more in Scurfy T cell-injected *Rag1*$^{-/-}$ mice, which lack the anti-inflammatory action of Treg. Angiotensin II increased MCP-1 expression in the vascular wall and PVAT, monocyte/macrophage infiltration and pro-inflammatory polarization in PVAT and the renal cortex, and T cell infiltration in the renal cortex only in Scurfy T cell-injected *Rag1*$^{-/-}$ mice. Wild-type Treg co-injection with vehicle or Scurfy T cells prevented or reduced the effects of angiotensin II. In conclusion, Treg counteracted angiotensin II-induced microvascular injury by modulating innate and adaptive immune responses. Aldosterone induced effects similar to Ang II, except that adoptive transfer of Treg from untreated mice did not lower BP and that hypertrophic remodeling was corrected [57], indicating that effects may be independent of hemodynamics. They could be mediated in part by the anti-inflammatory actions of IL-10, or by other mediators of Treg such as adenosine acting on adenosine A2A receptors, or the degradation of adenosine triphosphate (ATP) by CD73 and CD39 on Treg leading to reduction of ATP effects on the vasoconstrictor and pro-inflammatory P2X7 receptor [58].

7.4 γδ T Lymphocytes

Most T lymphocytes in the circulation have a T cell receptor (TCR) composed of two subunits, alpha and beta. A small percentage of circulating T lymphocytes (1–4%) are mostly CD4 and CD8 double negative and have gamma and delta subunits, the γδ T lymphocytes. They are much more abundant in tissues and are unconventional "innate-like" T cells that produce interferon gamma and IL-17 and are not MHC-restricted. They respond rapidly to public antigens such as nonprotein phosphoantigens, isoprenoid pyrophosphates, alkylamines, nonclassical MHC class I molecules, MHC class I chain-related proteins A and B, as well as heat shock protein-derived peptides, without antigen processing and MHC presentation [59]. γδ T lymphocytes may be a bridge between innate and adaptive immunity. They can prime T effector lymphocytes and contribute to activate adaptive immunity. We recently demonstrated that Ang II infusion was followed by increased number and activation of γδ T cells [60]. *Tcrδ* knockout mice or injection of γδ T cell-depleting antibodies prevented BP elevation in response to Ang II infusion and small artery endothelial dysfunction and activation of spleen and mesenteric artery PVAT $CD4^+$ and $CD8^+$ T cells. Thus, γδ T cells are at the initiation of inflammatory responses associated with the development of hypertension. Using multiple linear regression, a correlation existed between whole blood expression of the TCR gamma chain constant regions 1 and 2 that was used as an estimate of the frequency of blood γδ T cells and systolic BP in a human cohort with and without coronary artery disease and a full range of BPs [60], indicating a role of γδ T cells in the regulation of BP in humans.

7.5 T Cell Subtypes and the Microvasculature in Humans

Recent studies have examined the relationship in human essential hypertension between T cell subtypes and the microvasculature [61]. A significant inverse correlation was observed between total Treg (CD4+CD25HighCD127Low) and M/L ratio of subcutaneous small resistance arteries, while a direct correlation was observed between Th17 lymphocytes and M/L ratio of subcutaneous small resistance arteries. Total capillary density in the dorsum of the finger was significantly correlated with circulating effector memory cell lymphocytes. Also, significant direct correlations were observed between Th2 lymphocytes and basal or total capillary density on different sites examined. No significant correlations were observed between clinic systolic or diastolic BP and lymphocyte subpopulations. The balance between Th1/Treg and Th17/Treg did not correlate with either BP or capillary density or recruitment. An inverse correlation between recent thymus emigrant lymphocytes as well as naïve lymphocytes and W/L ratio of retinal arterioles was also demonstrated [61].

7.6 B Cells in Hypertension

Despite the evidence from Guzik et al. [47] that B cells do not participate in Ang II-induced hypertension, Chan et al. [62] demonstrated that B cells and IgGs are crucial for the development of Ang II-induced hypertension and blood vessel remodeling in mice. Ang II-dependent hypertension in mice was associated with B-cell activation in lymphoid organs, together with increased circulating IgG levels and increased IgG in the aortic wall. B-cell-deficient *BAFF-R*$^{-/-}$ mice were protected from the action of Ang II and vascular inflammation and dysfunction, immune cell infiltration, extracellular matrix changes, and stiffening of the aorta. Moreover, an anti-B-cell antibody prevented Ang II-induced increases in BP.

7.7 Role of Cytokines on the Microvasculature in Hypertension

The role of cytokines produced by T cells has been elucidated over recent years. The pro-inflammatory vascular effects of interferon gamma are well known for some time, contributing to the remodeling action of immune cells [63]. Mice deficient in the lymphocyte adaptor protein, LNK, have elevated numbers of CD4+ and particularly CD8+ T cells that produce IFN-γ [64]. They develop hypertension and vascular dysfunction in response to angiotensin II that is blunted in the presence of IFN-γ deficiency. Other cytokines have also been demonstrated to participate in pro-hypertensive actions and inducing microvascular injury. IL-6 gene deletion has been shown to be associated with attenuated hypertension [65]. A role for IL-22 has also been suggested [66]. Ang II infusion increased Th22 cells and IL-22 levels in mice. Recombinant IL-22 increased BP, induced inflammation, and worsened endothelial dysfunction. IL-22-neutralizing monoclonal antibody decreased BP, inflammatory responses, and endothelial dysfunction. The STAT3 inhibitor S31–201 reduced IL-22 increases in BP and endothelial dysfunction, suggesting that IL-22 contributed to BP elevation in Ang II-infused mice via the STAT3 pathway. As well, IL-23 may be involved as has been demonstrated in LNK$^{-/-}$ mice [64].

An important role for the CXCL1-CXCR2 axis in hypertension and vascular injury has been also recently demonstrated [67]. After 1 day of Ang II infusion, these authors showed that the *Cxcl1* was the most highly expressed chemokine in mouse aorta compared to other chemokines. Expression of *Cxcr2* in the aorta increased in Ang II-infused mice, associated to infiltration of leukocytes that expressed the CXCL1 receptor CXCR2. Knockout of *Cxcr2* or CXCR2 antagonism with SB265610, blunted Ang II-induced BP elevation, endothelial dysfunction and vascular remodeling, fibrosis, oxidative stress, inflammation, and macrophage and T cell infiltration. *Cxcr2* knockout also prevented DOCA/salt hypertension and vascular injury. Bone marrow transplantation studies demonstrated that CXCR2 on immune cells was responsible for the role played by CXCL1 in hypertension and vascular injury.

7.8 Conclusion

The participation of inflammation and immune mechanisms in vascular injury in hypertension and cardiovascular disease has increasingly been appreciated. As these different mechanisms that contribute to the remodeling of small arteries in hypertension, metabolic disease, and other cardiovascular diseases are identified, this will allow characterizing novel biomarkers and treatments for hypertension and other cardiovascular diseases, hopefully leading to improvement of patient outcomes.

Acknowledgments The work of the author was supported by Canadian Institutes of Health Research (CIHR) grants 37917, 82790, 102606, and 123465 and by the Canada Fund for Innovation and currently by the First Pilot Foundation Grant 143348 from the CIHR and by a Canada Research Chair (CRC) on Hypertension and Vascular Research from the CIHR/Government of Canada CRC Program.

References

1. Schiffrin EL. The vascular phenotypes in hypertension: relation to the natural history of hypertension. J Am Soc Hypertens. 2007;1:56–67.
2. Mitchell GF, Lacourcière Y, Ouellet JP, et al. Determinants of elevated pulse pressure in middle-aged and older subjects with uncomplicated systolic hypertension—the role of proximal aortic diameter and the aortic pressure-flow relationship. Circulation. 2003;108:1592–8.
3. Schiffrin EL. Vascular stiffening and arterial compliance—implications for systolic blood pressure. Am J Hypertens. 2004;17:39S–48S.
4. Heagerty AM, Aalkjaer C, Bund SJ, Korsgaard N, Mulvany MJ. Small artery structure in hypertension: dual processes of remodeling and growth. Hypertension. 1993;21:391–7.
5. Schiffrin EL, Deng LY, Larochelle P. Morphology of resistance arteries and comparison of effects of vasoconstrictors in mild essential hypertensive patients. Clin Invest Med. 1993;16:177–86.

6. Mulvany MJ, Baumbach GL, Aalkjaer C, et al. Vascular remodeling. Hypertension. 1996;28:505–6.
7. Schiffrin EL. Remodeling of resistance arteries in essential hypertension and effects of antihypertensive treatment. Am J Hypertens. 2004;17:1192–200.
8. Rizzoni D, Porteri E, Castellano M, et al. Vascular hypertrophy and remodeling in secondary hypertension. Hypertension. 1996;28:785–90.
9. Rizzoni D, Porteri E, Guelfi D, et al. Structural alterations in subcutaneous small arteries of normotensive and hypertensive patients with non-insulin-dependent diabetes mellitus. Circulation. 2001;103:1238–44.
10. Endemann D, Pu Q, De Ciuceis C, et al. Persistent remodeling of resistance arteries in type 2 diabetic patients on antihypertensive treatment. Hypertension. 2004;43:399–404.
11. Rizzoni D, Porteri E, Giustina A, et al. Acromegalic patients show the presence of hypertrophic remodeling of subcutaneous small resistance arteries. Hypertension. 2004;43:561–5.
12. Li JS, Larivière R, Schiffrin EL. Effect of a nonselective endothelin antagonist on vascular remodeling in deoxycorticosterone acetate-salt hypertensive rats: evidence for a role of endothelin in vascular hypertrophy. Hypertension. 1994;24:183–8.
13. Schiffrin EL, Larivière R, Li JS, Sventek P. Enhanced expression of the endothelin-1 gene in blood vessels of DOCA-salt hypertensive rats: correlation with vascular structure. J Vasc Res. 1996;33:235–48.
14. D'Uscio LV, Barton M, Shaw S, Moreau P, Lüscher TF. Structure and function of small arteries in salt-induced hypertension—effects of chronic endothelin-subtype-A-receptor blockade. Hypertension. 1997;30:905–11.
15. Schiffrin EL, Touyz RM. From bedside to bench to bedside: role of renin-angiotensin-aldosterone system in remodeling of resistance arteries in hypertension. Am J Physiol Heart Circ Physiol. 2004;287:H435–46.
16. Schiffrin EL, Larivière R, Li JS, Sventek P, Touyz RM. Deoxycorticosterone acetate plus salt induces overexpression of vascular endothelin-1 and severe vascular hypertrophy in spontaneously hypertensive rats. Hypertension. 1995;25(Part 2):769–73.
17. Schiffrin EL, Deng LY, Sventek P, Day R. Enhanced expression of endothelin-1 gene in resistance arteries in severe human essential hypertension. J Hypertens. 1997;15:57–63.
18. Amiri F, Virdis A, Neves MF, et al. Endothelium-restricted overexpression of human endothelin-1 causes vascular remodeling and endothelial dysfunction. Circulation. 2004;110:2233–40.
19. Coelho SC, Berillo O, Ouerd S, et al. Three-month endothelial human endothelin-1 overexpression causes blood pressure elevation and vascular and kidney injury. Hypertension. 2018;71:208–16.
20. Lee RMKW, Garfield RE, Forrest JB, Daniel EE. Morphometric study of structural changes in the mesenteric blood vessels of spontaneously hypertensive rats. Blood Vessels. 1983;20:57–71.
21. Mulvany MJ, Baandrup U, Gundersen HJ. Evidence for hyperplasia in mesenteric resistance vessels of spontaneously hypertensive rats using a three-dimensional disector. Circ Res. 1985;57:794–800.
22. Intengan HD, Deng LY, Li JS, Schiffrin EL. Mechanics and composition of human subcutaneous resistance arteries in essential hypertension. Hypertension. 1999;33:569–74.
23. Bakker ENTP, Van der Meulen ET, Van den Berg BM, et al. Inward remodeling follows chronic vasoconstriction in isolated resistance arteries. J Vasc Res. 2002;39:12–20.
24. Intengan HD, Thibault G, Li JS, Schiffrin EL. Resistance artery mechanics, structure, and extracellular components in spontaneously hypertensive rats—effects of angiotensin receptor antagonism and converting enzyme inhibition. Circulation. 1999;100:2267–75.
25. Pu Q, Neves MF, Virdis A, Touyz RM, Schiffrin EL. Endothelin antagonism on aldosterone-induced oxidative stress and vascular remodeling. Hypertension. 2003;42:49–55.
26. Neves MF, Virdis A, Schiffrin EL. Resistance artery mechanics and composition in angiotensin II-infused rats: effects of aldosterone antagonism. J Hypertens. 2003;21:189–98.
27. Bakker ENTP, Buus CL, Spaan JAE, et al. Small artery remodeling depends on tissue-type transglutaminase. Circ Res. 2005;96:119–26.

28. Brassard P, Amiri F, Thibault G, Schiffrin EL. Role of angiotensin type-1 and angiotensin type-2 receptors in the expression of vascular integrins in angiotensin II-infused rats. Hypertension. 2006;47:122–7.
29. Brassard P, Amiri F, Schiffrin EL. Combined angiotensin II type 1 and type 2 receptor blockade on vascular remodeling and matrix metalloproteinases in resistance arteries. Hypertension. 2005;46:598–606.
30. Barhoumi T, Mian MOR, Fraulob-Aquino JC, et al. Matrix metalloproteinase-2 knockout prevents angiotensin II-induced endothelial dysfunction and vascular remodeling, oxidative stress and inflammation. Cardiovasc Res. 2017;113:1752–62.
31. Park JB, Schiffrin EL. Small artery remodeling is the most prevalent (earliest?) form of target organ damage in mild essential hypertension. J Hypertens. 2001;19:921–30.
32. Intengan HD, Schiffrin EL. Vascular remodeling in hypertension—roles of apoptosis, inflammation, and fibrosis. Hypertension. 2001;38:581–7.
33. Levy B, Schiffrin EL, Mourad JJ, et al. Impaired tissue perfusion: a pathology common to hypertension, obesity and diabetes. Circulation. 2008;118:968–76.
34. Rizzoni D, Porteri E, Boari GE, et al. Prognostic significance of small-artery structure in hypertension. Circulation. 2003;108:2230–5.
35. Marchesi C, Ebrahimian T, Angulo O, Paradis P, Schiffrin EL. Endothelial NO synthase uncoupling and perivascular adipose oxidative stress and inflammation contribute to vascular dysfunction in a rodent model of metabolic syndrome. Hypertension. 2009;54:1384–92.
36. Deng LY, Li JS, Schiffrin EL. Endothelium-dependent relaxation of small arteries from essential hypertensive patients: mechanisms and comparison with normotensive subjects and with responses of vessels from spontaneously hypertensive rats. Clin Sci. 1995;88:611–22.
37. Endemann DH, Schiffrin EL. Endothelial dysfunction. J Am Soc Nephrol. 2004;15:1983–92.
38. Mason JC, Libby P. Cardiovascular disease in patients with chronic inflammation: mechanisms underlying premature cardiovascular events in rheumatologic conditions. Eur Heart J. 2015;36:482–9.
39. Armstrong AW, Harskamp CT, Armstrong EJ. The association between psoriasis and hypertension: a systematic review and meta-analysis of observational studies. J Hypertens. 2013;31:433–43.
40. Savoia C, Schiffrin EL. Vascular inflammation in hypertension and diabetes: molecular mechanisms and therapeutic interventions. Clin Sci. 2007;112:375–84.
41. Hoeppli RE, Wu D, Cook L, Levings MG. The environment of regulatory T cell biology: cytokines, metabolites, and the microbiome. Front Immunol. 2015;6:1–14.
42. Wilck N, Matus MG, Kearney SM, et al. Salt-responsive gut commensal modulates TH17 axis and disease. Nature. 2017;551:585–9.
43. Kirabo A, Fontana V, de Faria AP, et al. DC isoketal-modified proteins activate T cells and promote hypertension. J Clin Invest. 2014;124:4642–56.
44. De Ciuceis C, Amiri F, Brassard P, et al. Reduced vascular remodeling, endothelial dysfunction and oxidative stress in resistance arteries of angiotensin II-infused macrophage colony-stimulating factor-deficient mice: evidence for a role in inflammation in angiotensin-induced vascular injury. Arterioscler Thromb Vasc Biol. 2005;25:2106–13.
45. Ko EA, Amiri F, Pandey NR, Touyz RM, Schiffrin EL. Resistance artery remodeling in deoxycorticosterone acetate-salt hypertension is dependent on vascular inflammation: evidence from m-csf-deficient mice. Am J Physiol Heart Circ Physiol. 2007;292:H1789–95.
46. Wenzel P, Knorr M, Kossmann S, et al. Lysozyme M–positive monocytes mediate angiotensin II–induced arterial hypertension and vascular dysfunction. Circulation. 2011;124:1370–81.
47. Guzik TJ, Hoch NE, Brown KA, et al. Role of the T cell in the genesis of angiotensin II induced hypertension and vascular dysfunction. J Exp Med. 2007;204:2449–60.
48. Madhur M, Lob HE, McCann LA, et al. Interleukin 17 promotes angiotensin II–induced hypertension and vascular dysfunction. Hypertension. 2010;55:500–5.
49. Wu J, Thabet SR, Kirabo A, et al. Inflammation and mechanical stretch promote aortic stiffening in hypertension through activation of p38 mitogen-activated protein kinase. Circ Res. 2014;114:616–25.

50. Norlander AE, Saleh MA, Pandey AK, Itani HA, Wu J, Xiao L, Kang J, Dale BL, Goleva SB, Laroumanie F, Du L, Harrison DG, Madhur MS. A salt-sensing kinase in T lymphocytes, SGK1, drives hypertension and hypertensive end-organ damage. JCI Insight. 2017;2. https://doi.org/10.1172/jci.insight.92801.
51. Trott DW, Thabet SR, Kirabo A, et al. Oligoclonal CD8+ T cells play a critical role in the development of hypertension. Hypertension. 2014;64:1108–15.
52. UCSC Human Gene Sorter. http://genome.ucsc.edu/cgi-bin/hgNear. Accessed July 14, 2007.
53. Viel EC, Lemarié CA, Benkirane K, Paradis P, Schiffrin EL. Immune regulation and vascular inflammation in genetic hypertension. Am J Physiol Heart Circ Physiol. 2010;298:H938–44.
54. Leibowitz AA, Li MW, Paradis P, Schiffrin EL. Chromosome 2 plays a role in high salt diet-induced hypertension and vascular remodeling in Dahl salt-sensitive rats. Hypertension. 2011;58:e44.
55. Barhoumi T, Kasal DAB, Li MW, et al. T regulatory lymphocytes prevent angiotensin II-induced hypertension and vascular injury. Hypertension. 2011;57:469–76.
56. Mian MOR, Barhoumi T, Briet M, Paradis P, Schiffrin EL. Deficiency of T regulatory cells exaggerates angiotensin II-induced microvascular injury by enhancing immune responses. J Hypertens. 2016;34:97–108.
57. Kasal DA, Barhoumi T, Li MW, et al. Aldosterone-induced hypertension and vascular injury was attenuated by adoptive transfer of T-regulatory lymphocytes. Hypertension. 2012;59:324–30.
58. Stachon P, Heidenreich A, Merz J, et al. P2X7 deficiency blocks lesional inflammasome activity and ameliorates atherosclerosis in mice. Circulation. 2017;135:2524–33.
59. Godfrey DI, Uldrich AP, McCluskey J, Rossjohn J, Moody DB. The burgeoning family of unconventional T cells. Nat Immunol. 2015;16:1114–23.
60. Caillon A, Mian MOR, Fraulob-Aquino JC, et al. Gamma delta T cells mediate angiotensin II-induced hypertension and vascular injury. Circulation. 2017;135:2155–62.
61. De Ciuceis C, Rossini C, Airo P, et al. Relationship between different subpopulations of circulating CD4+ T-lymphocytes and microvascular structural alterations in humans. Am J Hypertens. 2017;30:51–60.
62. Chan CT, Sobey CG, Lieu M, et al. Obligatory role for B cells in the development of angiotensin II-dependent hypertension. Hypertension. 2015;66:1023–33.
63. Markó L, Kvakan H, Park JK, et al. Interferon-γ signaling inhibition ameliorates angiotensin II-induced cardiac damage. Hypertension. 2012;60:1430–6.
64. Saleh MA, McMaster WG, Wu J, et al. Lymphocyte adaptor protein LNK deficiency exacerbates hypertension and end-organ inflammation. J Clin Invest. 2015;125:1189–202.
65. Lee DL, Sturgis LC, Labazi H, Osborne JB Jr, Fleming C, Pollock JS, Manhiani M, Imig JD, Brands MW. Angiotensin II hypertension is attenuated in interleukin-6 knockout mice. Am J Physiol Heart Circ Physiol. 2006;290:H935–40.
66. Ye J, Ji Q, Liu J, et al. Interleukin 22 promotes blood pressure elevation and endothelial dysfunction in angiotensin II–treated mice. J Am Heart Assoc. 2017:e005875. https://doi.org/10.1161/JAHA.117.005875.
67. Wang L, Zhao XC, Cui W, et al. Genetic and pharmacologic inhibition of the chemokine receptor CXCR2 prevents experimental hypertension and vascular dysfunction. Circulation. 2016;134:1353–68.

8 Microvascular Endothelial Dysfunction in Hypertension

Agostino Virdis and Stefano Masi

8.1 Introduction

Endothelium represents the fundamental homeostatic tissue for the regulation of the vascular tone and structure, which is physiologically characterized by a balance between substances with vasodilating and antithrombogenic properties and substances with vasoconstricting and prothrombotic activities. In healthy conditions, NO modulates several pathways leading to protect the vascular wall against those mechanisms involved in favouring and promoting the atherosclerotic disease. Under disease conditions, including essential hypertension, the endothelium loses its protective role, becoming a pro-atherosclerotic structure [1]. The loss of the normal endothelial function is referred to as "endothelial dysfunction", characterized by impaired NO bioavailability. This can be determined by either a reduced production of NO by endothelial NO synthase (eNOS) or, more frequently, an increased breakdown by reactive oxygen species (ROS). Under such conditions, in addition to ROS, endothelial cells produce additional substances with vasoconstricting and prothrombotic activities, including endothelin-1 (ET-1) and prostanoids [2, 3] generically called endothelial-derived contracting factors (EDCFs). The aim of this review is to give a brief overview of the known mechanisms involved in the pathogenesis of endothelial dysfunction in the microcirculation of hypertensive patients.

A. Virdis (✉) · S. Masi
Department of Clinical and Experimental Medicine, University of Pisa, Pisa, Italy
e-mail: agostino.virdis@med.unipi.it

E. Agabiti-Rosei et al. (eds.), *Microcirculation in Cardiovascular Diseases*, Updates in Hypertension and Cardiovascular Protection,
https://doi.org/10.1007/978-3-030-47801-8_8

8.2 Endothelial Dysfunction in Hypertension: The Traditional Mechanisms

The principal EDCF is ET-1, generated by the vascular endothelium, which acts through specific ET_A and ET_B receptors. ET_A receptors are located on smooth muscle cells and promote growth and contraction. ET_B receptors are located on both endothelial and smooth muscle cells, with opposite effects. Activation of smooth muscle cell ET_B evokes contraction, whereas activation of endothelial ET_B induces relaxation [4, 5]. The overall biological effect of these activated receptors on vasculature derives from the balance between their protective or deleterious effects, and this delicate equilibrium likely explains why this physiological substance may shift to a pathological role in cardiovascular disease, including essential hypertension [6]. We previously observed that intra-arterial infusion of TAK-044, an ET_A/ET_B receptor antagonist, caused an increased vasodilation in the forearm microcirculation of hypertensive patients compared with healthy subjects. Moreover, vasoconstriction to L-NMMA was decreased in hypertensive patients compared with controls. In addition, vasodilation to TAK-044 and vasoconstriction to L-NMMA showed an inverse correlation [7]. These findings allowed to conclude that in essential hypertensive patients, endogenous ET-1 shows greater vasoconstrictor activity.

Beyond ET-1, the main EDCFs are represented by endoperoxides, deriving from the metabolism of arachidonic acid by COX activity into a range of bioactive prostanoids, including thromboxane A_2 or prostacyclin [2, 8].

The metabolism of oxygen by cells generates ROS. In healthy conditions, the rate and magnitude of oxidant formation is counterbalanced by the rate of oxidant elimination. An imbalance between pro-oxidants and antioxidants results in oxidative stress, which is the pathogenic outcome of oxidant overproduction. Vascular ROS derive primarily from NAD(P)H oxidase. Other important sources include COX, dysfunctional eNOS (uncoupled NOS), and xanthine oxidase [9]. The first experiments assessing the role of EDCFs on endothelial dysfunction in the forearm microcirculation of essential hypertensive patients demonstrated that intra-arterial administration of the COX inhibitor indomethacin improved the vasodilation to acetylcholine and restored the inhibitory effect of L-NMMA on that response, indicating that COX generates substances that reduce the availability of NO [10, 11]. Moreover, intra-arterial infusion of the ROS scavenger ascorbic acid evoked similar effect as indomethacin in these patients, with no further potentiation when the two compounds were coinfused [12]. In conjunction, these findings represent the demonstration that the COX pathway is a source of ROS in essential hypertension.

When does the EDCFs generation occur during the lifespan of a hypertensive patient? Important knowledge on this topic provided from vascular studies on the cross-relation between the hypertensive progress and ageing. The increasing age is the most powerful determinant of endothelial dysfunction and is accompanied by a progressive worsening of NO availability in resistance circulation [13–16]. The main mechanism responsible for age-related endothelial dysfunction in the peripheral microcirculation is a primary defect in the L-arginine-NO pathway. After the age of 60 years, along with a further impairment of the L-arginine-NO pathway, COX-dependent EDCF production becomes evident and significant [17]. Such age-related endothelial dysfunction is anticipated by hypertension, which therefore represents a condition of premature vascular ageing. Thus, while not effective in younger hypertensive patients (<30 years), in adult patients (31–45 years), indomethacin begins to show some effect, and in the oldest patients (46–60 and >60 years), COX-derived oxidative stress generation becomes the main determinant of endothelial dysfunction [17]. In conclusion, this study evidenced that ageing is an important factor altering endothelium-dependent vasodilation and that the mechanisms involved include a defect in the L-arginine-NO pathway and production of COX-dependent EDCFs. However, whereas in normotensive subjects, ageing mainly affects the formation of NO and EDCF production characterizes only old age, the presence of hypertension causes an earlier onset of altered L-arginine-NO pathway and also earlier formation of vasoconstrictor prostanoids.

8.3 Endothelial Dysfunction in Hypertension: The Emerging Conspirators

As described in the previous section, the first studies conducted in human hypertension documented the COX pathway actively interferes with NO availability and represents a source of ROS in the peripheral microcirculation. Until now, COX represents the unique pathway identified acting as a source of ROS in human hypertension. However, these pioneering investigations have left undetermined the question of which COX isoenzymes are effectively involved. Two different isoforms of COX are known to exist [18]. In most tissues, COX-1 is regarded as constitutively expressed to produce physiological prostanoids, while COX-2 is often induced by a number of stimuli, including inflammation or growth factors [19]. Nevertheless, COX-2 is also expressed constitutively in several organs. In particular, within the vasculature, endothelial and vascular smooth muscle cells do express both isoforms, with COX-1 being usually expressed at a higher extent than COX-2 [19]. Recently, in isolated small resistance arteries from essential hypertensive patients, we investigated the role of COX-1 and COX-2 and their cross-talks with NO and ROS. It was found that the blunted vascular response to acetylcholine, while not modified by the COX-1 inhibitor, was significantly improved by a selective COX-2 inhibitor, which also partly restored the inhibitory effect of L-NAME on acetylcholine [20]. In addition, a marked upregulation of COX-2 mainly in the vascular media layer was documented [20]. Of note, in these patients the intravascular superoxide excess was dramatically reduced by incubation with the selective COX-2 inhibitor and moderately blunted by the NAD(P)H inhibitor apocynin. These data provide evidence that, in small arteries isolated from essential hypertensive patients, an overexpression and increased activity of COX-2 occur, playing a major role in reducing NO availability. COX-2 represents a major source of ROS generation in essential hypertension. NAD(P)H oxidase participates also, although with a minor role, in promoting superoxide generation in these patients [21].

Ghrelin is a peptide recently discovered from human stomachs [22]. Although basically a gastric hormone, adjunctive growth hormone-independent properties have been recently attributed to ghrelin. Among others, a variety of cardiovascular properties, including vascular effects, are ascribable to this peptide. In particular, expression of ghrelin has been demonstrated in endothelial cells of human arteries and veins and in secretory vesicles of cultured endothelial cells [23], while its receptor GHS-R 1a has been stained within endothelial cells and vascular smooth muscle cells [24]. In vitro studies documented that ghrelin acutely stimulated increased production of NO in bovine aortic endothelial cells and in human aortic endothelial cells in a time- and dose-dependent manner, by using a signaling pathway that involves an increased expression of eNOS [25]. Ghrelin also demonstrated an anti-oxidant property, an activity dependent on the activation of cellular signaling pathways leading to inhibition of enzymes involved in ROS generation, including NAD(P)H oxidase [26]. The first evidence of the vascular effects of ghrelin in in vivo human studies comes from patients with metabolic syndrome. In the forearm microcirculation of these patients, intra-arterial infusion of human ghrelin was able to reverse endothelial dysfunction by increasing NO availability [27] and to exert a beneficial effect on NO and ET-1 imbalance [28]. More recently, we investigated the impact of exogenous ghrelin on endothelial dysfunction in human hypertension. Results revealed that acute ghrelin infusion improved endothelium-dependent vasodilation and restored the inhibition by L-NMMA in response to acetylcholine in the forearm resistance arterioles of essential hypertensive patients, thus representing the first demonstration in hypertensive patients of the ability of ghrelin to restore endothelial NO availability [29]. The antioxidant property of ghrelin was also assessed in our study, both in the forearm resistance circulation and in isolated small vessels. In the first vascular district, acute ghrelin infusion was able to induce a great reduction of markers of oxidative stress in our hypertensive population. In isolated small arteries from hypertensive patients, ghrelin dramatically reduced the enhanced superoxide generation. An identical result was obtained upon incubation with the NAD(P)H oxidase inhibitor gp91ds-tat, and no further superoxide reduction was seen when ghrelin and gp91 were simultaneously incubated (Fig. 8.1). In conjunction, these findings demonstrate, in two different microvascular districts of hypertensive patients, that the beneficial activity of ghrelin is related to its antioxidant property, an effect obtained through a marked inhibition of NAD(P)H oxidase activation, leading to an amelioration of NO activity.

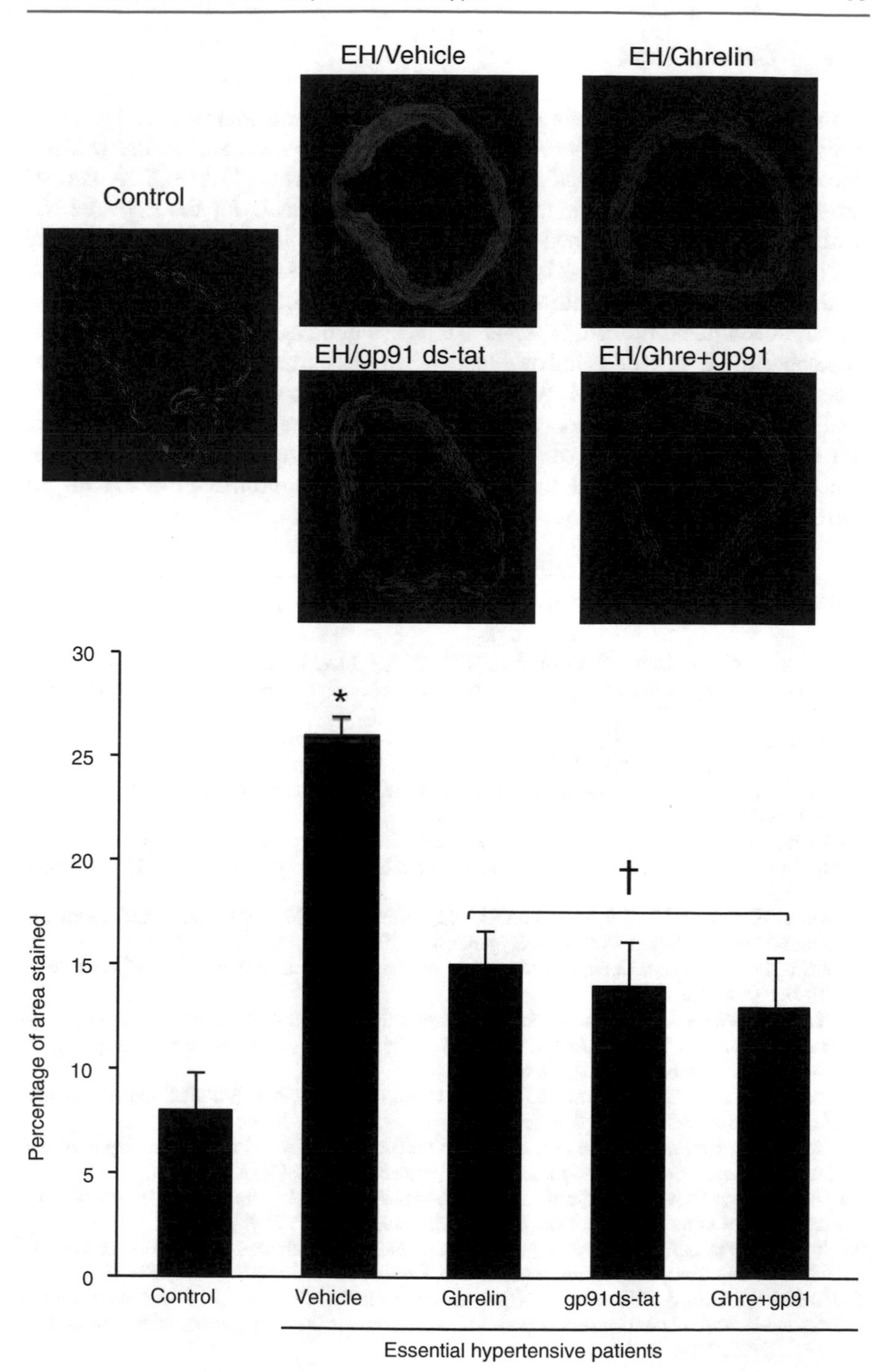

Fig. 8.1 Representative dihydroethidium staining (upper) and quantitative analysis of the red signal (lower) in small arteries from controls and essential hypertensive patients (EH) at baseline (vehicle) or after incubation with ghrelin, gp91 ds-tat, or both. $^*P < 0.001$ vs other groups; $^\dagger P < 0.05$ vs control. Modified from ref. 29

8.4 Conclusions

Essential hypertensive patients are characterized by an impaired vascular NO availability in the peripheral microcirculation. The first evidence from the last two decades indicates an increased production of EDCFs, which include COX-derived prostanoids and ROS, as the most important determinants for the impaired NO availability. More recent papers evidenced new mechanisms involved, recognized as "new conspirators". COX-2 is the isoform identified as a main source of intravascular ROS generation. In addition, important vascular protective properties by ghrelin have been demonstrated, in terms that its systemic reduction is involved in the pathophysiology of endothelial dysfunction, while a normalization of its levels may restore vascular homeostasis. When considering the importance of a reduced NO availability in the pathogenesis of atherosclerotic disease, the discovery of protective effects of ghrelin toward the vasculature is opening up many new research perspectives, thus highlighting the ghrelin system as a promising candidate for cardiovascular drug discovery.

References

1. Flammer AJ, Anderson T, Celermajer DS, Creager MA, Deanfield J, Ganz P, et al. The assessment of endothelial function: from research into clinical practice. Circulation. 2012;126(6):753–67.
2. Vanhoutte PM, Feletou M, Taddei S. Endothelium-dependent contractions in hypertension. Br J Pharmacol. 2005;144(4):449–58.
3. Virdis A, Ghiadoni L, Taddei S. Human endothelial dysfunction: EDCFs. Pflugers Arch. 2010;459(6):1015–23.
4. Cardillo C, Kilcoyne CM, Cannon RO, Panza JA. Interactions between nitric oxide and endothelin in the regulation of vascular tone of human resistance vessels in vivo. Hypertension. 2000;35(6):1237–41.
5. Dhaun N, Goddard J, Kohan DE, Pollock DM, Schiffrin EL, Webb DJ. Role of endothelin-1 in clinical hypertension: 20 years on. Hypertension. 2008;52(3):452–9.
6. Schiffrin EL. State-of-the-art lecture. Role of endothelin-1 in hypertension. Hypertension. 1999;34(4 Pt 2):876–81.
7. Taddei S, Virdis A, Ghiadoni L, Sudano I, Notari M, Salvetti A. Vasoconstriction to endogenous endothelin-1 is increased in the peripheral circulation of patients with essential hypertension. Circulation. 1999;100(16):1680–3.
8. Vanhoutte PM. Endothelium and control of vascular function. State of the art lecture. Hypertension. 1989;13(6 Pt 2):658–67.
9. Touyz RM. Reactive oxygen species, vascular oxidative stress, and redox signaling in hypertension: what is the clinical significance? Hypertension. 2004;44(3):248–52.
10. Taddei S, Virdis A, Mattei P, Salvetti A. Vasodilation to acetylcholine in primary and secondary forms of human hypertension. Hypertension. 1993;21(6 Pt 2):929–33.
11. Taddei S, Virdis A, Ghiadoni L, Magagna A, Salvetti A. Cyclooxygenase inhibition restores nitric oxide activity in essential hypertension. Hypertension. 1997;29(1 Pt 2):274–9.
12. Taddei S, Virdis A, Ghiadoni L, Magagna A, Salvetti A. Vitamin C improves endothelium-dependent vasodilation by restoring nitric oxide activity in essential hypertension. Circulation. 1998;97(22):2222–9.

13. Egashira K, Inou T, Hirooka Y, Kai H, Sugimachi M, Suzuki S, et al. Effects of age on endothelium-dependent vasodilation of resistance coronary artery by acetylcholine in humans. Circulation. 1993;88(1):77–81.
14. Zeiher AM, Drexler H, Saurbier B, Just H. Endothelium-mediated coronary blood flow modulation in humans. Effects of age, atherosclerosis, hypercholesterolemia, and hypertension. J Clin Invest. 1993;92(2):652–62.
15. Taddei S, Virdis A, Mattei P, Ghiadoni L, Fasolo CB, Sudano I, et al. Hypertension causes premature aging of endothelial function in humans. Hypertension. 1997;29(3):736–43.
16. Taddei S, Virdis A, Mattei P, Ghiadoni L, Gennari A, Fasolo CB, et al. Aging and endothelial function in normotensive subjects and patients with essential hypertension. Circulation. 1995;91(7):1981–7.
17. Taddei S, Virdis A, Ghiadoni L, Salvetti G, Bernini G, Magagna A, et al. Age-related reduction of NO availability and oxidative stress in humans. Hypertension. 2001;38(2):274–9.
18. Simmons DL, Botting RM, Hla T. Cyclooxygenase isozymes: the biology of prostaglandin synthesis and inhibition. Pharmacol Rev. 2004;56(3):387–437.
19. Feletou M, Huang Y, Vanhoutte PM. Endothelium-mediated control of vascular tone: COX-1 and COX-2 products. Br J Pharmacol. 2011;164(3):894–912.
20. Virdis A, Bacca A, Colucci R, Duranti E, Fornai M, Materazzi G, et al. Endothelial dysfunction in small arteries of essential hypertensive patients: role of cyclooxygenase-2 in oxidative stress generation. Hypertension. 2013;62(2):337–44.
21. Vanhoutte PM. One or two, does it matter as long as the Arterial Wall is Coxygenated? Hypertension. 2013;62:244–6.
22. Kojima M, Hosoda H, Date Y, Nakazato M, Matsuo H, Kangawa K. Ghrelin is a growth-hormone-releasing acylated peptide from stomach. Nature. 1999;402(6762):656–60.
23. Kleinz MJ, Maguire JJ, Skepper JN, Davenport AP. Functional and immunocytochemical evidence for a role of ghrelin and des-octanoyl ghrelin in the regulation of vascular tone in man. Cardiovasc Res. 2006;69(1):227–35.
24. Iglesias MJ, Pineiro R, Blanco M, Gallego R, Dieguez C, Gualillo O, et al. Growth hormone releasing peptide (ghrelin) is synthesized and secreted by cardiomyocytes. Cardiovasc Res. 2004;62(3):481–8.
25. Iantorno M, Chen H, Kim JA, Tesauro M, Lauro D, Cardillo C, et al. Ghrelin has novel vascular actions that mimic PI 3-kinase-dependent actions of insulin to stimulate production of NO from endothelial cells. Am J Physiol Endocrinol Metab. 2007;292(3):E756–64.
26. Kawczynska-Drozdz A, Olszanecki R, Jawien J, Brzozowski T, Pawlik WW, Korbut R, et al. Ghrelin inhibits vascular superoxide production in spontaneously hypertensive rats. Am J Hypertens. 2006;19(7):764–7.
27. Tesauro M, Schinzari F, Iantorno M, Rizza S, Melina D, Lauro D, et al. Ghrelin improves endothelial function in patients with metabolic syndrome. Circulation. 2005;112(19):2986–92.
28. Tesauro M, Schinzari F, Rovella V, Di Daniele N, Lauro D, Mores N, et al. Ghrelin restores the endothelin 1/nitric oxide balance in patients with obesity-related metabolic syndrome. Hypertension. 2009;54(5):995–1000.
29. Virdis A, Duranti E, Colucci R, Ippolito C, Tirotta E, Lorenzini G, et al. Ghrelin restores nitric oxide availability in resistance circulation of essential hypertensive patients: role of NAD(P)H oxidase. Eur Heart J. 2015;36(43):3023–30.

Interrelationships Between Micro- and Macrocirculation

9

Stéphane Laurent and Pierre Boutouyrie

9.1 Introduction

Large and small artery remodeling and functional changes are generally analyzed separately in pathophysiology. However, large and small arteries are strongly interconnected. The present chapter focuses on hypertension, particularly essential hypertension because this is the most dominant form, and proposes an integrated pathophysiological approach in order to better understand how large and small artery changes interact in pressure wave transmission; increase central pressure pulsatility; exaggerate cardiac, brain, and kidney damage; and lead to cardiovascular and renal complications. To better illustrate these complex interrelationships, we previously advanced the concept of "large/small artery cross-talk" [1]. The present chapter will first address the small artery remodeling and functional changes. They will be briefly summarized since they are extensively analyzed in several chapters. Second, we will discuss in more detail the large artery remodeling and functional changes [2]. And then, we will explain the large/small artery cross-talk, by which small artery alterations influence large artery phenotype and conversely large artery alterations influence small artery phenotype. Finally, we will apply this concept to explain the damage of target organs in hypertension, at the brain, kidney, and heart level.

S. Laurent (✉) · P. Boutouyrie
Department of Pharmacology, Assistance Publique-Hôpitaux de Paris, Hôpital Européen Georges Pompidou, Paris, France

Université Paris-Descartes, Paris, France

INSERM U 970, Paris, France
e-mail: stephane.laurent@egp.aphp.fr; pierre.boutouyrie@aphp.fr

E. Agabiti-Rosei et al. (eds.), *Microcirculation in Cardiovascular Diseases*, Updates in Hypertension and Cardiovascular Protection,
https://doi.org/10.1007/978-3-030-47801-8_9

9.2 Small Artery Alterations

Resistance arteries play a crucial role in the control of blood pressure. The main drop in hydrostatic pressure indeed occurs at their site. Peripheral resistance in small arteries (lumen diameter less than 350 μm) and arterioles (lumen diameter less than 100 μm) accounts for 45–50% of total peripheral resistance [3–6], whereas capillaries (about 7 μm lumen diameter) account for 23–30%. Poiseuille's law states that resistance is inversely proportional to the radius to the fourth power. Thus, slight alterations in arterial lumen, either functional or structural, result in significant changes in arterial resistance.

The major features of small resistance arteries in patients with essential hypertension are vasoconstriction, eutrophic remodeling with increased media-to-lumen ratio, increased myogenic tone, reduced distensibility, decreased vasodilation reserve, and rarefaction [2–9]. Most of structural changes observed in patients with essential hypertension are a consequence of inward eutrophic remodeling [2, 6, 9–11]. Inward eutrophic remodeling corresponds to a greater media thickness and a reduced lumen and external diameter, with increased media-to-lumen ratio, without any significant change in the total amount of wall tissue, as indicated by unchanged media cross-sectional area [5, 6, 9]. These structural changes are related to increased myogenic tone [7, 8, 12, 13].

In essential hypertension, the role of myogenic tone in the autoregulation of blood flow depends on the regional circulation [8]. For instance, in the renal circulation, myogenic tone primarily mediates the autoregulatory response, a mechanism which is responsible for protection against hypertensive injury [14]. If renal autoregulatory ability is impaired, even modest increases in systemic BP can be transmitted to the glomerular capillaries. Thus, impaired myogenic tone of the renal circulation, such as observed in vessels from hypertensive patients with type 2 diabetes, reduces the autoregulation capacity and increases the barotrauma due to high systolic BP, leading to glomerular injury. The specific features of the coronary and cerebral circulations, and their response to excessive pressure pulsatility, are described below.

Reduced distensibility of small resistance arteries has been variably observed in hypertensive humans and animals. The consequences of reduced distensibility on regional blood flow are complex. A reduced distensibility can contribute to narrow the lumen of small arteries at high BP levels, exaggerating the structural part of total peripheral resistance, thus reducing blood flow to target organs. However, according to Folkow [8], small artery geometric design and distensibility tend to be altered to an "ideal" extent, so that with ordinary changes in smooth muscle activity, a normal flow range is maintained, despite elevations in both perfusion and transmural pressures and in resistance.

Since blood circulation goes from the heart to large vessels and then to the small arteries, the latter are directly concerned by modification in hemodynamic conditions caused by alterations in large arteries' mechanical properties. In the next section, we will describe the changes in large artery properties in hypertension.

9.3 Large Artery Alterations

9.3.1 Large Artery Remodeling

In essential hypertension, large artery remodeling is characterized by an increase in intima-media thickness (IMT) (about +15 to 40%), a lumen enlargement of proximal elastic arteries, and no change in the lumen diameter of distal muscular arteries [15–20]. Wall thickening [21] allows compensation for the rise in BP and tends to normalize circumferential wall stress, according to the Lamé equation $\sigma_\theta = \frac{P \times R}{h}$, where stress is proportional to radius (R) and pressure (P) and inversely proportional to thickness (h).

The enlargement of proximal elastic large arteries with aging and elevated MBP has been extensively described in humans in studies using ultrasounds, particularly high-resolution echotracking systems [2, 15, 19, 22–24]. The enlargement of large arteries, such as the thoracic aorta and the common carotid artery, is generally attributed to the fracture of the load-bearing elastin fibers in response to the fatiguing effect of steady and pulsatile tensile stress. However, growth and apoptosis of vascular smooth muscle cells (VSMCs) could also be involved. Indeed, cyclic strain, which is a major determinant of gene expression, phenotype, and growth of VSMCs in vitro [25–27], could play a role as pulsatile load, exerting both a fatiguing effect and signaling growth on VSMCs. In untreated essential hypertensives, we showed [20], using high-resolution echotracking systems for measuring common carotid IMT and lumen diameter and applanation tonometry for measuring local pulse pressure (PP), that local PP was a strong independent determinant of lumen enlargement and wall thickening, whereas brachial PP and MBP were not. Thus, local pulsatile stress (PP) plays a more important role than steady stress (MBP) [2]. The enlargement of large proximal arteries can be viewed as a compensating phenomenon that enables to maintain a certain level of arterial compliance (expressed as $\Delta V/\Delta P$, i.e., the product of volumic distensibility $\Delta V/V.\Delta P$ and volume V) despite aortic stiffening [22].

In hypertensives, wall thickening, BP increase, and lumen enlargement, all conjugate to affect circumferential wall stress. At the site of hypertensive proximal elastic arteries, such as the common carotid artery, the intima-media thickening is insufficient to compensate for both the enlargement of internal diameter and the rise in BP: Circumferential wall stress is significantly increased compared to normotensives [28]. By contrast, at the site of hypertensive distal muscular arteries, such as the radial artery, the intima-media thickening compensates for the rise in BP (lumen diameter is not enlarged), and circumferential wall stress is not significantly increased compared to nomortensives [17].

9.3.2 Large Artery Stiffness

The wording "arterial stiffness" is a general term that refers to the loss of arterial compliance and/or changes in vessel wall properties. Compliance of large arteries, including the thoracic aorta that has the major role, represents their ability to dampen the pulsatility of ventricular ejection and to transform a pulsatile pressure (and flow) at the site of the ascending aorta into a continuous pressure (and flow) downstream at the site of arterioles, in order to lower the energy expenditure of organ perfusion [2]. Indeed, during ventricular contraction, part of the stroke volume is forwarded directly to the peripheral tissues and part of it is momentarily "stored" in the aorta and central arteries stretching the arterial walls and raising local blood pressure. During diastole, the stored energy recoils the aorta, squeezing the accumulated blood forward into the peripheral tissues, ensuring a continuous flow. The efficiency of this function depends on the stiffness and geometry of the arteries [22, 29]. When the stiffness is low, arterial wall opposes low resistance to distension and the pressure effect is minimized. In hypertension, the arterial system is rigid and necessitates high pressure to be stretched to the same degree as in normotensives. A larger proportion of the stroke volume will flow through the arterial system and peripheral tissues mainly during systole with two consequences: intermittent flow and pressure, excessive flow and pressure pulsatility at the site of distal small resistance and short capillary transit time with reduced metabolic exchanges, altogether damaging target organs. The pulsatility of blood pressure is further strongly influenced by the phenomenon of wave reflection [22, 29], as detailed below.

9.3.3 Hypertension and Arterial Stiffness

With age and hypertension, the repeated mechanical stress induced by the local pulsatility induces biomechanical fatigue of the wall components in the load-bearing media of elastic arteries. This is associated with a loss of the orderly arrangement of smooth muscle cells (SMCs) and extracellular matrix (ECM). Elastic fibers and laminae display thinning, splitting, fraying, and fragmentation [30, 31]. Degeneration of elastic fibers is associated with an increase in collagenous material and in ground substance and often with deposition of calcium together with degeneration of elastic fibers. These phenomenons occur in parallel with profound remodeling of the extracellular matrix. Indeed, any alteration in collagen and elastin production and molecular repair mechanisms can theoretically change arterial elasticity. The above mechanisms are variously associated in patients with essential hypertension, depending on the associated cardiovascular risk factors or disease. In secondary forms of hypertension, such as primary aldosteronism [32] or renovascular hypertension [33], arterial stiffness is mainly due either to the fibrotic effect of high aldosterone levels [32] or to the continuous activation of the renin-angiotensin system [33].

Regional and local arterial stiffness can be measured directly and noninvasively at various sites along the arterial tree in humans. The measurement of pulse wave velocity (PWV) is generally accepted as the most simple, noninvasive, robust, and reproducible method with which to determine regional arterial stiffness [34].

Carotid-femoral PWV (cfPWV), directly measured along the aortic and aortoiliac pathway, is the most clinically relevant, since the aorta is responsible for most of the pathophysiological effects of arterial stiffness. Carotid-femoral PWV is measured using the foot-to-foot velocity method and calculated as cfPWV = D (meters)/Δt (seconds), where D is the distance covered by the waves, usually assimilated to the surface distance between the two recording sites from various waveforms, and Δt is the time delay (or transit time). A large number of studies [35–39], including a collaborative study [40] in 16,867 subjects and patients, showed that age and BP were the main determinants of cfPWV and that at a given age, cfPWV was higher in hypertensives than in normotensives.

Local arterial stiffness of superficial arteries (carotid, brachial, radial, femoral) can be determined noninvasively in humans using high-resolution echotracking devices. They have been developed to measure internal diameter, stroke change in diameter, and intima-media thickness (IMT) with a very high precision [15–18, 41–43].

9.3.4 Wave Reflection and Transmission of Pulsatility to Small Arteries

In healthy subjects, the stiffness gradient between proximal elastic arteries and distal muscular arteries leads to an abrupt rise in impedance (impedance mismatch), generating reflection of the pressure wave that limits the transmission of pressure pulsatility to the small arteries of target organs (Fig. 9.1). Most of the reflected pulsatile energy that propagates backward (i.e., toward the heart) travels at low

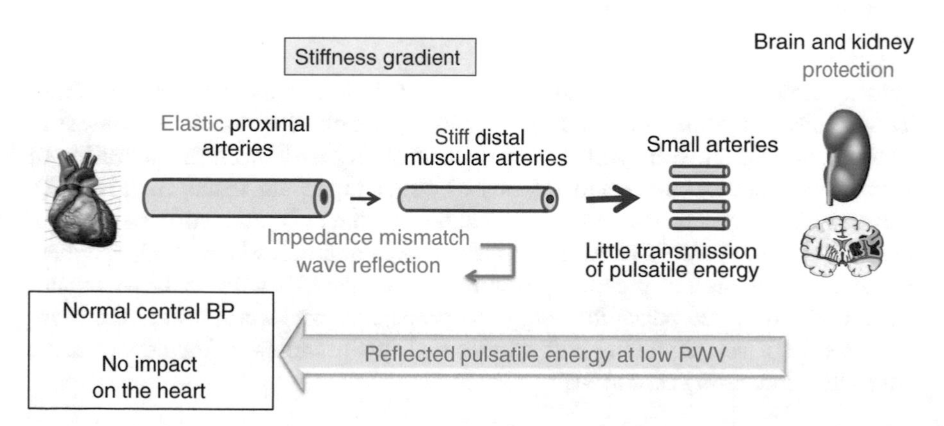

Fig. 9.1 Healthy aging, a means to retard brain, kidney, and heart damage. In healthy subjects, the stiffness gradient between proximal elastic arteries and distal muscular arteries leads to an impedance mismatch, generating pressure wave reflection that limits the transmission of pressure pulsatility to target organs. Most of the reflected pulsatile energy that propagates backward, toward the heart, travels at low velocity along elastic arteries and does not superimpose on incident pressure wave. Thus, central BP remains normal. With permission from Laurent S and Cunha P. Large vessels in hypertension: central blood pressure. *In* Hypertension in children and adolescents, Ed. Lurbe E and Wuehl E. Springer publisher 2018 [82]

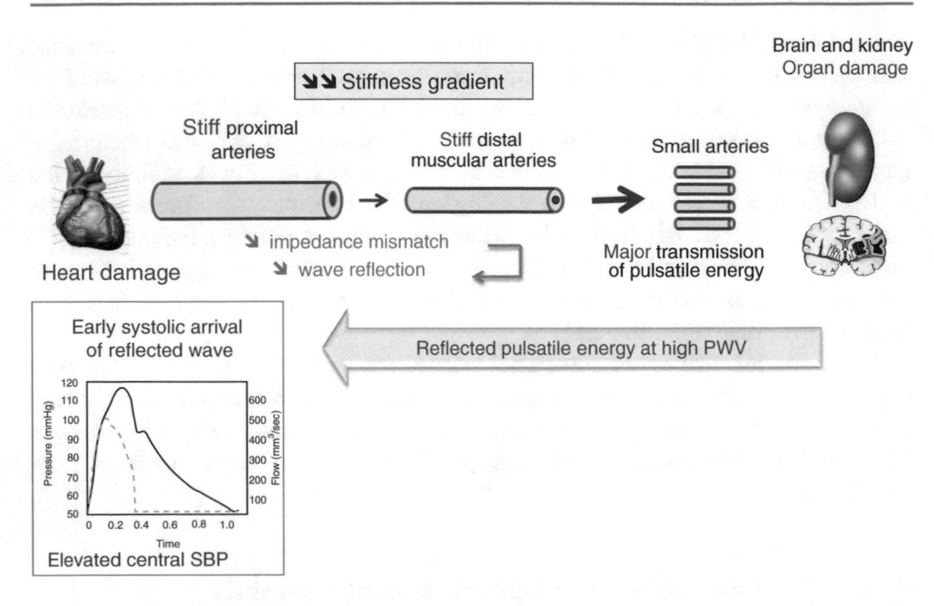

Fig. 9.2 Early damage of brain, kidney, and heart, with early vascular aging. When elastic proximal arteries, such as the aorta, stiffen with aging, they lose their ability to dampen the pulsatility of ventricular ejection, and small arteries of target organ are damaged. Because distal muscular arteries do not stiffen with age, the stiffness gradient between proximal elastic arteries and distal muscular arteries is reduced, and there is more transmission of pressure pulsatility toward small arteries of target organs. Most of the reflected pulsatile energy that propagates backward, toward the heart, travels at high velocity along stiff arteries and superimposes on incident pressure wave, thus increasing central SBP. With permission from Laurent S and Cunha P. Large vessels in hypertension: central blood pressure. *In* Hypertension in children and adolescents, Ed. Lurbe E and Wuehl E. Springer publisher 2018 [82]

velocity along elastic arteries and does not superimpose on incident pressure wave. Thus, central BP remains normal. By contrast, when elastic proximal arteries, such as the aorta, stiffen with aging, they lose their ability to dampen the pulsatility of ventricular ejection, and small arteries of target organ are damaged (Fig. 9.2). Because distal muscular arteries do not stiffen with age [2], the stiffness gradient between proximal elastic arteries and distal muscular arteries is reduced, and there is more transmission of pressure pulsatility toward small arteries of target organs. Most of the reflected pulsatile energy that propagates backward, toward the heart, travels at high velocity along stiff arteries and superimposes on incident pressure wave, thus increasing central SBP.

9.4 Large/Small Artery Cross-Talk in Hypertension

In essential hypertension, small and large artery alterations are closely interdependent. A temporal relationship is difficult to establish, and a cross-talk, by which small artery alterations influence larger artery phenotype and conversely large artery alterations influence small artery phenotype, described as a vicious circle is more likely than a linear sequence [1, 2].

9.4.1 Large/Small Artery Cross-Talk and Central Pulsatility

A cross-talk between small and large artery can be exemplified by the following vicious circle of small/large artery damages (Fig. 9.3), which can be described from any step. Each step refers to pathophysiological changes described above. When necessary, additional evidence is detailed. For instance, we can start the description from small artery damage:

Step 1: Vasoconstriction, impaired vasodilatation, increased wall-to-lumen ratio associated with reduced lumen diameter, and rarefaction of small arteries are major determinants of the increase in total peripheral resistance and mean BP in essential hypertension, as seen above.

Step 2: The higher mean BP in turn increases large artery stiffness. Indeed, when BP is raised, there is a progressive loading (uncurling) and recruitment of stiff collagen fibers and distensibility decreases. This recruitment can operate during a few seconds or within one cardiac cycle where arterial stiffness increases from diastole to systole.

Step 3: The increased large artery stiffness leads to high central systolic and pulse pressures, as seen above. In addition, structural alterations in small resistance

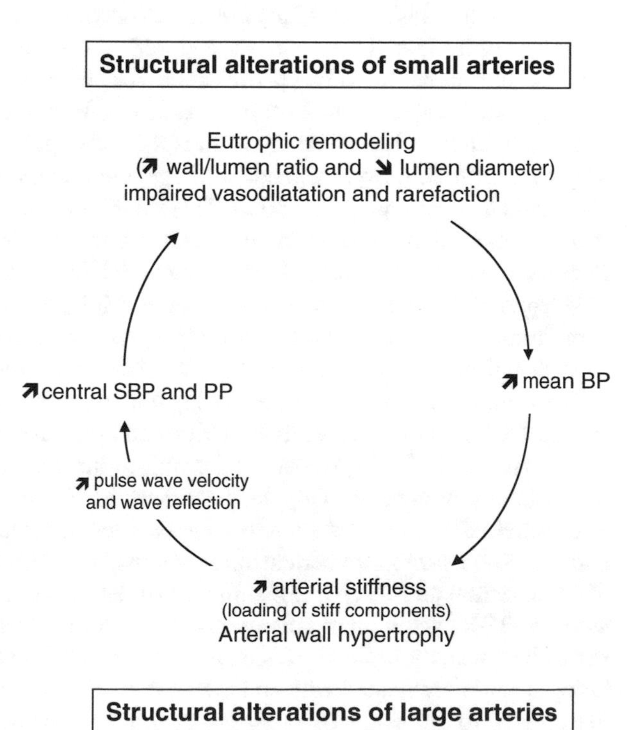

Fig. 9.3 Schematic representation of the large/small artery cross-talk: a vicious circle of aggravation between large and small artery alterations in essential hypertension

arteries contribute to increase the amplitude of wave reflection, which acts synergistically with the increased pulse wave velocity to ultimately raise central systolic and pulse pressures. These relationships are exemplified by the fact that, in hypertensive patients, media-to-lumen (M/L) ratio of subcutaneous small resistance arteries and carotid-femoral PWV are both independent determinants of central SBP [44]. Indeed Muiesan et al. [44] measured central BP, carotid-femoral pulse wave velocity, and M/L ratio of small resistance arteries originating from subcutaneous fat biopsies, in 73 patients with essential or secondary hypertension. M/L ratio was significantly related to brachial systolic and pulse pressures and to central systolic and pulse pressures. A positive correlation was observed between M/L ratio and carotid-femoral pulse wave velocity. This correlation remained statistically significant after adjustment for age and mean BP. M/L ratio was also associated with aortic augmentation index—an index of wave reflection—and this correlation remained statistically significant after adjustment for potential confounders. These findings strongly suggest that both M/L ratio of subcutaneous small resistance arteries and carotid-femoral PWV can act synergistically, or at least additionally, to increase central BP.

Step 4: High central BP pulsatility is correlated with damage of small resistance arteries, i.e., increased M/L ratio of subcutaneous small resistance arteries. This has initially been reported in hypertensive animals [45] and then in hypertensive patients with brachial pulse pressure [46]. In this clinical pioneering study, James et al. [46] analyzed the M/L ratio of small resistance arteries originating from subcutaneous fat biopsies and 24-h ambulatory BP in 32 individuals aged 70 years, 21 of whom were hypertensive and 11 normotensive. M/L ratio was higher in the hypertensive than the normotensive subjects and correlated with age, clinic systolic pressure, 24-h systolic pressure, and 24-h pulse pressure. Importantly, stepwise multivariate regression analysis identified clinic and 24-h pulse pressure as the only significant predictors of M/L ratio independent of age, other parameters of clinic blood pressure, and blood pressure variability. These findings thus confirmed those from animal models of hypertension in demonstrating the importance of pulse pressure in relation to small artery remodeling. More recently, as seen above, Muiesan et al. [44] reported that M/L ratio of small arteries originating from subcutaneous fat was significantly related to central systolic and pulse pressures.

Although most studies of small arteries have concerned subcutaneous fat originating from gluteal biopsy, an increasing number of studies relate to a noninvasive approach (i.e., in vivo rather than in vitro) and measure changes in the retinal arterioles. Salvetti et al. [47] measured the wall-to-lumen ratio (WLR) of retinal arterioles using scanning laser Doppler flowmetry in 295 subjects (mean age 54 years) and central BP with applanation tonometry. They observed a higher WLR in treated patients with essential hypertension in comparison with normotensive individuals. WLR of retinal arterioles was significantly related to clinic systolic and pulse pressures, to 24-h systolic and pulse pressures, and to central systolic and pulse pressures. However, multiple regression analysis revealed that only mean 24-h SBP was independently associated with an increased WLR of retinal arterioles. Further studies remain to be done in order to determine whether the lack of significant correlation between central BP and WLR of retinal arterioles is due to the variability of central BP measurement, the heterogeneity of the population, or the methods used for assessing retinal arteriole morphology. Indeed, the in vivo evaluation of retinal arterioles does not allow to completely separate structural from functional characteristics, whereas the more invasive micromyographic technique used to analyze subcutaneous small arteries clearly identifies the eutrophic or hypertrophic remodeling [47].

Steps 5 and 1: Increased media-to-lumen ratio of subcutaneous small resistance arteries, which is associated with reduced lumen diameter, represents the largest part of the structural aspect of increased total peripheral resistance, leading to a rise in mean BP, and so continues the vicious circle.

This pathophysiological approach can be completed at two levels:

(a) The hypertension-induced remodeling of the vasa vasorum contained in the periaortic fat of the adventitia can reduce the vasa vasorum flow and impair the nutrition of the outer layers of the thoracic aorta, leading to increased aortic stiffness [48].
(b) Not only is arterial stiffening a consequence of the rise in mean BP, but also arterial stiffening may be the determinant of elevated systolic BP on the long term. Indeed, longitudinal assessment of the temporal relationship between carotid [49] and aortic stiffness [39, 50] on the one side and incident hypertension on the other side suggests a precursor role of arterial stiffening in future altered systolic hemodynamic load.

9.4.2 Central Pressure Pulsatility, Small Artery Changes, and Target Organ Damage

As detailed above, the small/large artery cross-talk exerts a synergistic effect on target organs, mainly through the excess of pulsatile energy which is delivered to small arteries either as central peak BP (systolic) or as pulse pressure when upstream large arteries stiffen. As seen above, large artery stiffening leads to intermittent flow and pressure, i.e., excessive flow and pressure pulsatility at the site of distal small resistance and short capillary transit time with reduced metabolic exchanges, altogether damaging target organs. Recent findings are consistent with the hypothesis that central systolic and pulse pressures are the most damaging components of the blood pressure load on target organs. Their elevations have been found to exhibit a closer correlation with hypertensive target organ injury than either systolic and pulsatile components measured at the brachial artery or the steady component (mean BP) [2, 51, 52].

9.4.2.1 Cardiac Damage

Central SBP and PP rather than brachial are stronger determinants of LVH [53], systolic [54] and diastolic dysfunction [55], left atria dilatation, and AF [56]. Both large and small artery damages contribute to the pathophysiology of cardiovascular diseases. For instance, central SBP increases the load on the left ventricle, increasing myocardial oxygen demand. In addition, arterial stiffness is associated with left ventricular hypertrophy (LVH) [53], a known risk factor for coronary events [57]. The increase in central PP and the decrease in diastolic BP may directly cause subendocardial ischemia. Indeed, decreased large epicardial coronary vascular tree perfusion during diastole is a consequence of enhanced central pulsatility and leads to decreased coronary flow reserve [58]. Rarefaction and remodeling of intramyocardial coronary artery [54], left ventricular hypertrophy, and left ventricular diastolic dysfunction further contribute to microcirculatory flow reserve reduction, impaired tissue perfusion, and susceptibility in ischemia during high metabolic and oxygen demand [58]. Epicardial coronary atherosclerosis exaggerates the deleterious effects of above damages and contributes to the pathogenesis of ischemic heart disease.

Increased myogenic tone and arteriolar vasoconstriction contribute to the protection of downstream capillaries but also promote functional rarefaction, ultimately leading to structural rarefaction [59]. Capillary rarefaction, associated with small artery remodeling, impairs tissue perfusion and organ function via three main mechanisms [60]: the tissue nutritive role in response to variations in demand, the capillary protection against the potential damaging effect of blood pressure increase, and the resistance to flow perfusion. For instance, intramyocardial coronary rarefaction impairs tissue perfusion and increases susceptibility in ischemia during high metabolic and oxygen demand [59].

9.4.2.2 Brain Damage

Large/small artery damage can increase the risk of ischemic stroke, white matter lesions, lacunar infarcts, and cognitive decline through several mechanisms [61, 62]. An increasing body of evidence suggests that high pulse pressure transmitted into cerebral arteries can lead to small cerebral artery remodeling with progressive encroachment of the arterial lumen aimed at protecting the microcirculation from pulsatile stress. The cerebral circulation (together with the kidney) is particularly susceptible to pressure damage, since this is a torrential circulation with minimal vascular resistance; therefore, mean and pulse pressures are easily transmitted from the aorta to small cerebral (and renal) arteries [63]. Two studies reported that an increased arterial pulsatility due to large artery stiffening can be transmitted to cerebral small vessels and associated with white matter lesions [64, 65]. Indeed, carotid flow pulsatility index, measured either with transcranial Doppler in the middle cerebral artery or with extracranial Doppler in the internal carotid artery, and carotid pulse pressure and carotid-femoral pulse wave velocity were associated with increased risk for microvascular structural brain damage, such as silent subcortical infarcts or white matter lesions [64–66], and lower scores in various cognitive domains [65]. Carotid stiffness has also been used as proxy for middle-sized cerebral artery stiffness and reported to be associated with increasing large white matter hyperintensity volume, independently of vascular risk factors and carotid plaque [62].

In hypertension, the inward remodeling of small cerebral artery and associated increased myogenic tone impair vasomotor reactivity, limit the autoregulation of cerebral blood flow, and increase susceptibility to focal ischemia when blood pressure is transiently and/or acutely low [67]. Patients with exaggerated visit-to-visit variability of blood pressure, namely, SBP, are at increased risk of stroke [68], which suggests that repeated episodes of hypoperfusion and microvascular ischemia that are more likely in those patients could favor tissue damage and stroke.

Finally, a high central PP influences arterial remodeling not only at the site of the intracranial arteries but also in extracranial arteries, increasing carotid wall thickness, leading to the development of plaques [20, 69].

9.4.2.3 Renal Damage

We explained above how impaired myogenic tone of the renal circulation reduces the autoregulation capacity and increases the barotrauma due to high systolic BP, leading to glomerular injury. Thus, in hypertensive patients with type 2 diabetes, even modest increases in systemic BP are transformed into higher pulsatile energy transmitted to the kidney, with higher dissipation in the microcirculation, hyperfiltration, and glomerulosclerosis. Clinical investigation is consistent with this pathophysiological approach. Significant relationships have been demonstrated between brachial pulse pressure and either glomerular filtration rate (GFR) or microalbuminuria [70]; between arterial stiffness and either GFR [71, 72] or urinary albumin (126); between carotid stiffness and GFR [71, 72]; and between central PP and incident end-stage renal disease [73]. Although not all these relationships are independent of confounding factors, there is a large amount of evidence for linking the pulsatility of BP to renal damage.

9.4.2.4 Retinal Damage

Several studies reported significant relationships between the remodeling of retinal arterioles and arterial stiffness. Two aspects should be underlined before discussing their findings. First, as discussed above, although most studies of small arteries have concerned subcutaneous fat originating from gluteal biopsy, an increasing number of studies have used a noninvasive approach and measured the remodeling of retinal arterioles. Here, at the retinal level, small arteries not only represent an easier access to small resistive arteries but also allow the true localization of organ damage, i.e., at the site of the retinal arteriolar system. However, in contrast to the in vitro micromyographic technique using subcutaneous fat small arteries, this in vivo technique does not allow to completely separate structural from functional characteristics, and it is more difficult to clearly identify the eutrophic or hypertrophic remodeling. Second, during the last 20 years, the technologies for measuring retinal artery remodeling and arterial stiffness improved, thus increasing the level of evidence. It is important to keep it in mind when analyzing the results of these studies.

In 2004, Liao et al. [74] reported, in a population-based cross-sectional study involving 8031 men and women 45–64 years of age, a significant relationship between carotid arterial stiffness (high-resolution ultrasonic echotracking of the left common carotid artery) and retinal arteriolar narrowing (digitized retinal photographs). In 2007, Cheung et al. [75] measured, in the Multi-Ethnic Study of Atherosclerosis (MESA) study, a cross-sectional population-based study of 5731 participants (aged 45–84 years) who were free of clinical cardiovascular disease, retinal arteriolar calibers from retinal photographs. After adjusting for a number of confounding factors, reduced large artery compliance (determined from pressure wave analysis) was significantly associated with smaller retinal arteriolar caliber. In 2012, Da Silva et al. [76] investigated the relationship between aortic stiffness (cPWV) and retinal microvascular changes (retinal photographs) in 145 acute ischemic stroke patients and reported that retinal arteriolar caliber significantly decreased with increasing cPWV. In 2016, Aissopou et al. [77], using retinal vessel calibers measured from digital images in 181 individuals (mean age 54 years), reported that, after regression analysis and adjustment for confounders, both 24 h-ambulatory PWV (Mobil-O-Graph) and office cfPWV (SphygmoCor) were associated mutually independently with narrower retinal arterioles.

Several other studies reported significant relationships between the remodeling of retinal arterioles and central BP. As detailed above, Salvetti et al. [47] reported that the wall-to-lumen ratio (WLR) of retinal arterioles, measured using scanning laser Doppler flowmetry, was significantly related to clinic, 24-h, and central systolic and pulse pressures [47].

9.4.2.5 Cardiovascular, Cerebrovascular, and Renal Outcomes

Because the small/large artery cross-talk exerts a synergistic damaging effect on target organs, it is not surprising that arterial stiffness [35–38, 78], central systolic and pulse pressures [73, 79, 80], and media-to-lumen ratio of small resistance arteries [81] have independent predictive value for CV events and renal complications in hypertensive patients. Several reviews and book chapters [1, 6, 34, 51] have already addressed this issue.

9.5 Conclusion

We proposed an integrated pathophysiological approach that we named "large/small artery cross-talk" in order to better explain how large and small artery changes interact in pressure wave transmission; increase central pressure pulsatility; exaggerate cardiac, brain, and kidney damage; and lead to cardiovascular and renal complications. The cross-talk between the micro- and the macrocirculation promotes a vicious circle, which can begin either from large vessels or at the site of small arteries. Sequentially, increased resistance in small arteries increases mean BP and then increases arterial stiffness in the large elastic arteries, which concomitantly with more pressure wave reflections increases central SBP, ultimately leading to target organ damage. The increased central BP pulsatility is in turn a factor of small resistance artery damage. Lastly, the increase in the media-to-lumen ratio of small resistance arteries and the concomitant reduction in lumen diameter represent the most important part of the structural aspects of increased total peripheral resistance in essential hypertension, contributing to a rise in mean BP and thus prolonging the vicious circle.

9.6 Sources of Funding

This review was funded by INSERM, University Paris-Descartes, and Assistance Publique-Hôpitaux de Paris.

9.7 Disclosure

None.

References

1. Laurent S, Briet M, Boutouyrie P. Large/small artery cross talk and recent morbidity mortality trials in hypertension. Hypertension. 2009;54:388–92.
2. Laurent S, Boutouyrie P. The structural factor in hypertension: large and small artery alterations. Circ Res. 2015;116:1007–21.
3. Mulvany MJ, Aalkjaer C. Structure and function of small arteries. Physiol Rev. 1990;70:921–71.
4. Schiffrin EL. Reactivity of small blood vessels in hypertension: relation with structural changes. State of the art lecture. Hypertension. 1992;19:II1–9.
5. Mulvany MJ, Baumbach GL, Aalkjaer C, Heagerty AM, Korsgaard N, Schiffrin EL, Heistad DD. Vascular remodeling. Hypertension. 1996;28:505–6.
6. Rizzoni D, Agabiti-Rosei E. Structural abnormalities of small resistance arteries in essential hypertension. Intern Emerg Med. 2012;7:205–12.
7. Folkow B. Physiological aspects of primary hypertension. Physiol Rev. 1982;62:347–504.
8. Folkow B. Hypertensive structural changes in systemic precapillary resistance vessels: how important are they for in vivo haemodynamics? J Hypertens. 1995;13:1546–59.
9. Heagerty AM, Izzard AS. Small artery changes in hypertension. J Hypertens. 1995;13:1560–5.
10. Heagerty AM, Aalkjaaer C, Bund SJ, Korsgaard N, Mulvany MJ. Small artery structure in hypertension. Dual process of remodeling and growth. Hypertension. 1993;21:391–7.
11. Intengan HD, Deng LY, Li JS, Schiffrin EL. Mechanics and composition of human subcutaneous resistance arteries in essential hypertension. Hypertension. 1999;33:569–74.
12. Izzard AS, Rizzoni D, Agabiti-Rosei E, Heagerty AM. Small artery structure and hypertension: adaptive changes and target organ damage. J Hypertens. 2005;23:247–50.
13. Matrougui K, Levy BI, Henrion D. Tissue angiotensin II and endothelin-1 modulate differently the response to flow in mesenteric resistance arteries of normotensive and spontaneously hypertensive rats. Br J Pharmacol. 2000;130:521–6.
14. Bidani AK, Griffin KA, Williamson G, Wang X, Loutzenhiser R. Protective importance of the myogenic response in the renal circulation. Hypertension. 2009;54:393–8.
15. Laurent S, Hayoz D, Trazzi S, Boutouyrie P, Waeber B, Omboni S, Brunner H, Mancia G, Safar M. Isobaric compliance of the radial artery is increased in patients with essential hypertension. J Hypertens. 1993;11:89–98.
16. Benetos A, Laurent S, Hoeks AP, Boutouyrie PH, Safar ME. Arterial alterations with ageing and high blood pressure. A noninvasive study of carotid and femoral arteries. Arterioscler Thromb. 1993;13:90–7.
17. Laurent S, Girerd X, Mourad JJ, Lacolley P, Beck L, Boutouyrie P, Mignot JP, Safar M. Elastic modulus of the radial artery wall is not increased in patients with essential hypertension. Arterioscler Thromb. 1994;14:1223–31.

18. Laurent S, Caviezel B, Beck L, Girerd X, Billaud E, Boutouyrie P, Hoeks A, Safar M. Carotid artery distensibility and distending pressure in hypertensive humans. Hypertension. 1994;23:878–83.
19. Roman MJ, Saba PS, Pini R, Spitzer M, Pickering TG, Rosen S, Alderman M, Devereux R. Parallel cardiac and vascular adaptation in hypertension. Circulation. 1992;86:1909–18.
20. Boutouyrie P, Bussy C, Lacolley P, Girerd X, Laloux B, Laurent S. Local steady and pulse pressure and arterial remodeling. Circulation. 1999;100:1387–93.
21. Engelen L, Ferreira I, Stehouwer CD, Boutouyrie P, Laurent S, Reference Values for Arterial Measurements Collaboration. Reference intervals for common carotid intima-media thickness measured with echotracking: relation with risk factors. Eur Heart J. 2013;34:2368–80.
22. Nichols WW, O'Rourke MF. McDonald's blood flow in arteries; theoretical, experimental and clinical principles. 6th ed: Hodder Arnold; 2011. p. 755.
23. Lam CS, Xanthakis V, Sullivan LM, Lieb W, Aragam J, Redfield MM, Mitchell GF, Benjamin EJ, Vasan RS. Aortic root remodeling over the adult life course: longitudinal data from the Framingham Heart Study. Circulation. 2010;122:884–90.
24. Milan A, Tosello F, Caserta M, Naso D, Puglisi E, Magnino C, Comoglio C, Rabbia F, Mulatero P, Veglio F. Aortic size index enlargement is associated with central hemodynamics in essential hypertension. Hypertens Res. 2011;34:126–32.
25. Leung DY, Glagov S, Mathews MB. Cyclic stretching stimulates synthesis of matrix components by arterial smooth muscle cells in vitro. Science. 1976;191:475–7.
26. Reusch P, Wagdy H, Reusch R, Wilson E, Ives HE. Mechanical strain increases smooth muscle and decreases nonmuscle myosin expression in rat vascular smooth muscle cells. Circ Res. 1996;79:1046–53.
27. Lehoux S, Tedgui A. Cellular mechanics and gene expression in blood vessels. J Biomech. 2003;36:631–43.
28. Bussy C, Boutouyrie P, Lacolley P, Challande P, Laurent S. Intrinsic stiffness of the carotid artery wall material in essential hypertensives. Hypertension. 2000;35:1049–54.
29. Westerhof N, Sipkema P, van den Bos CG, Elzinga G. Forward and backward waves in the arterial system. Cardiovasc Res. 1972;6:648–56.
30. Lacolley P, Challande P, Osborne-Pellegrin M, Regnault V. Genetics and pathophysiology of arterial stiffness. Cardiovasc Res. 2009;81:637–48.
31. Lakatta EG. Cardiovascular regulatory mechanisms in advanced age. Physiol Rev. 1993;73:413–67.
32. Lacolley P, Labat C, Pujol A, Delcayre C, Benetos A, Safar M. Increased carotid wall elastic modulus and fibronectin in aldosterone-salt-treated rats: effects of eplerenone. Circulation. 2002;106:2848–53.
33. Zanchi A, Wiesel P, Aubert JF, Brunner HR, Hayoz D. Time course changes of the mechanical properties of the carotid artery in renal hypertensive rats. Hypertension. 1997;29:1199–203.
34. Laurent S, Cockcroft J, Van Bortel L, Boutouyrie P, Giannattasio C, Hayoz D, et al. Expert consensus document on arterial stiffness: methodological issues and clinical applications. Eur Heart J. 2006;27:2588–605.
35. Laurent S, Boutouyrie P, Asmar R, Gautier I, Laloux B, Guize L, Ducimetiere P, Benetos A. Aortic stiffness is an independent predictor of all-cause and cardiovascular mortality in hypertensive patients. Hypertension. 2001;37:1236–41.
36. Boutouyrie P, Tropeano AI, Asmar R, Gautier I, Benetos A, Lacolley P, Laurent S. Aortic stiffness is an independent predictor of primary coronary events in hypertensive patients: a longitudinal study. Hypertension. 2002;39:10–5.
37. Vlachopoulos C, Aznaouridis K, Stefanadis C. Prediction of cardiovascular events and all-cause mortality with arterial stiffness: a systematic review and meta-analysis. J Am Coll Cardiol. 2010;55:1318–27.
38. Karras A, Haymann JP, Bozec E, Metzger M, Jacquot C, Maruani G, Houillier P, Froissart M, Stengel B, Guardiola P, Laurent S, Boutouyrie P, Briet M, on behalf of The NephroTest study group. Large artery stiffening and remodeling are independently associated with all-cause mortality and cardiovascular events in chronic kidney disease. Hypertension. 2012;60:1451–7.

39. Najjar SS, Scuteri A, Shetty V, Wright JG, Muller DC, Fleg JL, et al. Pulse wave velocity is an independent predictor of the longitudinal increase in systolic blood pressure and of incident hypertension in the Baltimore longitudinal study of aging. J Am Coll Cardiol. 2008;51:1377–83.
40. Reference values for carotid-femoral pulse wave velocity in the reference values for arterial stiffness' collaboration database. Eur Heart J. 2010;31:2338–50.
41. Van Bortel L, Balkestein EJ, van der Heijden-Spek JJ, Vanmolkot FH, Staessen JA, Kragten JA, Vredeveld JW, Safar M, Stuijker-Boudier HA, Hoeks A. Non-invasive assessment of local arterial pulse pressure : comparison of applanation tonometry and echo-tracking. J Hypertens. 2001;19:1037–44.
42. Vermeersch SJ, Rietzschel ER, De Buyzere ML, De Bacquer D, De Backer G, Van Bortel LM, Gillebert TC, Verdonck PR, Segers P. Age and gender related patterns in carotid-femoral PWV and carotid and femoral stiffness in a large healthy, middle-aged population. J Hypertens. 2008;26:1411–9.
43. Giannattasio C, Salvi P, Valbusa F, Kearney-Schwartz A, Capra A, Amigoni M, Failla M, Boffi L, Madotto F, Benetos A, Mancia G. Simultaneous measurement of beat-to-beat carotid diameter and pressure changes to assess arterial mechanical properties. Hypertension. 2008;52:896–902.
44. Muiesan ML, Salvetti M, Rizzoni D, Paini A, Agabiti-Rosei C, Aggiusti C, Bertacchini F, Stassaldi D, Gavazzi A, Porteri E, De Ciuceis C, Agabiti-Rosei E. Pulsatile hemodynamics and microcirculation: evidence for a close relationship in hypertensive patients. Hypertension. 2013;61:130–6.
45. Christensen KL. Reducing pulse pressure in hypertension may normalize small artery structure. Hypertension. 1991;18:722–7.
46. James MA, Watt PAC, Potter JF, Thurston H, Swales JD. Pulse pressure and resistance artery structure in the elderly. Hypertension. 1995;26:301–6.
47. Salvetti M, Agabiti Rosei C, Paini A, Aggiusti C, Cancarini A, Duse S, Semeraro F, Rizzoni D, Agabiti Rosei E, Muiesan ML. Relationship of wall-to-lumen ratio of retinal arterioles with clinic and 24-hour blood pressure. Hypertension. 2014;63:1110–5.
48. Stefanadis C, Vlachopoulos C, Karayannacos P, Boudoulas H, Stratos C, Filippides T, et al. Effect of vasa vasorum flow on structure and function of the aorta in experimental animals. Circulation. 1995;91:2669–78.
49. Liao D, Arnett DK, Tyroler HA, Riley WA, Chambless LE, Szklo M, Heiss G. Arterial stiffness and the development of hypertension: the ARIC study. Hypertension. 1999;34:201–6.
50. Kaess BM, Rong J, Larson MG, Hamburg NM, Vita JA, Levy D, et al. Aortic stiffness, blood pressure progression, and incident hypertension. JAMA. 2012;308:875–81.
51. Roman MJ, Devereux RB. Association of central and peripheral blood pressures with intermediate cardiovascular phenotypes. Hypertension. 2014;63:1148–53.
52. Safar ME, O'Rourke MF. Handbook of hypertension, volume 23: arterial stiffness in hypertension: Elsevier; 2006. p. 598.
53. Boutouyrie P, Laurent S, Girerd X, Beck L, Abergel E, Safar M. Common carotid artery distensibility and patterns of left ventricular hypertrophy in hypertensive patients. Hypertension. 1995;25(part 1):651–9.
54. Laskey WK, Kussmaul WG. Arterial wave reflection in heart failure. Circulation. 1987;75:711–22.
55. Weber T, Auer J, O'Rourke MF, Punzengruber C, Kvas E, Eber B. Prolonged mechanical systole and increased arterial wave reflections in diastolic dysfunction. Heart. 2006;92:1616–22.
56. Mitchell GF, Vasan RS, Keyes MJ, Parise H, Wang TJ, Larson MG, D'Agostino RB Sr, Kannel WB, Levy D, Benjamin EJ. Pulse pressure and risk of new-onset atrial fibrillation. JAMA. 2007;297:709–15.
57. Koren MJ, Devereux RB, Casale PN, Savage DD, Laragh JH. Relation of left ventricular mass and geometry to morbidity and mortality in uncomplicated essential hypertension. Ann Intern Med. 1991;114:345–52.

58. Erdogan D, Yildirim I, Ciftci O, Ozer I, Caliskan M, Gullu H, Muderrisoglu H. Effects of normal blood pressure, prehypertension, and hypertension on coronary microvascular function. Circulation. 2007;115:593–599115.
59. Feihl F, Liaudet L, Levy BI, Waeber B. Hypertension and microvascular remodeling. Cardiovasc Res. 2008;78:274–85.
60. Yannoutsos A, Levy BI, Safar ME, Slama G, Blacher J. Pathophysiology of hypertension: interactions between macro and microvascular alterations through endothelial dysfunction. J Hypertens. 2014;32:216–24.
61. Tzourio C, Laurent S, Debette S. Is hypertension associated with an accelerated ageing of the brain? Hypertension. 2014;63:894–903.
62. Brisset M, Boutouyrie P, Pico F, Zhu Y, Zureik M, Schilling S, Dufouil C, Mazoyer B, Laurent S, Tzourio C, Debette S. Large-vessel correlates of cerebral small-vessel disease. Neurology. 2013;80:662–9.
63. O'Rourke MF, Safar ME. Relationship between aortic stiffening and microvascular disease in brain and kidney. Cause and logic of therapy. Hypertension. 2005;46:200–4.
64. Webb AJ, Simoni M, Mazzucco S, Kuker W, Schulz U, Rothwell PM. Increased cerebral arterial pulsatility in patients with leukoaraiosis: arterial stiffness enhances transmission of aortic pulsatility. Stroke. 2012;10:2631–6.
65. Mitchell GF, van Buchem MA, Sigurdsson S, Gotal JD, Jonsdottir MK, Kjartansson Ó, Garcia M, Aspelund T, Harris TB, Gudnason V, Launer LJ. Arterial stiffness, pressure and flow pulsatility and brain structure and function: the age, gene/environment Susceptibility-Reykjavik study. Brain. 2011;134:3398–407.
66. Jochemsen HM, Muller M, Bots ML, Scheltens P, Vincken KL, Mali WP, van der Graaf Y, Geerlings MI, SMART Study Group. Arterial stiffness and progression of structural brain changes: the SMART-MR study. Neurology. 2015;84:448–855.
67. Fernando MS, Simpson JE, Matthews F, Brayne C, Lewis CE, Barber R, Kalaria RN, Forster G, Esteves F, Wharton SB, Shaw PJ, O'Brien JT, Ince PG, MRC Cognitive Function and Ageing Neuropathology Study Group. White matter lesions in an unselected cohort of the elderly: molecular pathology suggests origin from chronic hypoperfusion injury. Stroke. 2006;37:1391–8.
68. Rothwell PM, Howard SC, Dolan E, O'Brien E, Dobson JE, Dahlöf B, Sever PS, Poulter NR. Prognostic significance of visit-to-visit variability, maximum systolic blood pressure, and episodic hypertension. Lancet. 2010;375:895–905.
69. Zureik M, Bureau JM, Temmar M, Adamopoulos C, Courbon D, Bean K, Touboul PJ, Benetos A, Ducimetière P. Echogenic carotid plaques are associated with aortic arterial stiffness in subjects with subclinical carotid atherosclerosis. Hypertension. 2003;41:519–27.
70. Fesler P, Safar ME, du Cailar G, Ribstein J, Mimran A. Pulse pressure is an independent determinant of renal function decline during treatment of essential hypertension. J Hypertens. 2007;25:1915–20.
71. Briet M, Bozec E, Laurent S, Fassot C, Jacquot C, Froissart M, Houillier P, Boutouyrie P. Arterial stiffness and enlargement in mild to moderate chronic kidney disease. Kidney Int. 2006;96:350–7.
72. Hermans MM, Henry R, Dekker JM, Kooman JP, Kostense PJ, Nijpels G, Heine RJ, Stehouwer CD. Estimated glomerular filtration rate and urinary albumin excretion are independently associated with greater arterial stiffness: the Hoorn Study. J Am Soc Nephrol. 2007;18:1942–52.
73. Briet M, Collin C, Karras A, Laurent S, Bozec E, Jacquot C, Stengel B, Houillier P, Froissart M, Boutouyrie P, for the Nephrotest Study Group. Maladaptive remodeling of large artery has a predictive value for chronic kidney disease progression. J Am Soc Nephrol. 2011;22:967–74.
74. Liao D, Wong TY, Klein R, Jones D, Hubbard L, Sharrett AR. Relationship between carotid artery stiffness and retinal arteriolar narrowing in healthy middle-aged persons. Stroke. 2004;35:837–42.
75. Cheung N, Islam FM, Jacobs DR Jr, Sharrett AR, Klein R, Polak JF, Cotch MF, Klein BE, Ouyang P, Wong TY. Arterial compliance and retinal vascular caliber in cerebrovascular disease. Ann Neurol. 2007;62:618–24.

76. De Silva DA, Woon FP, Manzano JJ, Liu EY, Chang HM, Chen C, Wang JJ, Mitchell P, Kingwell BA, Cameron JD, Lindley RI, Wong TY, Wong MC, behalf of the Multi-Centre Retinal Stroke Study Collaborative Group. The relationship between aortic stiffness and changes in retinal microvessels among Asian ischemic stroke patients. J Hum Hypertens. 2012;26:716–22.
77. Aissopou EK, Argyris AA, Nasothimiou EG, Konstantonis GD, Tampakis K, Tentolouris N, Papathanassiou M, Theodossiadis PG, Papaioannou TG, Stehouwer CD, Sfikakis PP, Protogerou AD. Ambulatory aortic stiffness is associated with narrow retinal arteriolar caliber in hypertensives: the SAFAR Study. Am J Hypertens. 2016;29:626–33.
78. Mitchell GF, Hwang SJ, Vasan RD, Larson MG, Pencina MJ, Hamburg NM, Vita JA, Levy D, Benjamin EJ. Arterial stiffness and cardiovascular events. The Framingham Heart Study. Circulation. 2010;121:505–11.
79. Vlachopoulos C, Aznaouridis K, O'Rourke MF, Safar ME, Baou K. Stefanadis C prediction of cardiovascular events and all-cause mortality with central haemodynamics: a systematic review and meta-analysis. Eur Heart J. 2010;31:1865–71.
80. Safar ME, Blacher J, Pannier B, Guerin A, Marchais SJ, Guyonvarc'h PM, London GM. Central pulse pressure and mortality in end-stage renal disease. Hypertension. 2002;39:735–8.
81. Mancia G, Fagard R, Narkiewicz K, et al. 2013 ESH/ESC Guidelines for the management of arterial hypertension: the task force for the management of arterial hypertension of the European Society of Hypertension (ESH) and of the European Society of Cardiology (ESC). Eur Heart J. 2013;34:2159–219.
82. Laurent S, Cunha P. Large vessels in hypertension: central blood pressure. In: Lurbe E, Wuehl E, editors. Hypertension in children and adolescents: Springer Publisher; 2018.

Alterations in Capillary and Microcirculatory Networks in Cardiovascular Diseases

10

Bernard I. Levy

10.1 Capillaries: Organization and Functions

The capillary beds, downstream from the terminal resistance arterioles, make up a dense network, with 10–15 capillaries branching from a single arteriole over a short distance in the skeletal muscle. The capillary network is therefore far denser than the arteriole network, with parallel branches some 25–30 μm apart. Capillaries are simple endothelial tubes surrounded by a basement membrane and pericytes. In striated muscles, capillary length can vary considerably with the mean length being somewhere between 200 and 500 μm [1]. Capillary diameter at the arterial end of the network is typically the same or even smaller than the size of unstressed erythrocytes, i.e., ≤7.2 μm, its mean value in humans. A typical capillary diameter increases gradually until it is approximately 20% greater by the time it reaches the venous end of the network. The presence, number, and size of endothelial fenestrations are strictly dependent on the specific organ, i.e., the capillary wall structure is specialized according to the function of each organ. Most capillaries are noncontractile although certain sinusoid capillaries in the liver respond actively to catecholamines [2].

In clinical practice, capillary density is usually measured on the skin, on the forearm, or more frequently on the upper surface of a phalanx. Nail capillaries are not commonly used to assess capillary density though are useful in the morphological diagnosis of systemic diseases such as scleroderma and autoimmune rheumatic diseases. Arteriole-venule anastomosis, where the luminal diameter is far larger than the red cell diameter (from 10 to 40 μm), enables the blood to short-circuit the capillary network [3]. This is necessary for the heat transfer from the skin to the

B. I. Levy (✉)
Inserm U970, PARCC & Vessels and Blood Institute, Paris, France
e-mail: bernard.levy@inserm.fr

E. Agabiti-Rosei et al. (eds.), *Microcirculation in Cardiovascular Diseases*, Updates in Hypertension and Cardiovascular Protection,
https://doi.org/10.1007/978-3-030-47801-8_10

ambient air, thus facilitating body thermoregulation: when the skin is exposed to cold temperatures, the arterioles feeding the epidermis undergo vasoconstriction, while the arteriovenous anastomoses vasodilate. Blood flow therefore bypasses the cutaneous surface and limits the cooling of the blood at the surface of the skin. When the body is exposed to high temperatures, the arteriovenous anastomoses are closed, and the whole cutaneous blood flow perfuses the epidermis, allowing maximum heat transfer from the core of the body to the ambient air. Figure 10.1 shows the microcirculatory network of the skin as visualized by scanning electron microscopy. Figure 10.2 is a diagram of the epidermal circulation corresponding to area "A" in Fig. 10.1.

10.1.1 Role of Capillaries: Oxygen Exchange

In addition to its role in determining vascular resistance to flow, which accounts for approximatively 15% of the total drop in pressure in resting skeletal muscle [5], the capillary network is the region of the vascular system that specializes in the transport of substances between the blood and tissues. Due to the large cross-sectional area of the whole capillary bed, blood flow velocity is lowest in these vessels, thus allowing more complete exchange of diffusible substances between the interstitium and plasma.

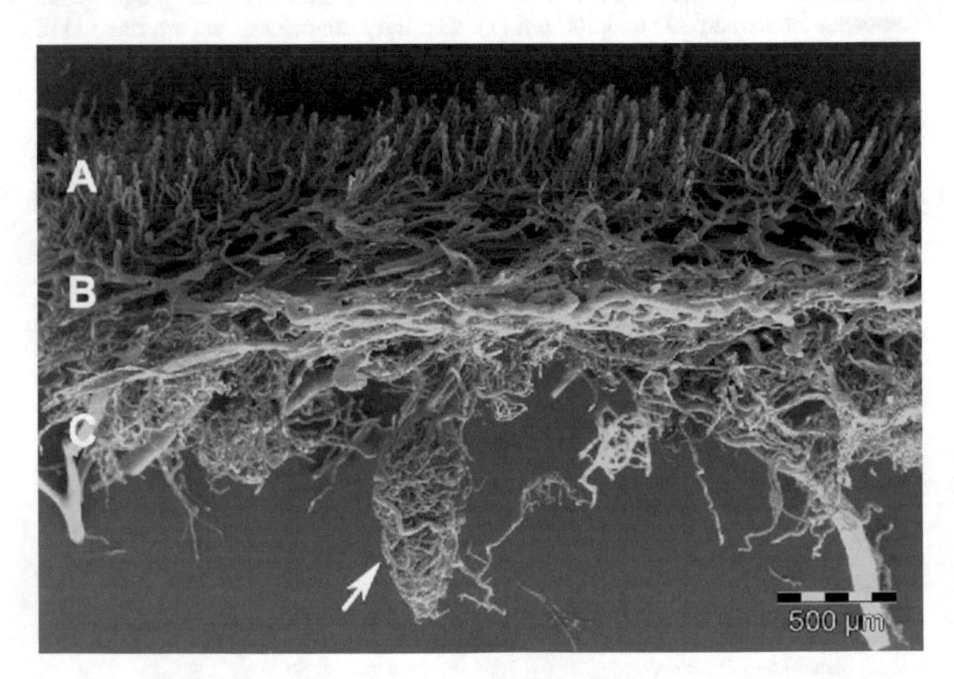

Fig. 10.1 The microvascular structure of the transversally sectioned human digit (thumb): (**a**) papillary layer, (**b**) dermal–epidermal junction, (**c**) hypodermal layer. Note the ovoid structure made up of the capillaries feeding a hair follicle (arrow). From Manelli A. et al. [4]

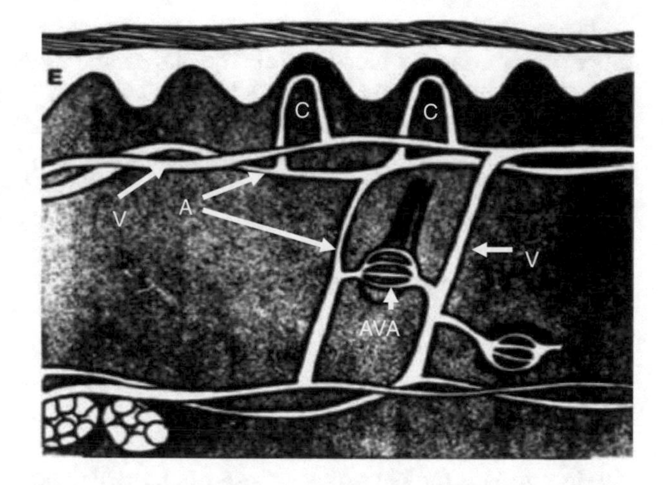

Fig. 10.2 Schematic drawing of microvascular organization in human skin. E, epidermis; A, arteriole; C, capillary; V, postcapillary venule; AVA, arteriovenous anastomosis. From Braverman IM [3]

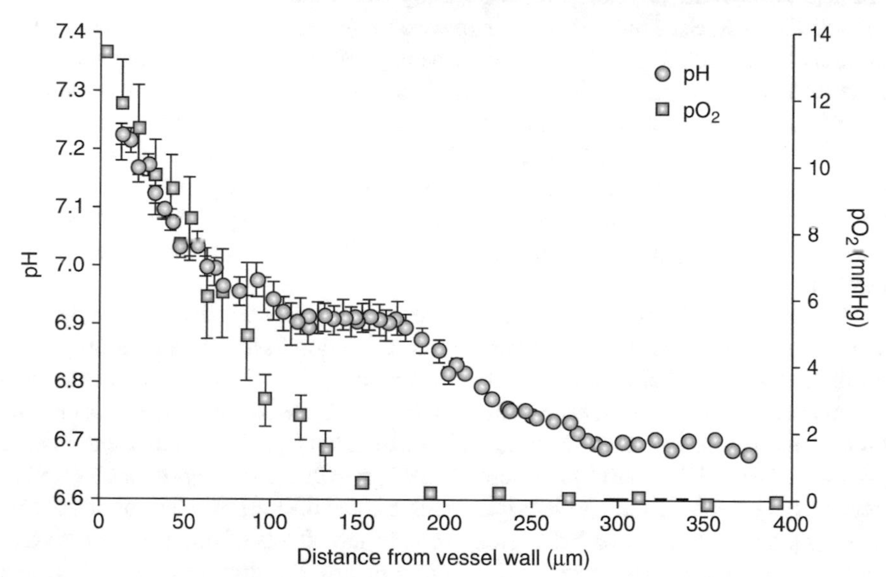

Fig. 10.3 Mean interstitial pH and pO_2 profiles measured at progressively greater distances from the nearest blood vessel (±SEM). Circles symbols, pH; squares symbols, pO_2. 0 μm on the x-axis corresponds to the center of the nearest vessel. The scale thereafter corresponds to the distance from the vessel wall. From Helmlinger G et al. [7]

It was long assumed that oxygen transfer only occurred in the capillaries. However, oxygen loss from blood to tissues occurs predominantly in the arteriolar network [6] and in such a way that the partial oxygen pressure (pO_2) in the capillary blood is no higher than 12–15 mmHg (Fig. 10.3). In highly metabolizing tissues, such as the myocardium and the brain, the volume flow per gram of tissue is higher, and the fraction of total oxygen loss is less in the arterioles and greater in the capillaries, likely related to the shorter blood transit time in the feeding arterioles and the denser capillary network.

Figure 10.3 shows the evolution of pO_2 (squares) and pH (circles) vs. the distance from the nearest blood vessel. The pO_2 decreases monotonically, and hypoxic values (>5 mmHg) were reached 70–80 μm away from the nearest vessel wall. Near anoxic values (0.0–0.5 mmHg) were consistently found ≥150 μm away from the vessel wall. It is noteworthy that (1) the capillary pO_2 is around 14 mmHg, i.e., much lower than the arterial value and lower than the levels generally presumed, and (2) it is virtually impossible for a cell to survive beyond 80–100 μm from the nearest capillary or arteriole. This physical characteristic of the distance of diffusion of O_2 in tissues implies that the intercapillary distance cannot be greater than 100 μm in any living tissue.

10.2 Angiogenesis

Angiogenesis is the physiological process by which new blood vessels form from pre-existing vessels. This is distinct from vasculogenesis, which is the de novo formation of endothelial cells from mesoderm cell precursors, and also from neovascularization, which corresponds to modifications to the existing vascular network to specifically increase blood flow to a hypoxic area. Neovascularization includes arterial and arteriolar remodeling, angiogenesis (creation of neo-capillaries), and remodeling of the venous component of the microcirculatory network. The first vessels in the developing embryo form through vasculogenesis; subsequently, angiogenesis is responsible for the growth of most, if not all, blood vessels during development and in disease [8].

Angiogenesis is a normal and vital process in growth and development, as well as in wound healing and in the formation of granulation tissue via migration, proliferation, and survival of endothelial cells. Vascular angiogenesis factors include different classes of molecules, each with a fundamental role in blood vessel formation. Numerous inducers of angiogenesis such as members of the vascular endothelial growth factor (VEGF) family, basic fibroblast growth factor, angiopoietin (ANG), hepatocyte growth factor, and hypoxia-inducible factor-1 all play an important role in angiogenesis. Interaction between and combined effects of different growth factors are essential in endothelial cell migration, proliferation, differentiation, and endothelial cell-cell communication, all of which ultimately lead to the formation of microvessels. VEGF plays a key role during angiogenesis and can therefore be considered a useful therapeutic target in clinical practice, not only to promote neovascularization under ischemic conditions but also to inhibit abnormal tumor angiogenesis in cancer (see Chap. 13 by Harry Struijker-Boudier). Figure 10.4 shows the main stages of angiogenesis: endothelial cell activation and permeability (A), endothelial proliferation and migration (B), and (C) sprouting and maturation of neo-capillaries.

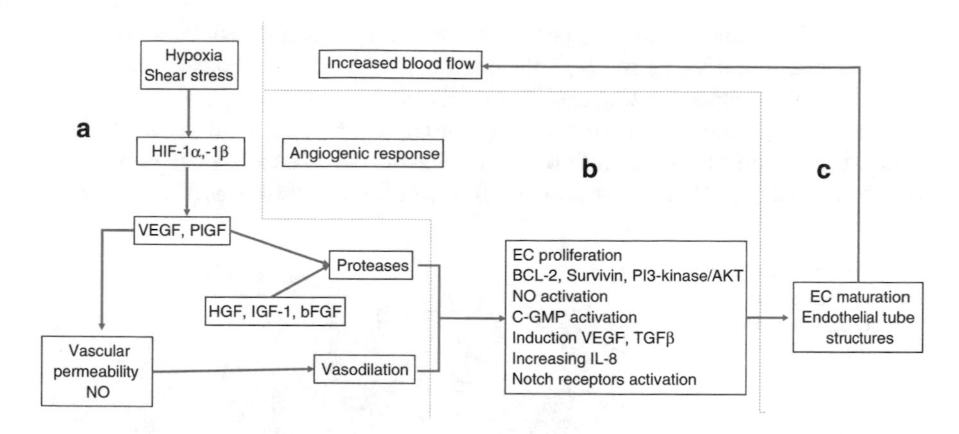

Fig. 10.4 Main stages of angiogenesis. **a**, endothelial cell activation and permeability, VEGF, and nitric oxide are important factors in the onset of angiogenesis, inducing the permeability and dilation of the vessel; **b**, endothelial cell migration and proliferation; proteases mobilize angiogenesis-related factors from the extracellular matrix pool; **c**, sprouting and maturation of endothelial cells. NO, nitric oxide; EC, endothelial cell; HIF-1, hypoxia-inducible factor-1; bFGF, basic fibroblast growth factor; HGF, hepatocyte growth factor

10.2.1 *Capillary Rarefaction*: Imbalance Between Angiogenesis and Capillary Regression

Microcirculatory networks are not stable definitive structures. Angiogenesis is initially a rapid process that decelerates over time until it reaches an equilibrium with capillary regression. Capillary "pruning" is the term used to describe the physiological regression of a subset of microvessels within a vasculature. Capillary density in a given tissue therefore depends on the balance between the formation of neocapillaries and the pruning of existing capillaries. Beyond vessel pruning, vascular networks may also undergo complete physiological regression. Under stable physiological conditions, pro- and anti-angiogenic factors have equipotent effects on vessel density, and the number of newly formed capillaries is approximately the same as the number of regressed capillaries, thus ensuring a stable capillary density.

Multiple signaling pathways are known to be regulators of vessel regression. VEGF/VEGFR2 signaling, non-canonical WNT signaling, and blood flow-induced signaling are all critical to vessel maintenance and play a role in the control of vessel regression. Canonical WNT signaling stabilizes the vascular network and promotes proliferation of endothelial cells. DLL4/Notch signaling assists in vessel regression by promoting vessel constriction and flow stasis. The outcome of ANG/TIE signaling during vessel remodeling is context-dependent. Whereas ANG1 supports endothelial cell survival, ANG2 disrupts the vascular network, driving it into regression in the absence of survival factor activity (e.g., VEGF).

Under circumstances that are common to all cardiovascular risk factors (arterial hypertension, diabetes, aging, etc.), the number of capillaries per tissue volume unit is reduced, thus characterizing capillary rarefaction.

Figure 10.5 shows a simplified diagram of the process of regression. (a–d) Following the primarily blood flow-driven selection of a branch for regression (a), the vessel constricts (b) until it occludes (c) and the blood flow ceases (d). Endothelial

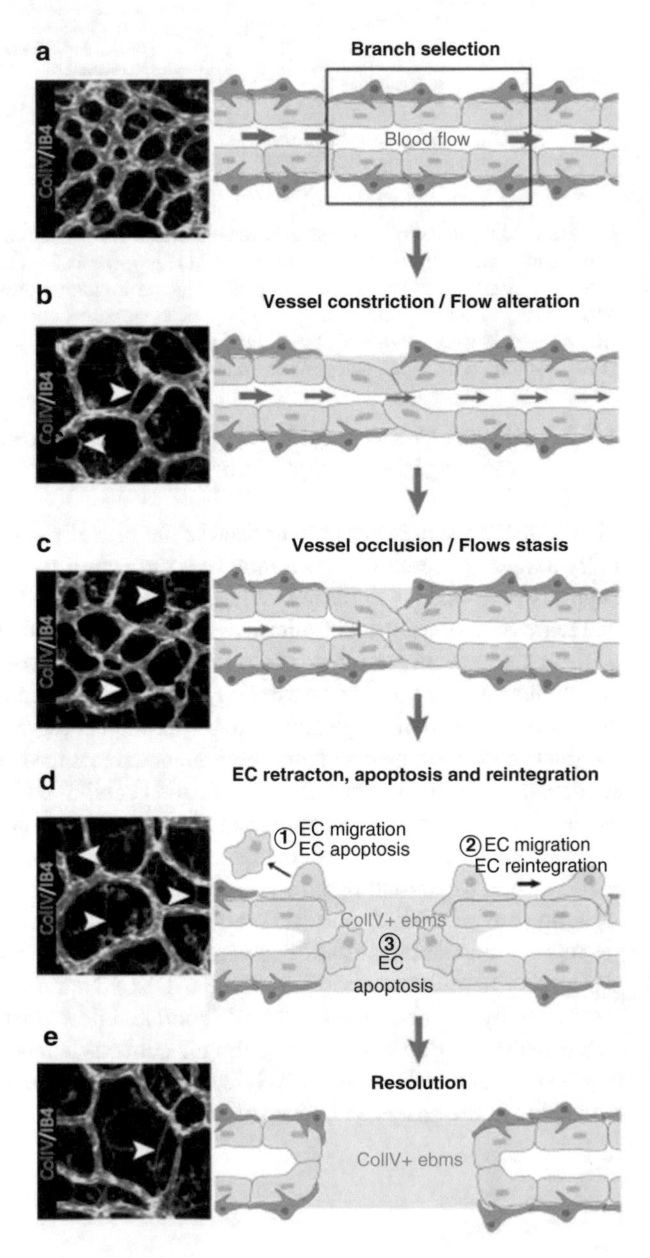

Fig. 10.5 Multistep nature of vessel pruning and regression. From Korn C et al. [9]

cells in regressing vessel segments may retract and undergo apoptosis, or they may migrate away to reintegrate elsewhere, leaving behind collagen IV+ empty basement membrane sleeves (ebms). Retracting endothelial cells are likely to disintegrate from the vascular network and undergo apoptosis due to detachment from the basement membrane. (e) In the final step of resolution, regression of the selected branch is completed, leaving behind a remodeled vascular network.

The global picture that emerges is that expansion and regression of the microcirculatory networks are not likely to be triggered by the same factors and are likely to have different expression patterns for pro- and anti-angiogenic regulators. The mechanisms that trigger or are responsible for initiating microvessel expansion or regression appear to be biologically constrained and dependent upon key angioregulatory peptides, much like a "brake" and an "accelerator" in a car that require separate input (forces) to set them in motion. In other words, it should no longer be assumed that the mechanisms behind capillary regression (and perhaps more importantly, capillary rarefaction associated with chronic disease) are controlled by simply turning off the factors that stimulate angiogenesis.

There is widespread speculation that diffuse systemic rarefaction may be one of the primary causes of essential hypertension [10]. On the other hand, rarefaction can also be a downstream consequence, as has been clearly shown by its presence in animal models of secondary hypertension [11].

10.3 Microvascular Rarefaction in Cardiovascular Diseases

A constant hallmark of both experimental and clinical hypertension has been that of microvascular rarefaction, defined as a reduced spatial density of microvascular networks [12–14]. Under physiological resting conditions, a substantial proportion of the microvascular networks of most organs remains closed, thus representing a flow reserve for adaptation to increased metabolic requirements. Defined as an abnormally low spatial density of microvessels, rarefaction can be functional, structural, or both.

Functional rarefaction designates the abnormal prevalence of anatomically present but non-perfused microvessels.

Structural rarefaction can be determined by quantitative histology or by in vivo detection of microvascular beds in conditions of maximal vasodilation and optimal perfusion [15].

Experimental studies have reported microvascular structural rarefaction in many experimental hypertensive models and tissues, including skeletal muscle [12, 13, 16, 17], intestine [18], and skin [19], though not in the brain [20]. And structural rarefaction has been detected as early as 4 weeks of age in spontaneously hypertensive rats [14].

Experimental data on myocardial microcirculation are a noteworthy separate entity because of the concomitant myocardial hypertrophy. Diminished myocardial capillary density (i.e., fewer vessels per square millimeter of cross-sectional area of tissue) has been widely studied in adult hypertensive animals [21–25]. Data appear to suggest that the microcirculation is unable to expand (through angiogenesis) sufficiently to keep up with the gradual increase in myocardial mass and myocyte dimensions. In line with this hypothesis, young spontaneously hypertensive rats (2.5 months, an age of fast growth in body dimensions and presumably more active angiogenesis than later in life) had normal capillary density in their already hypertrophic left ventricle, whereas rarefaction was detected in older animals (7 months) [26]. Consistent with these data, patients with left ventricular hypertrophy resulting from aortic stenosis only had myocardial capillary rarefaction when the valve defect was acquired in the adult age but not when it was congenital [27].

10.3.1 Rarefaction in Clinical Hypertension

In 1933, Ruedemann [28] reported an abnormally low number of small conjunctival vessels in hypertensive patients. This observation was later replicated using more sophisticated visualization equipment [29]. Serne et al. [30] used venous occlusion capillaroscopy to measure nail fold capillary density in nondiabetic patients with never-treated essential hypertension but also in healthy normotensive controls matched for age, sex, and lipid profile. The mean and SD of capillary counts of 52.5 ± 6.6/mm^2 in hypertensive individuals were significantly lower than in the controls (57.2 ± 8.6/mm^2, an approximately 10% difference). Antonios et al. [31] used an identical experimental design but conducted venous occlusion capillaroscopy on dorsal finger skin rather than the nail fold. Their results were analogous (never-treated hypertensive individuals, 73 ± 5 capillaries per mm^2; controls, 87 ± 7 capillaries per mm^2; 20% difference). Yet another group made similar findings after capillaroscopy on the forearm skin of hypertensive versus normotensive individuals [32]. There is some evidence that capillary rarefaction in the skin may antedate the clinical onset of essential hypertension. In two other studies by Antonios et al., dorsal finger venous occlusion capillaroscopy revealed an abnormally low capillary density in individuals with borderline hypertension and even in individuals with normotension but with a familial predisposition to the disease [33, 34]. Noon et al. also described microvascular rarefaction in dorsal finger and forearm skin of genetically predisposed normotensive individuals [10]. Finally, in hypertensive patients—but irrespective of antihypertensive therapy—we have shown that the Framingham cardiovascular risk score is negatively correlated to capillary density, as evaluated in the dorsal skin of the second phalanx of the third finger [35].

The cause–effect relationship between rarefaction and hypertension is still the subject of extensive discussion. The role played by mechanical forces in general, and elevated pressure in particular, remains uncertain. In a rat model of secondary hypertension induced by partial ligation of the abdominal aorta upstream from the renal arteries, Boegehold et al. [17] observed structural arteriolar rarefaction in hindquarter muscles. The fact that these vascular beds were not exposed to high pressure in this model suggested that pressure-independent mechanisms were involved. In contrast, in mice genetically deficient in endothelial NO synthase, the arteriolar rarefaction that develops in conjunction with hypertension was prevented in animals whose blood pressure was kept normal by chronic administration of the non-specific vasodilator, hydralazine [36].

The density of the microvascular network can diminish as a result of vessel destruction or insufficient angiogenesis. Prewitt et al. [15] used in vivo videomicroscopy to carefully investigate skeletal muscle microcirculation in spontaneously hypertensive rats at various stages of the disease. They concluded that nonneural factors (possibly related to the aforementioned hypersensitivity of vascular smooth muscle to vasoconstrictors) cause reversible closure of arterioles (functional rarefaction) and subsequently their anatomic disappearance. Endothelial cell apoptosis caused by oxidative stress plays a significant role in this process [37]. This oxidative stress is likely related to abnormal activation of membrane-bound reduced nicotinamide dinucleotide phosphate oxidase by angiotensin II.

One compelling argument in favor of the link between capillary rarefaction and the long-term control of angiogenesis and blood pressure is the crucial role played by NO and the renin–angiotensin system in both processes. The bioactivity of NO appears to be deficient in hypertension. NO is not only a vasorelaxant but is also required for adequate vascular budding in wound healing [38] and stimulates the expression of vascular growth factors, notably VEGF [39]. Impaired angiogenesis has been directly demonstrated in experimental hypertension induced by chronic pharmacological inhibition of NO synthesis [40]. The renin–angiotensin system plays an indisputable role in angiogenesis though likely in a complex fashion involving an interplay of antagonistic and context-dependent influences (Fig. 10.6). Pharmacological blockade of the renin–angiotensin system in experimental models of hypertension predictably affects rarefaction but not always in the same way.

10.3.2 Microvascular Rarefaction in Diabetes [52]

Diabetes mellitus, a major risk factor for cardiovascular disease, causes severe damage to the microvascular structure and function. High glucose concentrations have been shown to cause endothelial cell dysfunction—mediated, for example, by vasoconstrictor prostanoids [53]—and to impair NO activity [54, 55]. NO is involved in VEGF-induced angiogenesis [56]. Animal data are in favor of the pathophysiological concept of impaired angiogenesis and collateral vessel formation as consequences of diabetes.

Microvascular rarefaction has been described as a result of induced specialized anti-angiogenic programs predominantly mediated by angiostatin and endostatin [57]. Furthermore, following induction of type 1 diabetes, reduced levels of major angiogenic growth factors, including VEGF, were observed in the skeletal muscle capillaries of mice [58]. Increased tissue fibrosis and glycation in diabetes also favor the sequestration of VEGF by collagen fibers, thus contributing to the depleted

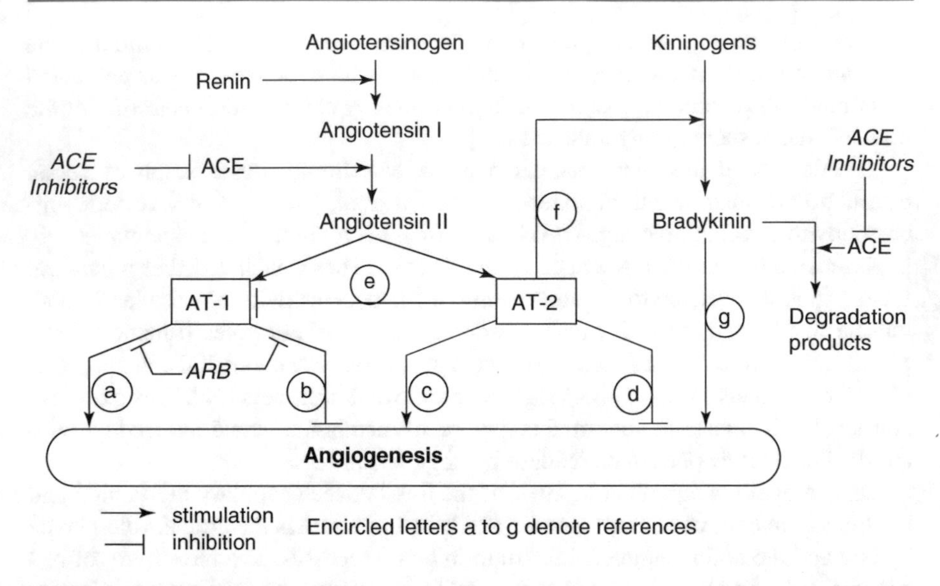

Fig. 10.6 Possible interactions between the renin–angiotensin system and angiogenesis. Although angiotensin II type 1 (AT-1) and type 2 (AT-2) receptors are known to modulate angiogenesis, conflicting data mean that, to date, details of when and how this occurs remain unclear. The figure includes the bradykinin pathway to show that AT-2 receptor stimulation may activate the production of bradykinin, with pro-angiogenic effects. We have omitted the complexities related to other peptides with potential influence on angiogenesis. Such peptides may be generated in the processing of either angiotensinogen by renin or angiotensin I by neutral endopeptidase. CE inhibitors not only reduce the levels of angiotensin II and thus the stimulation of both AT-1 and AT-2 receptors but also promote the bradykinin pathway by inhibiting enzymes responsible for bradykinin degradation. AT-1 receptor blockers (ARB) lead to a compensatory increase in the production of angiotensin II, thus shifting the balance in favor of AT-2 stimulation. Circled lowercase letters indicate references: a, [41–43]; b, [44–46]; c, [44]; d, [42, 47, 48]; e, [49]; f, [50]; and g, [51]. Redrawn from [14]

VEGF levels available for angiogenesis [REF]. In a diabetic mouse model of induced hindlimb ischemia, restoration of perfusion in the ischemic limb was impaired and capillary density reduced [59]. Goligorsky and colleagues demonstrated the occurrence of an angiogenic switch during the natural history of type 2 diabetes mellitus, from a pro-angiogenic phenotype in the early prediabetic stages to the profound inhibition of ex vivo angiogenesis and microvascular rarefaction in the kidneys of obese Zucker diabetic fatty rats [58]. Peritubular capillary rarefaction, a major hallmark of chronic kidney disease, predicts renal outcome in diabetic nephropathy [60]. There is evidence from animal models that endothelial dysfunction, oxidative stress, interstitial hypoxia due to arteriolar vasoconstriction, and pericyte detachment from the vasculature are all involved in this pathophysiological process [61].

Existing data on capillary density in patients with type 2 diabetes mellitus and overt end-organ damage vary considerably in their findings. While coronary collateral vessels have been seen to develop to a lesser extent in patients with type 1 or type 2 diabetes mellitus and advanced coronary atherosclerosis than in nondiabetic individuals [56], other authors have reported that capillary videomicroscopy of dorsal finger skin in patients with type 2 diabetes mellitus (50% of whom had evidence of diabetic retinopathy or microalbuminuria) did not reveal any difference in capillary density compared to healthy controls [62]. And conjunctival vessels of patients with type 1 or type 2 diabetes mellitus have also been shown to have fewer capillaries and postcapillaries than those of nondiabetic individuals [63]. There is a lack of human in vivo data on capillary rarefaction in the early stages of type 2 diabetes mellitus without overt end-organ damage. Increasing evidence points to the pivotal role played by capillary rarefaction—notably its effects on blood pressure and the blood flow—in the pathogenesis of end-organ damage, with a severe impact on peripheral vascular resistance (and consequently on systemic blood pressure), metabolisms [30, 64], and tissue ischemia [65]. The retinal circulation provides insight into the microcirculation as a whole [66] as it is thought to mirror the cerebrovascular circulation [67–69]. There is however a considerable mass of data on diabetic retinopathy, a widely recognized example of overt end-organ damage [70]. Jumar et al. recently reported that retinal capillary rarefaction was greater in patients in the early stages of type 2 diabetes than in healthy controls [71].

Moreover, impaired myogenic tone has been reported in experimental models of diabetes. The capillary beds are therefore less well protected from the effects of high blood pressure, thus favoring the occurrence of edemas. The lack of myogenic tone in pre-glomerular afferent arterioles of the kidney likely contributes to the proteinuria and kidney diseases associated with diabetes.

10.4 Conclusion

The capillary bed has a leading function within the vascular network: (1) it plays a role in determining vascular resistance to flow, which accounts for approximatively 15% of the total drop in blood pressure; (2) the capillary network is the region of the vascular system that specializes in the transport of substances between the blood and tissues; and (3) the capillary network is affected by the diseases frequently associated with elevated cardiovascular risk, i.e., arterial hypertension and diabetes mellitus. Future studies of the capillary bed should focus on our currently incomplete understanding of the precise mechanisms of rarefaction, its actual significance in elevating hemodynamic peripheral resistance and blood pressure, and its role in the onset of end-organ damage. Finally, the effects of antihypertensive and antidiabetic drugs on capillary structure and functions should unquestionably be the subject of further investigation.

References

1. Smaje L, Zweifach BW, Intaglietta M. Micropressures and capillary filtration coefficients in single vessels of the cremaster muscle of the rat. Microvasc Res. 1970;2:96–110.
2. McCuskey RS, Reilly FD. Hepatic microvasculature: dynamic structure and its regulation. Semin Liver Dis. 1993;13:1–12.
3. Braverman IM. Ultrastructure and organization of the cutaneous microvasculature in normal and pathologic states. J Invest Dermatol. 1989;93:2S–9S.
4. Manelli A, Sangiorgi S, Ronga M, Reguzzoni M, Bini A, Raspanti M. Plexiform vascular structures in the human digital dermal layer: a SEM—corrosion casting morphological study. Eur J Morphol. 2005;42:173–7.
5. Froneck K, Zweifach BW. Microvascular pressure distribution in skeletal muscle and the effect of vasodilatation. Am J Phys. 1975;228:791–6.
6. Dulling BR, Berne RM. Longitudinal gradients in periarteriolar oxygen tension: possible mechanism for the participation of oxygen in local regulation of blood flow. Circ Res. 1970;27:669–78.
7. Helmlinger G, Yuan F, Dellian M, Jain RK. Interstitial pH and pO2 gradients in solid tumors in vivo: high-resolution measurements reveal a lack of correlation. Nat Med. 1997;3:177–82.
8. Flamme I, Frölich T, Risau. Molecular mechanisms of vasculogenesis and embryonic angiogenesis. J Cell Physiol. 1997;173:206–10.
9. Korn C, Augustin HG. Mechanisms of vessel pruning and regression. Dev Cell. 2015;34:5–17.
10. Noon JP, Walker BR, Webb DJ, Shore AC, Holton DW, Edwards HV, Watt GC. Impaired microvascular dilatation and capillary rarefaction in young adults with a predisposition to high blood pressure. J Clin Invest. 1997;99:1873–9.
11. Meirelles Pereira LM, Mandarim-de-Lacerda CA. Effect of antihypertensive drugs on the myocardial microvessels in rats with nitric oxide blockade. Pathol Res Pract. 2000;196:305–11.
12. Struijker Boudier HA, le Noble JL, Messing MW, Huijberts MS, le Noble FA, van Essen H. The microcirculation and hypertension. J Hypertens. 1992;10(suppl):S147–56.
13. Levy BI, Ambrosio G, Pries AR, Struijker-Boudier HAJ. Microcirculation in hypertension–a new target for treatment? Circulation. 2001;104:735–40.
14. Feihl F, Liaudet L, Waeber B, Levy BI. Hypertension: a disease of the microcirculation? Hypertension. 2006;48:1012–7.
15. Prewitt RL, Chen II, Dowell R. Development of microvascular rarefaction in the spontaneously hypertensive rat. Am J Phys. 1982;243:H243–51.
16. Chen II, Prewitt RL, Dowell RF. Microvascular rarefaction in spontaneously hypertensive rat cremaster muscle. Am J Phys. 1981;241:H306–10.
17. Boegehold MA, Johnson MD, Overbeck HW. Pressure-independent arteriolar rarefaction in hypertension. Am J Phys. 1991;261:H83–7.
18. Henrich H, Hertel R, Assmann R. Structural differences in the mesentery microcirculation between normotensive and spontaneously hypertensive rats. Pflugers Arch. 1978;375:153–9.
19. Haack DW, Schaffer JJ, Simpson JG. Comparisons of cutaneous microvessels from spontaneously hypertensive, normotensive Wistar-Kyoto, and normal Wistar rats. Proc Soc Exp Biol Med. 1980;164:453–8.
20. Lin SZ, Sposito N, Pettersen S, Rybacki L, McKenna E, Pettigrew K, Fenstermacher J. Cerebral capillary bed structure of normotensive and chronically hypertensive rats. Microvasc Res. 1990;40:341–57.
21. Rakusan K, Cicutti N, Kazda S, Turek Z. Effect of nifedipine on coronary capillary geometry in normotensive and hypertensive rats. Hypertension. 1994;24:205–11.
22. Sabri A, Samuel JL, Marotte F, Poitevin P, Rappaport L, Levy BI. Microvasculature in angiotensin II-dependent cardiac hypertrophy in the rat. Hypertension. 1998;32:371–5.
23. Kobayashi N, Kobayashi K, Hara K, Higashi T, Yanaka H, Yagi S, Matsuoka H. Benidipine stimulates nitric oxide synthase and improves coronary circulation in hypertensive rats. Am J Hypertens. 1999;12:483–91.

24. Rakusan K, Cicutti N, Maurin A, Guez D, Schiavi P. The effect of treatment with low dose ACE inhibitor and/or diuretic on coronary microvasculature in stroke-prone spontaneously hypertensive rats. Microvasc Res. 2000;59:243–54.
25. Levy BI, Duriez M, Samuel JL. Coronary microvasculature alteration in hypertensive rats. Effect of treatment with a diuretic and an ACE inhibitor. Am J Hypertens. 2001;14:7–13.
26. Tomanek RJ, Searls JC, Lachenbruch PA. Quantitative changes in the capillary bed during developing, peak, and stabilized cardiac hypertrophy in the spontaneously hypertensive rat. Circ Res. 1982;51:295–304.
27. Rakusan K, Flanagan MF, Geva T, Southern J, Van Praagh R. Morphometry of human coronary capillaries during normal growth and the effect of age in left ventricular pressure-overload hypertrophy. Circulation. 1992;86:38–46.
28. Ruedemann AD. Conjunctival vessels. J Am Med Assoc. 1933;101:1477–81.
29. Sullivan JM, Prewitt RL, Josephs JA. Attenuation of the microcirculation in young patients with high-output borderline hypertension. Hypertension. 1983;5:844–51.
30. Serne EH, Gans ROB, ter Maaten JC, Tangelder GJ, Donker AJM, Stehouwer CDA. Impaired skin capillary recruitment in essential hypertension is caused by both functional and structural capillary rarefaction. Hypertension. 2001;38:238–42.
31. Antonios TFT, Singer DRJ, Markandu ND, Mortimer PS, MacGregor GA. Structural skin capillary rarefaction in essential hypertension. Hypertension. 1999;33:998–1001.
32. Prasad A, Dunnill GS, Mortimer PS, MacGregor GA. Capillary rarefaction in the forearm skin in essential hypertension. J Hypertens. 1995;13:265–8.
33. Antonios TFT, Singer DRJ, Markandu ND, Mortimer PS, MacGregor GA. Rarefaction of skin capillaries in borderline essential hypertension suggests an early structural abnormality. Hypertension. 1999;34:655–8.
34. Antonios TFT, Rattray FM, Singer DRJ, Markandu ND, Mortimer PS, MacGregor GA. Rarefaction of skin capillaries in normotensive offspring of individuals with essential hypertension. Heart. 2003;89:175–8.
35. Debbabi H, Uzan L, Mourad JJ, Safar M, Levy BI, Tibirica E. Increased skin capillary density in treated essential hypertensive patients. Am J Hypertens. 2006;19:477–83.
36. Kubis N, Besnard S, Silvestre JS, Feletou M, Huang PL, Levy BI, Tedgui A. Decreased arteriolar density in endothelial nitric oxide synthase knockout mice is due to hypertension, not to the constitutive defect in endothelial nitric oxide synthase enzyme. J Hypertens. 2002;20:273–80.
37. Kobayashi N, DeLano FA, Schmid-Schönbein GW. Oxidative stress promotes endothelial cell apoptosis and loss of microvessels in the spontaneously hypertensive rats. Arterioscler Thromb Vasc Biol. 2005;25:2114–21.
38. Lee PC, Salyapongse AN, Bragdon GA, Shears LL 2nd, Watkins SC, Edington HD, Billiar TR. Impaired wound healing and angiogenesis in eNOS-deficient mice. Am J Phys. 1999;277:H1600–8.
39. Dulak J, Jozkowicz A, Dembinska-Kiec A, Guevara I, Zdzienicka A, Zmudzinska-Grochot D, Florek I, Wojtowicz A, Szuba A, Cooke JP. Nitric oxide induces the synthesis of vascular endothelial growth factor by rat vascular smooth muscle cells. Arterioscler Thromb Vasc Biol. 2000;20:659–66.
40. Kiefer FN, Misteli H, Kalak N, Tschudin K, Fingerle J, Van der Kooij M, Stumm M, Sumanovski LT, Sieber CC, Battegay EJ. Inhibition of NO biosynthesis, but not elevated blood pressure, reduces angiogenesis in rat models of secondary hypertension. Blood Press. 2002;11:116–24.
41. Sasaki K, Murohara T, Ikeda H, Sugaya T, Shimada T, Shintani S, Imaizumi T. Evidence for the importance of angiotensin II type 1 receptor in ischemia-induced angiogenesis. J Clin Invest. 2002;109:603–11.
42. Munzenmaier DH, Greene AS. Opposing actions of angiotensin II on microvascular growth and arterial blood pressure. Hypertension. 1996;27:760–5.
43. Emanueli C, Salis MB, Stacca T, Pinna A, Gaspa L, Madeddu P. Angiotensin AT(1) receptor signalling modulates reparative angiogenesis induced by limb ischaemia. Br J Pharmacol. 2002;135:87–92.

44. Walther T, Menrad A, Orzechowski HD, Siemeister G, Paul M, Schirner M. Differential regulation of in vivo angiogenesis by angiotensin II receptors. FASEB J. 2003;17:2061–7.
45. de Boer RA, Pinto YM, Suurmeijer AJ, Pokharel S, Scholtens E, Humler M, Saavedra JM, Boomsma F, van Gilst WH, van Veldhuisen DJ. Increased expression of cardiac angiotensin II type 1 (AT(1)) receptors decreases myocardial microvessel density after experimental myocardial infarction. Cardiovasc Res. 2003;57:434–42.
46. Forder JP, Munzenmaier DH, Greene AS. Angiogenic protection from focal ischemia with angiotensin II type 1 receptor blockade in the rat. Am J Phys. 2005;288:H1989–96.
47. Silvestre JS, Tamarat R, Senbonmatsu T, Icchiki T, Ebrahimian T, Iglarz M, Besnard S, Duriez M, Inagami T, Levy BI. Antiangiogenic effect of angiotensin II type 2 receptor in ischemia-induced angiogenesis in mice hindlimb. Circ Res. 2002;90:1072–9.
48. Benndorf R, Boger RH, Ergun S, Steenpass A, Wieland T. Angiotensin II type 2 receptor inhibits vascular endothelial growth factor-induced migration and in vitro tube formation of human endothelial cells. Circ Res. 2003;93:438–47.
49. AbdAlla S, Lother H, Abdel-tawab AM, Quitterer U. The angiotensin II AT2 receptor is an AT1 receptor antagonist. J Biol Chem. 2001;276:39721–6.
50. Tsutsumi Y, Matsubara H, Masaki H, Kurihara H, Murasawa S, Takai S, Miyazaki M, Nozawa Y, Ozono R, Nakagawa K, Miwa T, Kawada N, Mori Y, Shibasaki Y, Tanaka Y, Fujiyama S, Koyama Y, Fujiyama A, Takahashi H, Iwasaka T. Angiotensin II type 2 receptor overexpression activates the vascular kinin system and causes vasodilation. J Clin Invest. 1999;104:925–35.
51. Silvestre JS, Bergaya S, Tamarat R, Duriez M, Boulanger CM, Levy BI. Proangiogenic effect of angiotensin-converting enzyme inhibition is mediated by the bradykinin B(2) receptor pathway. Circ Res. 2001;89:678–83.
52. Silvestre JS, Smadja DM, Lévy BI. Postischemic revascularization: from cellular and molecular mechanisms to clinical applications. Physiol Rev. 2013;93:1743–802.
53. Tesfamariam B, Brown ML, Deykin D, Cohen RA. Elevated glucose promotes generation of endothelium-derived vasoconstrictor prostanoids in rabbit aorta. J Clin Invest. 1990;85:929–32.
54. Pieper GM, Peltier BA. Amelioration by L-arginine of a dysfunctional arginine/nitric oxide pathway in diabetic endothelium. J Cardiovasc Pharmacol. 1995;25:397–403.
55. Abaci A, Oguzhan A, Kahraman S, Eryol NK, Unal S, Arinc H, et al. Effect of diabetes mellitus on formation of coronary collateral vessels. Circulation. 1999;99:2239–42.
56. Parenti A, Morbidelli L, Cui XL, Douglas JG, Hood JD, Granger HJ, et al. Nitric oxide is an upstream signal of vascular endothelial growth factor-induced extracellular signal-regulated kinase1/2 activation in postcapillary endothelium. J Biol Chem. 1998;273(7):4220–6.
57. Goligorsky MS. Microvascular rarefaction: the decline and fall of blood vessels. Organogenesis. 2010;6:1–10.
58. Kivela R, Silvennoinen M, Touvra AM, Lehti TM, Kainulainen H, Vihko V. Effects of experimental type 1 diabetes and exercise training on angiogenic gene expression and capillarization in skeletal muscle. FASEB J. 2006;20:1570–2.
59. Ebrahimian TG, Heymes C, You D, Blanc-Brude O, Mees B, Waeckel L, Duriez M, Vilar J, Brandes RP, Levy BI, Shah AM, Silvestre JS. NADPH oxidase-derived overproduction of reactive oxygen species impairs postischemic neovascularization in mice with type 1 diabetes. Am J Pathol. 2006;169:719–1728.
60. Kida Y, Tchao BN, Yamaguchi I. Peritubular capillary rarefaction: a new therapeutic target in chronic kidney disease. Pediatr Nephrol. 2014;29:333–42.
61. Howangyin KY, Silvestre JS. Diabetes mellitus and ischemic diseases: molecular mechanisms of vascular repair dysfunction. Arterioscler Thromb Vasc Biol. 2014;34:1126–35.
62. Jaap AJ, Shore AC, Stockman AJ, Tooke JE. Skin capillary density in subjects with impaired glucose tolerance and patients with type 2 diabetes. Diabet Med. 1996;13:160–1604.
63. Fenton BM, Zweifach BW, Worthen DM. Quantitative morphometry of conjunctival microcirculation in diabetes mellitus. Microvasc Res. 1979;18:153–1566.

64. Clark MG, Barrett EJ, Wallis MG, Vincent MA, Rattigan S. The microvasculature in insulin resistance and type 2 diabetes. Semin Vasc Med. 2002;2:21–31.
65. Aellen J, Dabiri A, Heim A, Liaudet L, Burnier M, Ruiz J, et al. Preserved capillary density of dorsal finger skin in treated hypertensive patients with or without type 2 diabetes. Microcirculation. 2012;19:554–62.
66. Rizzoni D, Porteri E, Duse S, De Ciuceis C, Agabiti-Rosei C, La Boria E, et al. Relationship between media-to-lumen ratio of subcutaneous small arteries and wall-to-lumen ratio of retinal arterioles evaluated noninvasively by scanning laser Doppler flowmetry. J Hypertens. 2012;30:1169–75.
67. Kaiser SE, Sanjuliani AF, Estato V, Gomes MB, Tibirica E. Antihypertensive treatment improves microvascular rarefaction and reactivity in low-risk hypertensive individuals. Microcirculation. 2013;20:703–16.
68. Tomassoni D, Mancinelli G, Mignini F, Sabbatini M, Amenta F. Quantitative image analysis of choroid and retinal vasculature in SHR: a model of cerebrovascular hypertensive changes? Clin Exp Hypertens. 2002;24:741–52.
69. Muiesan ML, Salvetti M, Rizzoni D, Paini A, Agabiti-Rosei C, Aggiusti C, et al. Resistant hypertension and target organ damage. Hypertens Res. 2013;36:485–91.
70. Lorenzi M, Gerhardinger C. Early cellular and molecular changes induced by diabetes in the retina. Diabetologia. 2001;44:791–804.
71. Jumar A, Harazny JM, Ott C, Friedrich S, Kistner I, Striepe K, Schmieder RE. Retinal capillary rarefaction in patients with type 2 diabetes mellitus. PLoS One. 2016;11:e0162608.

Microvascular Alterations in Obesity

11

Gino Seravalle and Guido Grassi

Obesity is a worldwide health problem which finds in changes in food composition and in sedentary lifestyle two of the main pathophysiological factors [1–3]. Epidemiological studies have demonstrated that obesity is associated with increased mortality mostly due to augmented risk of cardiovascular death [4]. Not only the volume of adipose quantity but also quality of fat tissue is very important for the link to cardiovascular (CV) risk [5, 6]. Several clinical studies using adiposity measures such as waist circumference and waist-to-hip ratio as markers of central obesity as well as cross-sectional abdominal imaging have established clear links between overall fat burden and visceral fat and systemic cardiometabolic disease [5–11]. The structural and functional abnormalities in microcirculation that are present just in the early phases of this pathophysiological condition represent one of the main factors contributing to the progression of the CV risk. The aim of this chapter is to review the alterations in microcirculation associated with weight increase and abdominal fatty deposition and the mechanisms responsible.

11.1 Resistance Arteries

The microcirculation includes small arteries (diameter ranging from 100 to 300 μm), arterioles (diameter < 100 μm), and capillaries (diameter < 10 μm) [12, 13] (Fig. 11.1). Small arteries contribute to about 30–50% of precapillary blood

G. Seravalle
Cardiology Department, S. Luca Hospital, IRCCS Istituto Auxologico Italiano, Milan, Italy

G. Grassi (✉)
Clinica Medica, Department of Medicine and Surgery, University Milano-Bicocca, Milan, Italy

IRCCS Multimedica, Sesto San Giovanni, Milan, Italy
e-mail: guido.grassi@unimib.it

E. Agabiti-Rosei et al. (eds.), *Microcirculation in Cardiovascular Diseases*, Updates in Hypertension and Cardiovascular Protection,
https://doi.org/10.1007/978-3-030-47801-8_11

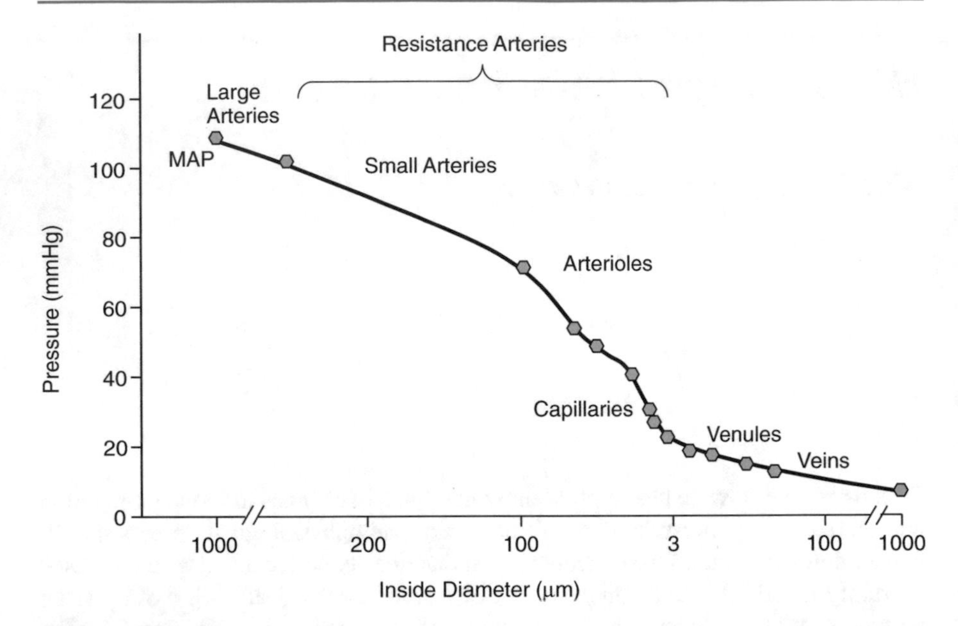

Fig. 11.1 Relation between inside diameter and blood pressure drop in the vascular bed. Modified from ref. 12

pressure drop, and an additional 30% occurs at the arteriolar level, with a different distribution of resistance among different vascular beds [13]. Nowadays, a definition based on arterial vessel physiology rather than diameter or structure has been proposed, depending on the response of the isolated vessel to increased internal pressure. Thus, all vessels that respond to increasing pressure by a myogenic reduction in lumen diameter would be considered part of the microcirculation [14]. The main functions of microcirculation are to avoid large fluctuations in hydrostatic pressure at the capillary level to preserve exchanges and to contribute to overall peripheral resistance by reducing the hydrostatic pressure.

11.2 How to Measure

The methods for the evaluation of microvascular structure and function in humans are relatively few.

Histology is invasive, and artifacts introduced by fixation, staining, and dehydration represent the main disadvantage of this procedure.

The plethysmographic technique is based on the occlusion of the brachial artery of the dominant arm, through the inflation of a sphygmomanometric bladder up to 300 mm Hg for 13 min associated with a dynamic exercise. The arterial occlusion is rapidly removed, while venous occlusion is maintained, and arterial flow is measured every 10 s for 3 min by a mercury strain gauge, which evaluates the increased volume of the forearm. In the absence of venous backward flow, the increased forearm volume is proportional to the arterial flow. The mean blood pressure divided by maximum arterial flow allows to determine the minimum vascular resistance [15]. It has been shown the maximum postischemic flow is correlated with the ratio between wall thickness and internal lumen in resistance arterioles [16]. The advantage of this technique is that it is inexpensive and is easy to access and after a period of personnel training shows a good reproducibility.

Two widely used miniinvasive techniques are represented by wire and pressurized micromyography that allows a reliable evaluation of structural characteristics of vascular wall and functional response to stimuli. The wire micromyography was developed in the 1970s [17] (Fig. 11.2). In humans small arteries are obtained from biopsies of subcutaneous tissue from the gluteal or anterior abdominal region. Small artery (diameter 100–300 μm) segments, free of periadventitial fat tissue, are cannulated with stainless steel wires, resulting in a ring preparation that is mounted in a micromyograph. Basically the passive tension is recorded, and then contraction responses by various substances added to the organ bath could be collected. Microscope observation allows to calculate in relaxed condition the total wall thickness; the relative thickness of the tunica adventitia, media, and intima; and the

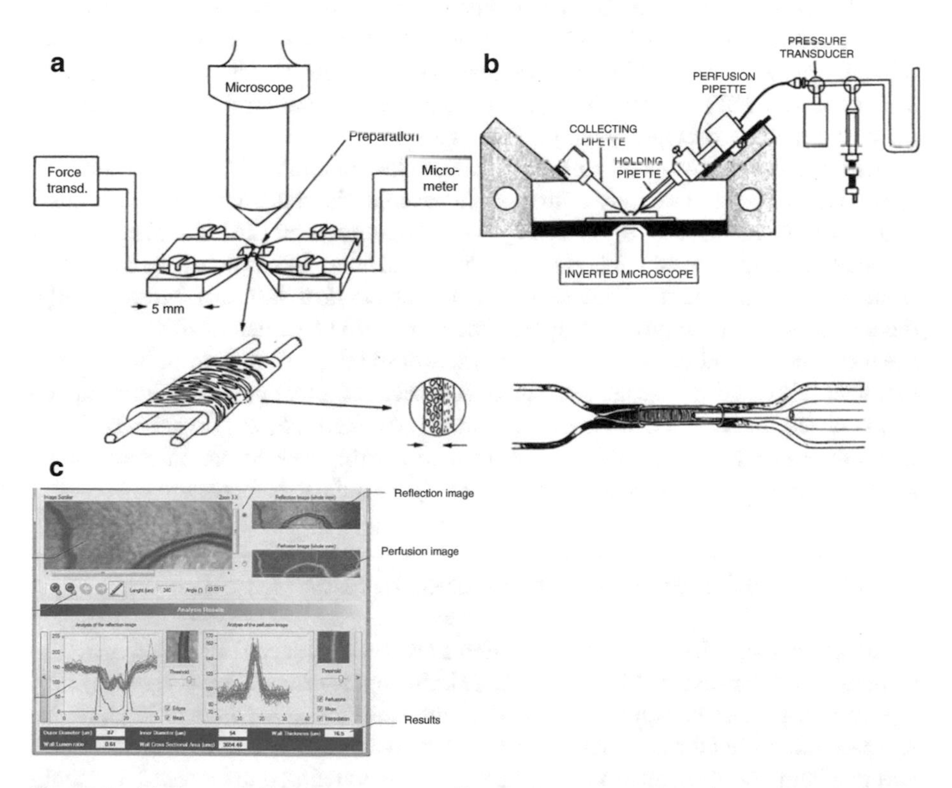

Fig. 11.2 (**a**) Cartoon of a subcutaneous small resistance artery mounted on a wire micromyograph as a ring preparation; (**b**) cartoon of vessel mounted on a pressurized micromyography; (**c**) example of small retinal artery morphology by scanner laser Doppler flowmetry and automatic full-field perfusion imaging analysis. Modified from refs. 12 and 26

internal diameter and cross-sectional area. The pressurized micromyography (Fig. 11.2) differs from the previous technique for the fact that isolated vessels are mounted in a pressurized myograph chamber and slipped into two glass microcannulae, connected to a perfusion system that allows a constant intraluminal pressure of 60 mm Hg [18–20]. Morphology of the vessels is evaluated by a computer-assisted video analyzer at a constant pressure or constant flow. This method allows a better evaluation of functional responses during pharmacological stimuli. The disadvantage of these two methods are related to the availability of adequate tissue obtained during surgical procedure or by a small and well-tolerated biopsy of subcutaneous fat tissue and to the need of well-trained personnel. These two methods, as regards the measurement of the media-to-lumen ratio, are currently considered, due to sensitivity, specificity, and accuracy, the gold standard for the evaluation of structural alterations of small resistance arteries in humans.

Another noninvasive method which evaluates the absolute number of perfused vessels that contribute to total vascular resistance is the calculation of microvascular rarefaction [21]. This is obtained by capillaroscopy or intravital videomicroscopy that allows a high-contrast image or video, with or without intravenous administration of fluorescent dyes, useful to assess capillary morphology, density, flow velocity, and red cell column width [22, 23]. Capillary density is defined as the number of capillaries per unit of skin area. The method cannot directly show capillaries that are not perfused at rest; thus, inducing a venous congestion for few minutes allows the recruitment of previously not perfused vessels.

As regards the new techniques to investigate the microcirculation, a noninvasive approach consists in the study of the retinal vascular district that is a microvascular bed that may be directly evaluated by an ophthalmoscope [24]. By means of an automated computerized method is possible to calculate the ratio between the arteriolar and venular external diameter in circular segments of the retina. More recently, the evolution of this approach proposed the association between confocal measurement of the external diameter of retinal arterioles and an evaluation of the internal diameter with a laser Doppler technique. A dedicated software makes the comparison between the reflected and perfused images obtained [25, 26] (Fig. 11.2). These methods are easy to use, and intraobserver and interobserver variation coefficients, as indices of the reproducibility of the method, are quite satisfactory.

11.3 Small Resistance Arteries and Obesity

Structural changes in the microcirculation have been observed in hypertension at different vascular levels [14, 16, 19, 21, 27]. Eutrophic remodeling (Fig. 11.3) may represent a protective and a physiological response to hemodynamic load, to the increased neurohumoral activity, and to the activation of reactive oxygen species and mediators of inflammation. A hypertrophic remodeling due to vascular smooth muscle cell hypertrophy (volume increase) or hyperplasia (cell number increase) is the result of the presence of humoral growth factors (angiotensin II, insulin/insulin-like growth factor 1). It has been established that visceral obesity is characterized by

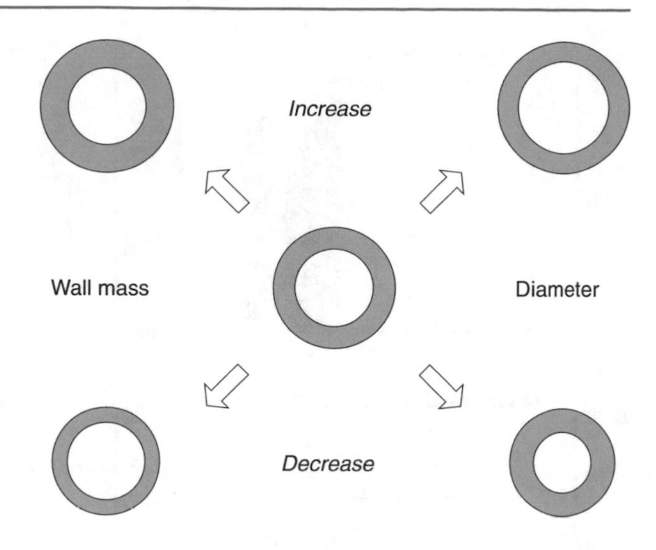

Fig. 11.3 Scheme of arterial remodeling

an insulin resistance state, by a production of reactive oxygen species (ROS), by a chronic inflammation with increased production of adipokines and cytokines (tumor necrosis factor-α, interleukin-1, interleukin-6, free fatty acids, leptin, resistin), and by an altered coagulation/fibrinolysis [28–30]. An alteration in large- and medium-sized artery structure and function in human obesity has been established. This condition is characterized by an increase in arterial stiffness [31, 32], a reduction in arterial compliance and distensibility [33, 34], and an impairment of endothelial function [35, 36]. More recently these aspects have been addressed by our group in small resistance arteries by making use of the pressurized micromyographic technique and employing biopsies of subcutaneous tissue from the anterior abdominal region [37]. Compared to lean individuals, for superimposable blood pressure, the media thickness, media cross-sectional area, and media-to-lumen ratio values of resistance arteries were markedly and significantly greater in obese individuals demonstrating for the first time a hypertrophic process probably secondary to vascular smooth muscle cell growth (Fig. 11.4). Acetylcholine-induced relaxation was impaired in obese subjects, whereas endothelium-independent vasorelaxation was similar in both groups. In the population as a whole, a significant correlation was found between media-to-lumen ratio and body mass index ($r = 0.81$, $p < 0.0001$), waist circumference ($r = 0.84$, $p < 0.0001$), plasma insulin levels ($r = 0.69$, $p < 0.0001$), and HOMA-IR values ($r = 0.68$, $p < 0.0001$). An inverse significant correlation was also observed between maximal percent vasorelaxation to acetylcholine and the same parameters [37]. In multivariate analysis the only independent predictor of abnormalities in media-to-lumen ratio as well as in maximal vasorelaxation to acetylcholine was waist circumference. The data of this study provide evidence that in severe obesity, at variance from what has been reported in the large- and medium-sized arteries, wall stiffness in the microcirculation is not increased, presumably because the tunica media content of collagen/elastin is not markedly altered. This could also suggest that increased collagen and fibronectin deposition

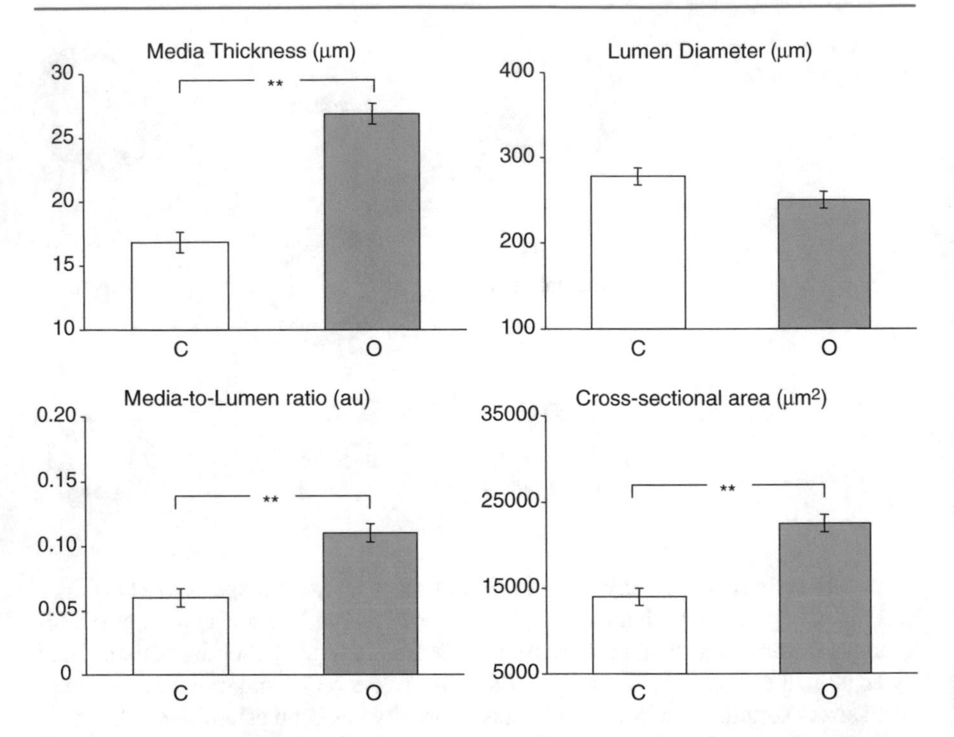

Fig. 11.4 Average values of media thickness, lumen diameter, media-to-lumen ratio, and cross-sectional area of the subcutaneous small arteries in lean normotensive controls (C) and in obese individuals (O). Data are shown as mean ± SEM. Asterisks ($**p < 0.01$) refer to statistical significance between groups. Modified from ref. 37

and collagen/elastin ratio which characterized small arteries of hypertensive individuals is a consequence of elevated blood pressure [18, 38–40]. Evaluation of obese subjects 1 year after bariatric surgery allowed to observe a marked reduction in body mass index together with a metabolic and inflammatory profile [41]. A statistically significant reduction in media-to-lumen ratio, media thickness, and media cross-sectional area was also observed, and this was accompanied by a clear shift from hypertrophic to eutrophic remodeling. Structure changes were also accompanied by an improvement in endothelial function as evaluated by vasodilation to acetylcholine [41].

11.4 Microvascular Dysfunction: Mechanisms Involved

In the previously mentioned study by Grassi and coworkers [37], a direct relationship was observed between arteriolar structure and plasma insulin levels and HOMA index, thus suggesting an important role of humoral factors such as insulin resistance in the development of vascular structural alterations. Insulin has two actions on the arterial vasculature to promote the delivery of insulin and glucose to skeletal muscles. First, it has been demonstrated the insulin's ability to vasodilate resistance vessels and consequently increase total skeletal muscle blood flow and insulin-mediated glucose uptake [42–44]. Second, it has been demonstrated an insulin-mediated capillary recruitment effect [45–47]. An inhibition of insulin-mediated microvascular effects causes a concomitant 30–40% reduction in glucose disposal [48]. This indicates a functional coupling between insulin-induced effects on muscle microvascular perfusion and glucose uptake, and it appears that microvascular dysfunction precedes and even predicts the development of insulin resistance [49].

Several are the mechanisms involved in the development of obesity-associated microvascular dysfunction.

In addition to its vasodilator actions, insulin also has vasoconstrictor effects that are mainly mediated by endothelin-1 (ET-1), and the net effect is on the balance between these two effects. Obesity is also associated with an increased production of ROS that limit the bioavailability of nitric oxide (NO). It has been demonstrated that obese individuals show an imbalance between NO and ET-1 production with a vasoreactivity shifted from vasodilation to vasoconstriction [50].

The second aspect is linked with adipose tissue and visceral adipocytes whose functions are not merely as a passive storage depot but as a highly active endocrine organ. In obesity there is an enhanced production of free fatty acids (FFAs), angiotensinogen, leptin, resistin, and several inflammatory cytokines such as tumor necrosis factor (TNF)-α and interleukin-6, whereas the production of adiponectin and anti-inflammatory adipokine is reduced [51–53]. FFAs impair basal and insulin-mediated effects on microvascular function blunting the PI3-kinase pathways, increasing ROS production, and increasing the release of ET-1 which has vasoconstrictor properties [54]. On the other side, TNF-α impairs insulin-mediated effect on microvascular function by impairing the balance between endothelial-derived vasodilator and vasoconstrictor substances; it downregulates the expression of eNOS and upregulates ET-1 expression in endothelial cells [55, 56].

11.5 News from the Eyes

The development of new noninvasive approaches for the evaluation of microvascular damage has been focused on the retinal vascular district due to the direct simple approach to this microvascular bed [24–26]. This method with its technical improvement during the recent years, quantifying the same structural characteristics observed with invasive methods, allowed to investigate a great number of subjects, at different ages and with different pathophysiological conditions. This allowed to observe that subjects with retinal arteriolar narrowing and venular widening were independently associated with an increased risk of hypertension [57]. The abnormalities observed in retinal circulation have shown to be also associated with the presence of diabetes, obesity, metabolic disorders, and cerebral diseases [58–61]. As regards the relation between obesity, microvascular alterations, and cardiovascular risk, a recent meta-analysis [62] investigated this aspect in about 44,000 obese individuals. Retinal images, with evaluation of central retinal arteriolar and venular equivalent (CRAE and CRVE) and their ratio, were compared with body mass index

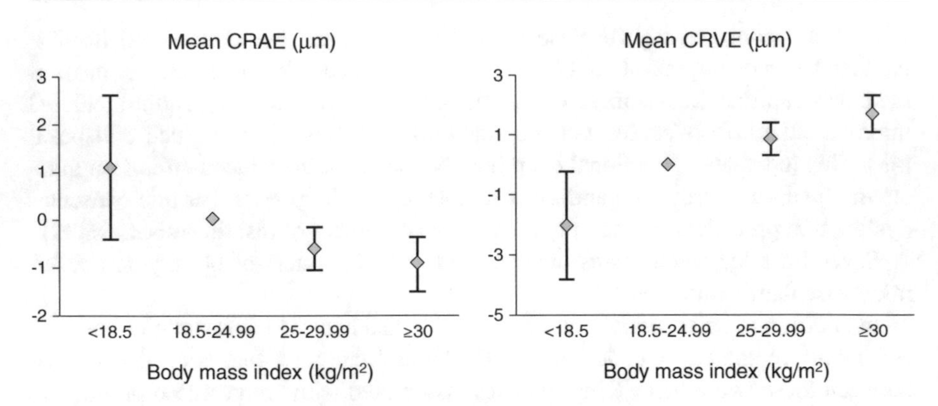

Fig. 11.5 Adjusted adult mean retinal arterial (CRAE, left) and venular (CRVE, right) values across body mass index (BMI) categories. Modified from ref. 62

categories both in children and adults. The meta-analysis found that higher BMI levels were associated with narrower arteriolar and wider venular caliber, independent of conventional cardiovascular risk factors (Fig. 11.5). A significant heterogeneity was observed in the pooled data for children, but this heterogeneity disappeared when correlated with increase in BMI values, thus suggesting that alterations are independent by ages and more related to mechanisms associated with arterial dysfunction, that is, metabolic, inflammatory, and neuroregulatory ones.

Acknowledgment None.

References

1. Poirier P, Giles TD, Bray GA, et al. Obesity and cardiovascular disease: pathophysiology, evaluation, and effect of weight loss: an update of the 1997 American Heart Association Scientific Statement on Obesity Committee of the Council on Nutrition, Physical activity, and Metabolism. Circulation. 2006;113:898–918.
2. Church TS, Thomas DM, Tudor-Locke C, et al. Trends over 5 decades in U.S. occupation-related physical activity and their associations with obesity. PLoS One. 2011;6:e19657.
3. Taubes G. The science of obesity: what do we really know about what makes us fat? An essay by Gary Taubes. BMJ. 2013;346:f1050.
4. Prospective Studies Collaboration. Body mass index and cause-specific mortality in 900000 adults: collaborative analyses of 57 prospective studies. Lancet. 2009;373:1083–96.
5. Rosenquist KJ, Massaro JM, Pedley A, et al. Fat quality and incident cardiovascular disease, all-cause mortality, and cancer mortality. J Clin Endocrinol Metab. 2015;100:227–34.
6. Abraham TM, Pedely A, Massaro JM, Hoffmann U, Fox CS. Association between visceral and subcutaneous adipose depots and incident cardiovascular disease risk factors. Circulation. 2015;132:1639–47.
7. Pou KM, Massaro JM, Hoffmann U, et al. Visceral and subcutaneous adipose tissue volumes are cross-sectionally related to markers of inflammation and oxidative stress: the Framingham Heart Study. Circulation. 2007;116:1234–41.
8. Sahakyan KR, Somers VK, Rodriguez-Escudero JP, et al. Normal-weight central obesity: implications for total and cardiovascular mortality. Ann Intern Med. 2015;163:827–35.

9. Shah RV, Murthy VL, Abbasi SA, et al. Visceral adiposity and the risk of metabolic syndrome across body mass index: the MESA study. JACC Cardiovasc Imaging. 2014;7:1221–35.
10. Grassi G, Seravalle G, Cattaneo BM, et al. Sympathetic activation in obese normotensive subjects. Hypertension. 1995;25:560–3.
11. Grassi G, Dell'Oro R, Facchini A, et al. Effect of central and peripheral body fat distribution on sympathetic and baroreflex function in obese normotensives. J Hypertens. 2004;22:2363–9.
12. Mulvany MJ, Aalkjaer C. Structure and function of small arteries. Physiol Rev. 1990;70:921–71.
13. Virdis A, Savoia C, Grassi C, et al. Evaluation of microvascular structure in humans: a "state of the art" document of the Working Group on Macrovascular and Microvascular Alterations of the Italian Society of Arterial Hypertension. J Hypertens. 2014;32:2120–9.
14. Levy BI, Ambrosio G, Pries AR, Struijker-Boudier HA. Microcirculation in hypertension: a new target for treatment? Circulation. 2001;104:735–40.
15. Pedrinelli R, Taddei S, Spessot M, Salvetti A. Maximal postischemic forearm vasodilation in human hypertension: a reassessment of the method. J Hypertens. 1987;5(suppl 5):S431–3.
16. Agabiti-Rosei E, Rizzoni D, Castellano M, et al. Media:lumen ratio in human small resistance arteries is related to forearm minimal vascular resistance. J Hypertens. 1995;13:341–7.
17. Mulvany MJ, Halpern W. Contractile properties of small resistance vessels in spontaneously hypertensive and normotensive rats. Circ Res. 1977;41:19–26.
18. Schiffrin EL, Hayoz D. How to assess vascular remodelling in small and medium-sized muscular arteries in humans. J Hypertens. 1997;15:571–84.
19. Schiffrin EL. Remodelling of resistance arteries in essential hypertension and effects of antihypertensive treatment. Am J Hypertens. 2004;17:1192–200.
20. Lew MJ. Wall stress and wall to lumen ratios differ between pressurized and myograph-mounted arteries. In: Mulvany MJ, Aalkjaer C, Heagerty AM, Nyborg NCB, Strandgaard S, editors. Resistance arteries: structure and function. Amsterdam, The Netherlands: Excerpta Medica; 1991. p. 353–6.
21. Struijker-Boudier HA, Agabiti-Rosei E, Bruneval P, et al. Evaluation of the microcirculation in hypertension and cardiovascular disease. Eur Heart J. 2007;28:2834–40.
22. Antonios TF, Singer DR, Markandu ND, Mortimer PS, MacGregor GA. Structural skin capillary rarefaction in essential hypertension. Hypertension. 1999;33:998–1001.
23. Shore AC. Capillaroscopy and the measurement of capillary pressure. Br J Clin Pharmacol. 2000;50:501–13.
24. Wong TY, Klein R, Sharrett AR, et al. Retinal arteriolar narrowing and risk of coronary heart disease in men and women. The atherosclerosis risk in communities study. JAMA. 2002;287:1153–9.
25. Harazny JM, Ritt M, Baleanu D, et al. Increased wall:lumen ratio of retinal arterioles in male patients with a history of a cerebrovascular event. Hypertension. 2007;50:623–9.
26. Rizzoni D, Porteri E, Duse S, et al. Relationship between media to lumen ratio of subcutaneous small arteries and wall to lumen ratio of retinal arterioles evaluated non invasively by scanning laser Doppler flowmetry. J Hypertens. 2012;30:1169–75.
27. Heagerty AM. Predicting hypertension complications from small artery structure. J Hypertens. 2007;25:939–40.
28. Iyer A, Fairlie DP, Prins JB, Hammock BD, Brown L. Inflammatory lipid mediators in adipocyte function and obesity. Nat Rev Endocrinol. 2010;6:71–82.
29. Karalis KP, Giannogonas P, Kodela E, Koutmani Y, Zoumakis M, Teli T. Mechanisms of obesity and related pathology: linking immune responses to metabolic stress. FEBS J. 2009;276:5747–54.
30. Hopps E, Noto D, Caimi G, Averna MR. A novel component of the metabolic syndrome: the oxidative stress. Nutr Metab Cardiovasc Dis. 2010;20:72–7.
31. Ferreira I, Twisk JW, van Mechelen E, et al. Current and adolescent body fatness and fat distribution: relationships with carotid intima-media thickness and large artery stiffness at the age of 36 years. J Hypertens. 2004;22:145–55.
32. Zebekakis PE, Nawrot T, Thijs L, et al. Obesity is associated with increased arterial stiffness from adolescence until old age. J Hypertens. 2005;23:1839–46.

33. Giltay EJ, Lambert J, Elbers JM, et al. Arterial compliance and distensibility are modulated by body composition in both men and women but by insulin sensitivity only in women. Diabetologia. 1999;42:214–21.
34. Mangoni AA, Giannattasio C, Brunani A, et al. Radial artery compliance in young, obese, normotensive subjects. Hypertension. 1995;26:984–8.
35. Steinberg HO, Chaker H, Leaming R, et al. Obesity/insulin resistance is associated with endothelial dysfunction. Implications for the syndrome of insulin resistance. J Clin Invest. 1996;97:2601–10.
36. Levy BI, Schiffrin EL, Mourad JJ, et al. Impaired tissue perfusion: a pathology common to hypertension, obesity, and diabetes mellitus. Circulation. 2008;118:968–76.
37. Grassi G, Seravalle G, Scopelliti F, et al. Structural and functional alterations of subcutaneous small resistance arteries in severe human obesity. Obesity. 2009;18:92–8.
38. Integan HD, Deng LY, Li JS, Schiffrin EL. Mechanics and composition of human subcutaneous resistance arteries in essential hypertension. Hypertension. 1999;33:569–74.
39. Integan HD, Schiffrin EL. Structure and mechanical properties of resistance arteries in hypertension: role of adhesion molecules and extracellular matrix determinants. Hypertension. 2000;36:312–8.
40. Savoia C, Touyz RM, Amiri F, Schiffrin EL. Selective mineralocorticoid receptor blocker eplerenone reduces resistance artery stiffness in hypertensive patients. Hypertension. 2008;51:432–9.
41. De Ciuceis C, Porteri E, Rizzoni D, et al. Effects of weight loss on structural and functional alterations of subcutaneous small arteries in obese patients. Hypertension. 2011;58:29–36.
42. Laakso M, Edelman SV, Brechtel G, Baron AD. Decreased effect of insulin to stimulate skeletal muscle blood flow in obese man. A novel mechanism for insulin resistance. J Clin Invest. 1990;85:1844–52.
43. Baron AD. Cardiovascular actions of insulin in humans. Implications for insulin sensitivity and vascular tone. Bailliere Clin Endocrinol Metab. 1993;7:961–87.
44. Baron AD, Steinberg H, Brechtel G, Johnson A. Skeletal muscle blood flow independently modulates insulin-mediated glucose uptake. Am J Physiol Endocrinol Metab. 1994;266:E248–53.
45. Vincent MA, Dawson D, Clark AD, et al. Skeletal muscle microvascular recruitment by physiological hyperinsulinemia precedes increases in total blood flow. Diabetes. 2002;51:42–8.
46. Kim JA, Montagnani M, Koh KK, Quon MJ. Reciprocal relationships between insulin resistance and endothelial dysfunction: molecular and pathophysiological mechanisms. Circulation. 2006;113:1888–904.
47. Clerk LH, Vincent MA, Jahn LA, Liu Z, Lindner JR, Barrett EJ. Obesity blunts insulin-mediated microvascular recruitment in human forearm muscle. Diabetes. 2006;55:1436–42.
48. Vincent MA, Barrett EJ, Lindner JR, Clark MG, Rattigan S. Inhibiting NOS blocks microvascular recruitment and blunts muscle glucose uptake in response to insulin. Am J Physiol Endocrinol Metab. 2003;285:E123–9.
49. Meigs JB, Hu FB, Rifai N, Manson JE. Biomarkers of endothelial dysfunction and risk of type 2 diabetes mellitus. JAMA. 2004;291:1978–86.
50. Cardillo C, Campia U, Iantorno M, Panza JA. Enhanced vascular activity of endogenous endothelin-1 in obese hypertensive patients. Hypertension. 2004;43:36–40.
51. Hotamisligil GS, Arner P, Caro JF, Atkinson RL, Spiegelman BM. Increased adipose tissue expression of tumor necrosis factor-alpha in human obesity and insulin resistance. J Clin Invest. 1995;95:2409–15.
52. Williams IL, Wheatcroft SB, Shah AM, Kearney MT. Obesity, atherosclerosis, and the vascular endothelium: mechanisms of reduced nitric oxide bioavailability in obese humans. Int J Obes Relat Metab Disord. 2002;26:754–64.
53. Arita Y, Kihara S, Ouchi N, et al. Paradoxical decrease of an adipose-specific protein, adiponectin, in obesity. Biochem Biophys Res Commun. 1999;257:79–83.
54. Piatti PM, Monti LD, Conti M, et al. Hypertriglyceridemia and hyperinsulinemia are potent inducers of endothelin-1 release in humans. Diabetes. 1996;45:316–21.

55. Rask-Madsen C, King GL. Mechanisms of disease: endothelial dysfunction in insulin resistance and diabetes. Nat Clin Pract Endocrinol Metab. 2007;3:46–56.
56. Mohamed F, Monge JC, Gordon A, et al. Lack of role for nitric oxide in the selective destabilization of endothelial NO synthase mRNA by tumor necrosis factor-aloha. Arterioscler Thromb Vasc Biol. 1995;15:52–7.
57. Ding J, Wai KL, McGeechan K, Meta-Eye Study Group, et al. Retinal vascular caliber and the development of hypertension: a meta-analysis of individual participant data. J Hypertens. 2014;32:207–15.
58. Klein R, Klein BE, Moss SE, Wong TY, Sharrett AR. Retinal vascular caliber in persons with type 2 diabetes: the Wisconsin Epidemiological Study of Diabetic Retinopathy. Ophthalmology. 2006;113:1488–98.
59. McGeechan K, Liew G, Macaskill P, et al. Prediction of incident stroke events based on retinal vessel caliber: a systematic review and individual-participant meta-analysis. Am J Epidemiol. 2009;170:1323–32.
60. Mimoun L, Massin P, Steg G. Retinal microvascularisation abnormalities and cardiovascular risk. Arch Cardiovasc Dis. 2009;102:449–56.
61. Al-Fiadh AH, Wong T, Kawasaky R, et al. Usefulness of retinal microvascular endothelial dysfunction as a predictor of coronary artery disease. Am J Cardiol. 2015;115:609–13.
62. Boillot A, Zoungas S, Mitchell P, for the META-EYE Study Group, et al. Obesity and the microvasculature: a systematic review and meta-analysis. PLoS One. 2013;8:e52708. https://doi.org/10.1371/journal.pone.0052708.

12 Microvascular Alterations in Diabetes: Focus on Small Resistance Arteries

Carolina De Ciuceis

12.1 Introduction

Diabetes mellitus and impaired glycaemic control are well-known risk factors for cardiovascular disease and are associated with high mortality and health cost [1, 2]. Both type 2 and type 1 diabetes are associated with microvascular and macrovascular complications affecting several organs including the muscles, skin, heart, brain, kidneys and lower limbs, which are responsible for poor quality of life and clinical outcomes of the disease.

Alterations in the microcirculation involve small resistance arteries (diameter between 100 and 300 microns), arterioles (diameter below 100 microns), capillaries and postcapillary venules and lead to morbidity and mortality of diabetes [1, 2]. Damage of the small vessels can lead to blindness (retinopathy), end-stage renal failure (kidney disease), diabetic cardiomyopathy and peripheral neuropathy.

Whereas a lot of data have been reported about alterations of capillary network, less is known about arterioles and small resistance arteries in humans. Indeed, until a few years ago the assessment of resistance arteries required an invasive approach to obtain the small vessels, such as gluteal biopsy, which was not suitable for general use. In the last years, new techniques have been developed which allow a non-invasive and reproducible approach for studying the retinal arterioles morphology, such as confocal laser Doppler flowmetry and especially adaptive optics. These techniques were recently validated [3] and could provide further knowledge in the field.

C. De Ciuceis (✉)
University of Brescia, Brescia, Italy

E. Agabiti-Rosei et al. (eds.), *Microcirculation in Cardiovascular Diseases*, Updates in Hypertension and Cardiovascular Protection,
https://doi.org/10.1007/978-3-030-47801-8_12

12.2 Mechanisms Involved in the Genesis of Diabetic Microvascular Alterations

Mechanisms leading to the diabetic microangiopathy are multiple and still largely unclear. They include both hemodynamic and metabolic factors, as well as growth factors, and involve several defects affecting blood cells interaction with the vessel wall, reactivity of the vessels and their anatomical structure [4].

12.2.1 Hemodynamic Factors

From a physiological point of view, the microvascular bed is responsible for oxygen and micronutrient delivery and removal of metabolic waste products. The regulation of the blood flow occurs according to local factors (metabolic needs, transmural pressure) which modulate the constriction of the arterioles and venules through different mechanisms, such as arteriolar myogenic response, venoarteriolar reflex and precapillary arteriolar vasomotion [5], in order to avoid capillary hypertension and increased permeability. Arteriolar vasomotion is blunted in both experimental and clinical diabetes, and this may be caused by hyperinsulinaemia [6]. In addition, diabetic patients, mainly those with neuropathy, showed impaired slow-wave vasomotion, a defect appearing very early and correlated with sympathetic dysfunction [7]. In small resistance arteries, blunted myogenic response seems to be involved in vascular hypertrophy.

12.2.2 Metabolic Factors

Among several metabolic factors, hyperglycaemia is considered one of the main mechanisms setting the stage for the onset of microvascular diseases. It causes an overproduction of reactive oxygen species (ROS) which leads to non-enzymatic glycation of intracellular and extracellular proteins forming advanced glycation end products (AGEs). AGEs in turn induce consequent changes in enzyme activity, protein cross-linking, molecule binding, proteolysis susceptibility, endocytosis and immunogenicity [8]. All these changes result in alterations in different physiological processes bringing about decreased bioavailability of nitrogen oxide and endothelial dysfunction and the development of chronic diabetic complications [8, 9]. Collagen cross links derived by extracellular protein advanced glycation may contribute to impairment of the mechanics of small vessels and vascular distensibility which may lately induce changes in the vascular structure [10]. Glycation of proteins involves also haemoglobin, with reduction of its affinity to oxygen and consequent induction of micro-ischemia and endothelial damage.

Furthermore, an important consequent of hyperglycaemia is an increased flux of glucose and other sugars through the polyol pathway with enhanced conversion of glucose to sorbitol and its oxidation to fructose. This results in exacerbation of ROS production leading to damage in endothelium and other vascular cells [8].

High concentrations of intracellular glucose increase also the production of the second messenger diacylglycerol in endothelial cells, which activates the diacylglycerol/protein kinase C pathway and causes the expression of different pro-inflammatory genes and growth factors (such as TGFβ, collagen, fibronectin, PAI-1, VEGF) possibly linked by the activation of NF-κB [8].

In addition, hyperglycaemia determines an increased synthesis of O-linked acetylglucosamine through the hexosamine pathway, with subsequent modification of transcription factors that regulate activity and expression of PAI-1 and TGFβ [8, 9].

All these different pathways altered by hyperglycaemia may ultimately induce changes in vascular function and structure.

12.2.3 Neurohumoral and Growth Factors

Direct effect of neurohumoral and growth factors such as angiotensin II [11], insulin or insulin-like growth factor-1 [12, 13] on vascular smooth muscle cell growth may also play a role in structural and functional alterations of small resistance arteries. Type 2 diabetes is characterized by insulin resistance, with high levels of circulating insulin, and it is frequently associated with arterial hypertension [14].

At the cellular level, insulin works via two pathways. The phosphatidylinositol 3-kinase (PI3K)-dependent signalling pathway is involved in nitric oxide (NO) production and vasodilation in vascular endothelium [15, 16]. Another signalling pathway is the mitogen-activated protein kinase (MAPK)-dependent insulin signalling pathway that regulates the secretion of vasoconstrictor endothelin-1 (ET-1) and intracellular adhesion molecules 1, as well as vascular cell adhesion molecule 1 expression in vascular endothelium [15, 16].

Insulin is able to induce a vasodilating effect in the microcirculation (which is, at least in part, endothelium dependent) in normal patients [17, 18]. However, such signalling pathways seem to be impaired in an insulin resistance state such as type 2 diabetes. Insulin resistance attenuates the physiological increase in eNOS activation and suppresses normal NO production because of a selective deficiency of phosphatidylinositol 3-kinase (PI3K)/Akt pathway, whereas signalling via the MAPK pathway is unaffected, resulting in blunted vasodilation and peripheral tissue blood [19]. Moreover, insulin mediates activation of MAPK and promotes endothelin-1 expression and cellular proliferation [20], which is crucial for structural alterations of small vessels.

Indeed, it has been demonstrated that hyperinsulinaemia induces an increase in the media-to-lumen ratio of small intramyocardial arterioles of spontaneously hypertensive rats [12]. Similarly, long-term insulin infusion into one femoral artery in the dog caused vascular hypertrophy only in the ipsilateral side [13].

12.3 Structural Alterations in Small Resistance Arteries in Diabetes Mellitus

Different from hypertensive disease, there are few studies, and non-univocal, addressing the structural alterations of small resistance arteries in diabetes. Indeed, the observation of changes in small artery structure, commonly called remodelling, in diabetic patients is more recent (Fig. 12.1). In one study [21], no difference in subcutaneous small artery structure was observed between control subjects and patients affected by type 1 diabetes. In another study, forearm minimal vascular resistance (an indirect index of resistance artery structure) was greater in type 2 diabetes patients than in normotensive controls [22]. By using a wire myograph system, a direct, reliable and well-assessed technique, Rizzoni et al. [23] first demonstrated the presence of vascular remodelling (Fig. 12.2) in small resistance arteries from type 2 diabetic patients, as shown in hypertensive patients, regardless of the presence of hypertension. Similar results were obtained by Schofield et al. who reported vascular remodelling, as well as abnormal myogenic responses of small resistance arteries, in type 2 diabetic patients [24].

Importantly, different from resistance arteries of patients with essential hypertension alone, which show inward eutrophic remodelling with increased media-to-lumen ratio [25–27] but no significant change in the total amount of tissue within the vascular wall [28, 29], type 2 diabetic patients present hypertrophic remodelling of small arteries with an increased media-to-lumen ratio and media cross-sectional

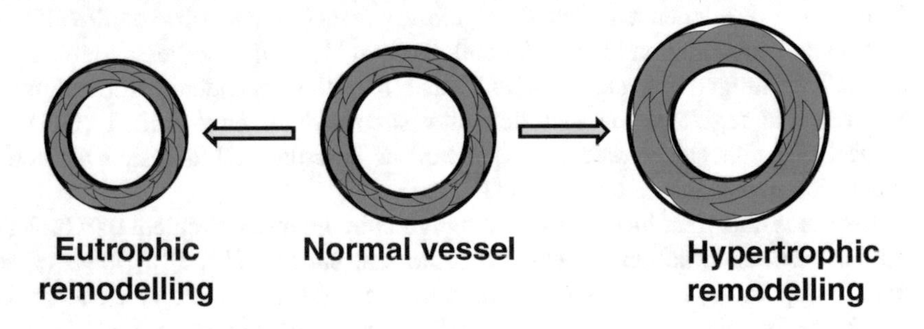

↑ internal and external diameters

↑ media to lumen ratio
no change in the number or size of smooth muscle cells

↑ media cross sectional area

↑ media to lumen ratio and external diameter

Fig. 12.1 Different remodelling processes in small resistance arteries from hypertensive and diabetic patients

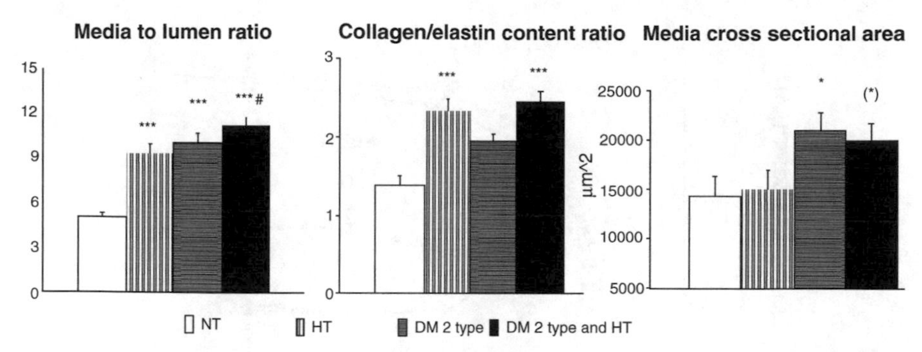

Fig. 12.2 Structural alterations of small resistance arteries in the presence of diabetes and/or hypertension. Media-to-lumen ratio (left panel), collagen-to-elastin ratio (measured with electronic microscope) (central panel) and media cross-sectional area (right panel) in subcutaneous small resistance arteries from normotensive patients (NT), essential hypertensive patients (HT), normotensive patients with type 2 diabetes mellitus (DM 2 type) and hypertensive diabetic patients (DM2 type and HT) (from ref. [92] modified)

area because of vascular smooth muscle cell hypertrophy or hyperplasia [23, 24] (Figs. 12.1 and 12.2). Hypertrophic remodelling was previously demonstrated in patients with renovascular hypertension in whom a pronounced activation of the renin–angiotensin system is present [29, 30], thus indicating that this system may represent an important pathological mechanism associated with the microvascular structural alterations also in diabetic patients. In addition, the impaired myogenic response observed in small arteries of diabetic patients compared with hypertensive or control patients [24] may play a role in the development of hypertrophic remodelling (Fig. 12.3). Indeed, whereas sustained myogenic constriction is crucial for the development of inward eutrophic remodelling, an impaired myogenic response of small arteries may induce wall stress increase, thus representing a stimulus promoting vascular hypertrophy [24]. This process is mediated by both haemodynamic changes and alterations in the inflammatory and redox microenvironment leading to dual processes of remodelling and growth [28, 30, 31].

As stated in previous section, insulin or other related substances like insulin-like growth factor-1 may also have a role in promoting vascular cell growth [32]. Accordingly, a significant correlation between circulating levels of insulin and media-to-lumen ratio of subcutaneous small arteries was observed in diabetic patients [23].

Furthermore, patients with type 2 diabetes present changes in the vascular extracellular matrix, as suggested by the observation of increased collagen-to-elastin ratio in their small arteries [23] (Fig. 12.2), which may have a relevant role in the process of vascular remodelling [33]. Vascular remodelling has been demonstrated also in hypertensive patients with type 1 diabetes [34].

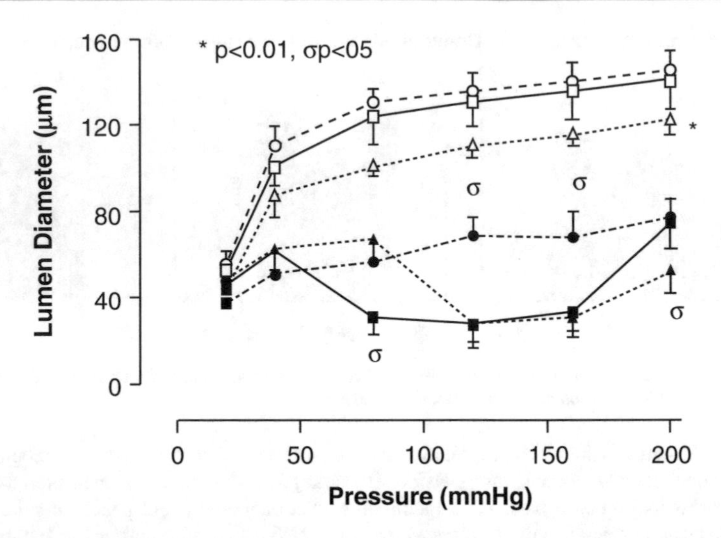

Fig. 12.3 Myogenic response in subcutaneous small resistance arteries in patients with type 2 diabetes mellitus. Passive pressure–lumen diameter relations for arteries from control subjects (□), patients with EH (Δ) and patients with type 2D ± H (○). *P 0.05, ANOVA, vs control. Active pressure–lumen diameter relations for arteries from control subjects (■), patients with EH (●) and patients with type 2D ± H (▲). ∂ P < 0.05 vs control vessels (from ref. [24])

Interestingly, the simultaneous presence of diabetes and hypertension seems to be associated with a significant higher media-to-lumen ratio compared to patients with either diabetes or hypertension only, thus suggesting that clustering of risk factors may have synergistic detrimental effects on the vasculature [23, 24].

From a clinical point of view, in hypertension, the structural alterations in subcutaneous small arteries due to inward eutrophic remodelling result in decreased vasodilator reserves and changes in the distensibility of arterioles [35] which may be of great clinical importance when they occur in some districts such as the coronary vasculature. In agreement with this, coronary flow reserve and minimum coronary resistance correlate significantly with the media-to-lumen ratio and internal diameter of subcutaneous small arteries of hypertensive patients [36]. Similarly, an impaired microvascular hyperaemic response (which may reflect an altered flow reserve) has been observed in children with diabetes mellitus [37] as well as in adult patients with type 2 diabetes [38]. However, it has been recently demonstrated that, in the presence of good blood pressure control, no differences in vascular structure of coronary arteries from diabetic and/or hypertensive patients were observed compared to non-diabetic normotensive patients [39]. This is in contrast to that observed in subcutaneous resistance arteries [23, 24].

The importance of structural alterations has been further emphasized by the direct correlation reported between media-to-lumen ratio of subcutaneous resistance arteries and major cardiovascular events in different populations of hypertensive patients, including also patients with diabetes mellitus [40–42]. Moreover, for the same values of internal diameter, those patients with a greater media cross-sectional area (and therefore with hypertrophic remodelling), such as that observed in diabetic patients, present a greater incidence of cardiovascular events [40, 43, 44] meaning a worse prognosis. Hypertrophic remodelling might indeed represent a 'less physiological' adaptation to increased blood pressure levels, thus playing a role in the development of target organ damage, both in hypertension and mainly in diabetes. This may be suggested also by the observation that, despite effective antihypertensive treatment, resistance arteries from hypertensive diabetic patients showed persisted marked remodelling, even greater than non-diabetic untreated hypertensive patients [45]. Thus, alterations in the microcirculation may play a crucial role in the development of organ damage not only in hypertension but also in diabetes.

12.4 Endothelial Dysfunction

The vascular endothelium plays a pivotal role in maintaining vascular protective conditions and controlling smooth muscle tone. Specifically, control is managed through the release of potent vasodilators, such as NO. Most of the arterial vasculature is the nutritive network of smaller resistance vessels (arterioles and capillaries) which have a dominant influence on local blood flow and are directly connected with other tissues. Also, resistance arteries are submitted to chemical and neurohormonal stimulation and exposed to the continuous effects of shear stress generated by the bloodstream [46].

In patients with type 2 diabetes, microvascular and macrovascular dysfunction frequently coexist and are thought to be mediated in part by reduced NO bioavailability [47] and increased ROS production resulting in progressive tissue damage [48]. The mechanisms underlying such dysfunction include an impaired production and/or bioavailability of endogenous NO·. NO· is scavenged by ROS including superoxide ($\cdot O_2^-$) and hydroxyl radicals, and endothelial NO synthase may also become uncoupled under conditions of increased oxidative stress [49, 50]. Damaged endothelial cells are also crucial for the release of vasoconstrictor instead of vasodilator factors as well as for production of other substances which regulate platelet aggregation and leucocyte extravasation contributing to the inflammatory and thrombotic vascular lesions.

A wide body of evidence shows an impairment of the endothelial function, as evaluated by the vasodilator response to acetylcholine, in large and small arteries of patients with type 1 diabetes [21, 51–54] as well as of those with type 2 diabetes [23, 24, 55–57].

Particularly, subcutaneous small resistance arteries of type 2 diabetic hypertensive or normotensive patients show the presence of an impaired dilatation to acetylcholine and bradykinin [23, 24, 57]. However, inhibitors of NO synthase (i.e. L-NMMA) were only able to partially block the vasodilator effect of acetylcholine or bradykinin which was conversely obtain with ouabain (a blocker of ATP-dependent sodium-potassium exchanger), thus suggesting that production of both NO and endothelium-derived hyperpolarizing factor may be involved [23, 57].

Furthermore, differently from what observed for the structural alterations, the simultaneous presence of hypertension and diabetes does not appear to further worsen endothelial function of small arteries [23, 57]. In fact, if both hypertension and diabetes mellitus may induce oxidative stress and reduction of NO bioavailability, with subsequent endothelial dysfunction, through a similar pathogenic mechanism, no additive deleterious effect may be observed when the two conditions occur together.

As already reported, it has been also proposed that insulin and insulin resistance may be involved in the genesis of endothelial dysfunction in diabetes [17, 18].

In addition, glucotoxicity, lipotoxicity and inflammation are often present in an insulin resistance state [58]. Various cytokines produced in inflammation selectively downregulate the PI3K signalling pathway [15, 16] with subsequent reduced expression of NO and increased release of vasoconstrictor ET-1, collectively resulting in endothelial vasoconstriction. Hyperinsulinaemia itself, as a result of insulin resistance, may have detrimental effects on microvascular function also in the prediabetic state [59]. On the other hand, microvascular structural alterations may contribute to an impaired delivery of insulin to skeletal muscles [60].

As previously mentioned, in type 2 as well as in type 1 diabetes, increased glucose levels are probably also responsible for increased oxidative stress and decreased NO bioavailability which cause endothelial dysfunction. Hyperglycaemia can impact multiple pathways, including direct uncoupling of endothelial NO synthase, increased ROS generation and activation of the $\cdot O_2^-$ generating enzyme NADPH oxidase, ultimately leading to reduced synthesis, bioavailability and efficacy of NO· [50].

Interestingly, it has been suggested that the severity of endothelial dysfunction reported in type 2 diabetes mellitus is attributable to abnormal lipid profile [22, 61]. Although the complete mechanism remains undetermined, enhanced oxidative stress appears to be crucial for the endothelial dysfunction caused by hypercholesterolaemia [62]. Studies in vitro and in humans have shown that hypercholesterolaemia induces angiotensin II type 1 (AT1) receptor overexpression [63, 64] and that AT1 receptor activation is a predominant source of free radical release in the vascular wall [65].

Therefore, together with structural alterations, microvascular dysfunction may, at least in part, explain the deleterious consequences of diabetes in terms of increased incidence of cardiovascular events.

12.5 Effect of Treatment

Data about the effect of treatment on structural and functional alterations in the microcirculation of patients with diabetes mellitus are relatively scarce. It is known that a tight haemodynamic and metabolic control is associated with a lower incidence of microvascular disease and, in general, of clinical end points related to microvascular disease in type 2 diabetic patients [66]. In those patients, treatment with ACE inhibitors [67] and AT1 receptor blockers [68] significantly improves both macrovascular and microvascular end points, including nephropathy [69–71], retinopathy [72] and neuropathy [73], thus confirming the importance of renin–angiotensin system blocking.

Among various antihypertensive agents, angiotensin-converting enzyme (ACE) inhibitors, AT1 receptor blockers and calcium channel blockers are more effective, despite similar blood pressure decrease, in reducing the media-to-lumen ratio and regress structural or functional alterations of subcutaneous small resistance arteries in essential hypertensive patients [74–80].

Similarly, 1-year treatment with the ACE inhibitor enalapril or with the AT1 receptor blocker candesartan was equally effective in improving subcutaneous small artery structure (Fig. 12.4) in hypertensive diabetic patients [81]. However, candesartan was more effective than enalapril in reducing collagen content in the vasculature [81], probably because of a more pronounced stimulation of the local production of metalloproteinase 9 (a collagen-degrading enzyme) from candesartan (Fig. 12.4). Furthermore, blockade of AT1 receptor with valsartan, added on top to ACE inhibitor and/or calcium channel blocker, provides added value in terms of regression of small artery media-to-lumen ratio in hypertensive diabetic patients, compared to the beta blocker atenolol [82]. Beyond blood pressure reduction, AT1 receptor blocker may exert beneficial actions on vascular remodelling acting as a vasodilator by blocking AT1 receptor and stimulating AT2 receptor which could reduce vascular tone and elicit antigrowth and proapoptotic effects [83]. In addition, the direct renin

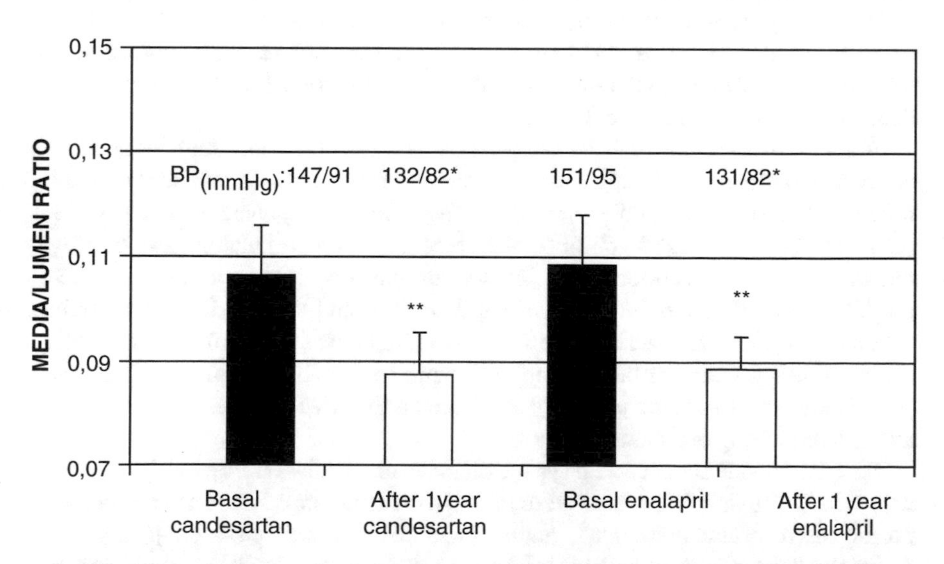

Fig. 12.4 Effects of 1-year treatment with candesartan or enalapril on subcutaneous small resistance artery structure in diabetic hypertensives. Media-to-lumen ratio in subcutaneous small resistance arteries from hypertensive patients with NIDDM, before and after 1-year treatment with the ACE inhibitor enalapril or the angiotensin II receptor blocker candesartan. BP, blood pressure. $^{**}P < 0.01$ versus basal (from ref. [92])

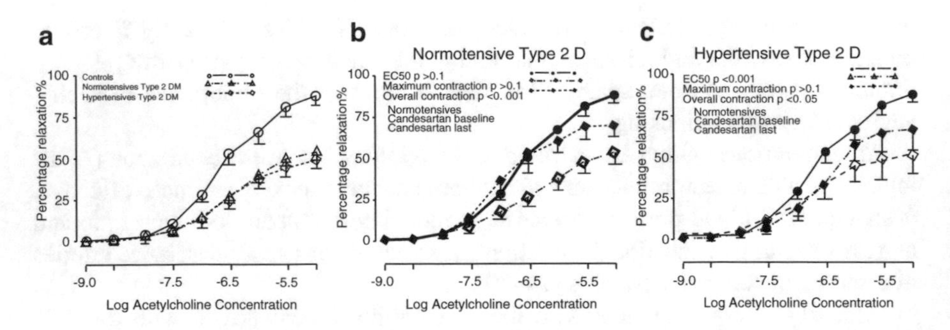

Fig. 12.5 Dilation of small arteries comparing control subjects with normotensive and hypertensive patients with type 2 diabetes mellitus treated with the AT1 receptor blocker candesartan (**a**) and response assessed by change in EC50, maximal relaxation and overall relaxation to candesartan in normotensive (**b**) and hypertensive (**c**) type 2 DM exposed to increasing concentrations of acetylcholine after pre-constriction with norepinephrine (from ref. [61])

inhibitor aliskiren was also proved to be equally effective to ramipril in correcting the retinal arteriolar remodelling in hypertensive diabetic patients [84]. Particularly, aliskiren might have some modest advantages compared with ramipril in terms of reduction in media-to-lumen ratio of subcutaneous small resistance arteries [84].

Both ACE inhibitors and AT1 receptor blockers improve also endothelial dysfunction of resistance arteries from patients with hypertension and/or diabetes [61] related either to hemodynamic actions or reduction of oxidative stress (Fig. 12.5).

It has been proposed that drugs that may stimulate PPAR-α or PPAR-γ receptors (such as fibrates of glitazones) may be useful in terms of vascular protection and regression of structural alterations in the microcirculation [85, 86]. However, no data are presently available in human beings.

Among different anti-diabetes drugs, the direct effects of anti-diabetes agents on vascular structure and function have been studied using different microvascular models in short-term studies [87–89]. The effects of glucagon-like peptide-1 (GLP-1)-based or DPP-4 inhibitor therapies on microvasculature are heterogeneous. Short-term treatment with the sodium–glucose cotransporter 2 (SGLT-2) inhibitor dapagliflozin reduced retinal capillary flow and improved structural retinal arteriole remodelling (arteriolar wall cross-sectional area and wall-to-lumen ratio), evaluated using scanning laser Doppler flowmetry, in patients with type 2 diabetes [90]. However, no data are presently available on the effects of these drugs on small artery remodelling or function in humans.

Few data about patients with type 1 diabetes are available. Previously, in type 1 diabetes mellitus, no differences were found in the structure of small arteries between normotensive normoalbuminuric patients and control participants [21]. In another study, small arteries from patients with type 1 diabetes develop hypertrophic growth in response to elevated blood pressure, similar to that seen in type 2 diabetes [34]. Interestingly, in patients with type 1 diabetes, metabolic control enables to reverse hypertrophic remodelling of subcutaneous small arteries to a more favourable structural profile of eutrophic remodelling in response to hypertension [34].

As already reported, it is more difficult to obtain a complete regression of small resistance artery structural alterations in patients with type 2 diabetes, despite the prolonged use of combination antihypertensive therapy and consistent blood pressure reductions [45, 81, 82]. It was recently demonstrated that on-treatment subcutaneous small artery structure, as evaluated by the media-to-lumen ratio, identifies individuals still at increased cardiovascular risk despite long-term blood pressure normalization and may represent an additional therapeutical target to prevent cardiovascular events [91] in hypertensive non-diabetic patients. However, it is not presently known whether a regression of vascular alterations in diabetic patients may be associated with a real protection from cardiovascular events [92].

12.6 Conclusions

Small arteries of patients with type 2 diabetes mellitus demonstrate both functional and structural alterations that include hypertrophic remodelling and deficient endothelium-dependent relaxation. In type 2 diabetes, humoral factors such as chronic hyperglycaemia, hyperinsulinaemia, oxidative stress and chronic low-grade inflammation might be involved in the small artery hypertrophic remodelling processes, as well as in endothelial dysfunction. Structural alterations of small arteries are associated with an increased cardiovascular risk in hypertensive and diabetic patients, and hypertrophic remodelling seems to be associated with an even worse prognosis.

Data about the effect of therapy on microvascular structure in diabetic patients are still scarce even though renin–angiotensin system blockade seems to be effective. However, persistent and severe abnormalities observed in diabetic patients suggest that other treatment strategies must be superimposed to improve the microvascular damage and high cardiovascular risk in the case of diabetes especially when associated with hypertension.

References

1. American Diabetes Association. Diagnosis and classification of diabetes mellitus. Diabetes Care. 2013;36(Suppl 1):S67–74.
2. American Diabetes Association. Standards of medical care in diabetes-2013. Diabetes Care. 2013;36(Suppl 1):S11–66.
3. De Ciuceis C, Agabiti Rosei C, Caletti S, Trapletti V, Coschignano MA, Tiberio GAM, Duse S, Docchio F, Pasinetti S, Zambonardi F, Semeraro F, Porteri E, Solaini L, Sansoni G, Pileri P, Rossini C, Mittempergher F, Portolani N, Ministrini S, Agabiti-Rosei E, Rizzoni D. Comparison between invasive and noninvasive techniques of evaluation of microvascular structural alterations. J Hypertens. 2018;36(5):1154–16.
4. Kollros PR. Microvascular disease in diabetes mellitus. J Cardiovasc Risk. 1997;4:70–5.
5. Wiernsperger NF. In defence of microvascular constriction in diabetes. Clin Hemorheol Microcirc. 2001;25:55–62.
6. Wiernsperger NF, Bouskela E. Microcirculation in insulin resistance and diabetes: more than just a complication. Diabetes Metab. 2003;29:6S77–87.

7. Meyer MF, Rose CJ, Hülsmann JO, Schatz H, Pfohl M. Impaired 0.1-Hz vasomotion assessed by laser Doppler anemometry as an early index of peripheral sympathetic neuropathy in diabetes. Microvasc Res. 2003;65:88–95.
8. Madonna R, De Caterina R. Cellular and molecular mechanisms of vascular injury in diabetes–part I: pathways of vascular disease in diabetes. Vasc Pharmacol. 2011;54:68–74.
9. Sena CM, Pereira AM, Seiça R. Endothelial dysfunction—a major mediator of diabetic vascular disease. Biochim Biophys Acta. 2013;1832:2216–31.
10. Aronson D. Cross-linking of glycated collagen in the pathogenesis of arterial and myocardial stiffening of aging and diabetes. J Hypertens. 2003;21:3–12.
11. Schelling P, Fischer H, Ganten D. Angiotensin and cell growth: a link to cardiovascular hypertrophy? J Hypertens. 1991;9(1):3–15.
12. Zimlichman R, Zaidel L, Nofech-Mozes S, Shkedy A, Matas Z, Shahar C, Eliahou HE. Hyperinsulinemia induces myocardial infarctions and arteriolar medial hypertrophy in spontaneously hypertensive rats. Am J Hypertens. 1997;10:646–53.
13. Cruz AB, Amatuzio DS, Grande F, Hay L. Effect of intra-arterial insulin on tissue cholesterol and fatty acids in alloxan-diabetic dogs. Circ Res. 1961;9:39–43.
14. Bhanot S, McNeill JH. Insulin and hypertension: a causal relationship? Cardiovasc Res. 1996;31:212–21.
15. Arcaro G, Cretti A, Balzano S, Lechi A, Muggeo M, Bonora E, Bonadonna RC. Insulin causes endothelial dysfunction in humans: sites and mechanisms. Circulation. 2002;105:576–82.
16. DeFronzo RA. Insulin resistance, lipotoxicity, type 2 diabetes and atherosclerosis: the missing links. The Claude Bernard Lecture 2009. Diabetologia. 2010;53:1270–87.
17. Rossi M, Cupisti A, Ricco R, Santoro G, Pentimone F, Carpi A. Skin vasoreactivity to insulin iontophoresis is reduced in elderly subjects and is absent in treated non-insulin-dependent diabetes patients. Biomed Pharmacother. 2004;58:560–5.
18. Clerk LH, Vincent MA, Lindner JR, Clark MG, Rattigan S, Barrett EJ. The vasodilatory actions of insulin on resistance and terminal arterioles and their impact on muscle glucose uptake. Diabetes Metab Res Rev. 2004;20:3–12.
19. Tabit CE, Chung WB, Hamburg NM, Vita JA. Endothelial dysfunction in diabetes mellitus: molecular mechanisms and clinical implications. Rev Endocr Metab Disord. 2010;11:61–74.
20. Roberts AC, Porter KE. Cellular and molecular mechanisms of endothelial dysfunction in diabetes. Diab Vasc Dis Res. 2013;10:472–82.
21. McNally PG, Watt PAC, Rimmer T, Burden AC, Hearnshaw JR, Thurston H. Impaired contraction and endothelium-dependent relaxation in isolated resistance vessels from patients with insulin-dependent diabetes mellitus. Clin Sci. 1994;87:31–6.
22. Longhini CS, Scorzoni D, Bazzanini F, Manservigi D, Fratti D, Gilli G, Musacci GF. The structural arteriolar changes in diabetes mellitus and essential hypertension. Eur Heart J. 1997;18:1135–40.
23. Rizzoni D, Porteri E, Guelfi D, Muiesan ML, Valentini U, Cimino A, Girelli A, Rodella L, Bianchi R, Sleiman I, Rosei EA. Structural alterations in subcutaneous small arteries of normotensive and hypertensive patients with non–insulin-dependent diabetes mellitus. Circulation. 2001;103:1238–44.
24. Schofield I, Malik R, Izzard A, Austin C, Heagerty A. Vascular structural and functional changes in type 2 diabetes mellitus: evidence for the roles of abnormal myogenic responsiveness and dyslipidemia. Circulation. 2002;106:3037–43.
25. Mulvany MJ, Aalkjaer C. Structure and function of small arteries. Physiol Rev. 1990;70:921–61.
26. Aalkjaer C, Heagerty AM, Petersen KK, Swales JD, Mulvany MJ. Evidence for increased media thickness, increased neuronal amine uptake, and depressed excitation–contraction coupling in isolated resistance vessels from essential hypertensives. Circ Res. 1987;61:181–6.
27. Agabiti-Rosei E, Rizzoni D, Castellano M, Porteri E, Zulli R, Muiesan ML, Bettoni G, Salvetti M, Muiesan P, Giulini SM. Media: lumen ratio in human small resistance arteries is related to forearm minimal vascular resistance. J Hypertens. 1995;13:341–7.
28. Heagerty AM, Aalkjaer C, Bund SJ, Korsgaard N, Mulvany MJ. Small artery structure in hypertension. Dual processes of remodeling and growth. Hypertension. 1993;21:391–7.

29. Rizzoni D, Porteri E, Castellano M, Bettoni G, Muiesan ML, Muiesan P, Giulini SM, Agabiti-Rosei E. Vascular hypertrophy and remodeling in secondary hypertension. Hypertension. 1996;28:785–90.
30. Rizzoni D, Porteri E, Guefi D, Piccoli A, Castellano M, Pasini G, Muiesan ML, Mulvany MJ, Agabiti-Rosei E. Cellular hypertrophy in subcutaneous small arteries of patients with renovascular hypertension. Hypertension. 2000;35:931–5.
31. Khavandi K, Greenstein AS, Sonoyama K, Withers S, Price A, Malik RA, Heagerty AM. Myogenic tone and small artery remodelling: insight into diabetic nephropathy. Nephrol Dial Transplant. 2009;24(2):361–9.
32. King GL, Goodman AD, Buzney S, Moses A, Kahn CR. Receptors and growth promoting effects of insulin and insulin-like growth factors on cells from bovine retinal capillaries and aorta. J Clin Invest. 1985;75:1028–36.
33. Intengan HD, Deng LY, Li JS, Schiffrin EL. Mechanics and composition of human subcutaneous resistance arteries in essential hypertension. Hypertension. 1999;33(part II):569–74.
34. Greenstein AS, Price A, Sonoyama K, Paisley A, Khavandi K, Withers S, Shaw L, Paniagua O, Malik RA, Heagerty AM. Eutrophic remodeling of small arteries in type 1 diabetes mellitus is enabled by metabolic control: a 10-year follow-up study. Hypertension. 2009;54(1):134–41.
35. Laurent S, Boutouyrie P. The structural factor of hypertension: large and small artery alterations. Circ Res. 2015;116:1007–21.
36. Rizzoni D, Palombo C, Porteri E, Muiesan ML, Kozakova M, La Canna G, Nardi M, Guelfi D, Salvetti M, Morizzo C, Vittone F, Rosei EA. Relationships between coronary flow vasodilator capacity and small artery remodelling in hypertensive patients. J Hypertens. 2003;21:625–31.
37. Shore AC, Price KJ, Sandeman DD, Green EM, Tripp JH, Tooke JE. Impaired microvascular hyperhaemic response in children with diabetes mellitus. Diabet Med. 1991;8:619–23.
38. Strain WD, Chaturvedi N, Nihoyannopoulos P, Bulpitt CJ, Rajkumar C, Shore AC. Differences in the association between type 2 diabetes and impaired microvascular function among Europeans and African Caribbeans. Diabetologia. 2005;48:2269–77.
39. Lynch FM, Izzard AS, Austin C, Prendergast B, Keenan D, Malik RA, Heagerty AM. Effects of diabetes and hypertension on structure and distensibility of human small coronary arteries. J Hypertens. 2012;30:384–9.
40. Rizzoni D, Porteri E, Boari GE, De Ciuceis C, Sleiman I, Muiesan ML, Castellano M, Miclini M, Agabiti-cRosei E. Prognostic significance of small-artery structure in hypertension. Circulation. 2003;108:2230–5.
41. De Ciuceis C, Porteri E, Rizzoni D, Rizzardi N, Paiardi S, Boari GE, Miclini M, Zani F, Muiesan ML, Donato F, Salvetti M, Castellano M, Tiberio GA, Giulini SM, Agabiti RE. Structural alterations of subcutaneous small-resistance arteries may predict major cardiovascular events in patients with hypertension. Am J Hypertens. 2007;20:846–52.
42. Mathiassen ON, Buus NH, Sihm I, Thybo NK, Mørn B, Schroeder AP, Thygesen K, Aalkjaer C, Lederballe O, Mulvany MJ, Christensen KL. Small artery structure is an independent predictor of cardiovascular events in essential hypertension. J Hypertens. 2007;25:1021–6.
43. Izzard AS, Rizzoni D, Agabiti-Rosei E, Heagerty AM. Small artery structure and hypertension: adaptive changes and target organ damage. J Hypertens. 2005;23:247–50.
44. Heagerty AM. Predicting hypertension complications from small artery structure. J Hypertens. 2007;25:939–40.
45. Endemann DH, Pu Q, De Ciuceis C, Savoia C, Virdis A, Neves MF, Touyz RM, Schiffrin EL. Persistent remodeling of resistance arteries in type 2 diabetic patients on antihypertensive treatment. Hypertension. 2004;43:399–404.
46. Roustit M, Cracowski JL. Assessment of endothelial and neurovascular function in human skin microcirculation. Trends Pharmacol Sci. 2013;34:373–84.
47. Krentz AJ, Clough G, Byrne CD. Interactions between microvascular and macrovascular disease in diabetes: pathophysiology and therapeutic implications. Diabetes Obes Metab. 2007;9:781–91.

48. De Vriese AS, Verbeuren TJ, Van de Voorde J, Lameire NH, Vanhoutte PM. Endothelial dysfunction in diabetes. Br J Pharmacol. 2000;130:963–74.
49. Gryglewski RJ, Palmer RM, Moncada S. Superoxide anion is involved in the breakdown of endothelium-derived vascular relaxing factor. Nature. 1986;320:454–6.
50. Hink U, Li H, Mollnau H, Oelze M, Matheis E, Hartmann M, Skatchkov M, Thaiss F, Stahl RA, Warnholtz A, Meinertz T, Griendling K, Harrison DG, Forstermann U, Munzel T. Mechanisms underlying endothelial dysfunction in diabetes mellitus. Circ Res. 2001;88:E14–22.
51. Johnstone MT, Creager SJ, Scales KM, Cusco JA, Lee BK, Creager MA. Impaired endothelium-dependent vasodilation in patients with insulin dependent diabetes mellitus. Circulation. 1994;88:2510–6.
52. Clarkson P, Celermajer DS, Donald AE, Sampson M, Sorensen KE, Adams M, Yue DK, Betteridge DJ, Deanfield JE. Impaired vascular reactivity in insulin-dependent diabetes mellitus is related to disease duration and low density lipoprotein cholesterol levels. J Am Coll Cardiol. 1996;28:573–9.
53. Zenere BM, Arcaro G, Saggiani F, Rossi L, Muggeo M, Lechi A. Noninvasive detection of functional alterations of the arterial wall in IDDM patients with and without microalbuminuria. Diabetes Care. 1995;18:975–82.
54. Lambert J, Aarsen M, Donker AJ, Stehouwer CD. Endothelium-dependent and –independent vasodilation of large arteries in normoalbuminuric insulin-dependent diabetes mellitus. Arterioscler Thromb Vasc Biol. 1996;16:705–11.
55. Goodfellow J, Ramsey MW, Luddington LA, Jones CJ, Coates PA, Dunstan F, Lewis MJ, Owens DR, Henderson AH. Endothelium and inelastic arteries: an early marker of vascular dysfunction in non-insulin dependent diabetes. BMJ. 1996;312:744–5.
56. Parving HH, Nielsen FS, Bang LE, Smidt UM, Svendsen TL, Chen JW, Gall MA, Rossing P. Macro-microangiopathy and endothelial dysfunction in NIDDM patients with and without diabetic nephropathy. Diabetologia. 1996;39:1590–7.
57. Rizzoni D, Porteri E, Guelfi D, Muiesan ML, Piccoli A, Valentini U, Cimino A, Girelli A, Salvetti M, De Ciuceis C, Tiberio GA, Giulini SM, Sleiman I, Monteduro C, Rosei EA. Endothelial dysfunction in small resistance arteries of patients with non-insulin dependent diabetes mellitus. J Hypertens. 2001;19:913–9.
58. Potenza MA, Addabbo F, Montagnani M. Vascular actions of insulin with implications for endothelial dysfunction. Am J Physiol Endocrinol Metab. 2009;297:E568–77.
59. Jaap AJ, Shore AC, Tooke JE. Relationship of insulin resistance to microvascular dysfunction in subjects with fasting hyperglycaemia. Diabetologia. 1997;40:238–43.
60. Vincent MA, Clerk LH, Rattigan S, Clark MG, Barrett EJ. Active role for the vasculature in the delivery of insulin to skeletal muscle. Clin Exp Pharmacol Physiol. 2005;32:302–7.
61. Malik RA, Schofield IJ, Izzard A, Austin C, Bermann G, Heagerty AM. Effects of angiotensin type-1 receptor antagonism on small artery function in patients with type 2 diabetes mellitus. Hypertension. 2005;45:264–9.
62. Harrison DG. Endothelial function and oxidative stress. Clin Cardiol. 1997;20:II-11–7.
63. Nickenig G, Sackinidis A, Michaelsen F, Böhm M, Seewald S, Vener H. Upregulation of vascular angiotensin II receptor gene expression by low-density lipoprotein in vascular smooth muscle cells. Circulation. 1997;95:473–8.
64. Nickenig G, Bäumer AT, Temur Y, Kebben D, Jockenhövel F, Böhm M. Statin-sensitive dysregulated AT1 receptor function and density in hypercholesterolemic men. Circulation. 1999;100:2131–4.
65. Griendling KK, Minieri CA, Ollerenshaw JD, Alexander RW. Angiotensin II stimulates NADH and NADPH oxidase activity in cultured vascular smooth muscle cells. Circ Res. 1994;74:1141–8.
66. Adler AI, Stratton IM, Neil HA, Yudkin JS, Matthews DR, Cull CA, Wright AD, Turner RC, Holman RR. Association of systolic blood pressure with macrovascular and microvascular complications of type 2 diabetes (UKPDS 36): prospective observational study. BMJ. 2000;321:412–9.

67. Heart Outcomes Prevention Evaluation Study Investigators. Effects of ramipril on cardiovascular and microvascular outcomes in people with diabetes mellitus: results of the HOPE study and MICRO-HOPE substudy. Lancet. 2000;355:253–9.
68. Lindholm LH, Ibsen H, Dahlof B, Devereux B, Beevers G, de Faire U, Fyhrquist F, Julius S, Kjeldsen SE, Kristiansson K, Lederballe-Pedersen O, Nieminen MS, Omvik P, Oparil S, Wedel H, Aurup P, Edelman J, Snapinn S, The LIFE Study Group. Cardiovascular morbidity and mortality in patients with diabetes in the losartan intervention for endpoint reduction in hypertension study (LIFE): a randomised trial against atenolol. Lancet. 2002;359:1004–10.
69. Parving HH, Lehnert H, Brochner-Mortensen J, Gomis R, Andersen S, Arner P, Irbesartan in Patients with Type 2 Diabetes and Microalbuminuria Study Group. The effect of irbesartan on the development of diabetic nephropathy in patients with type 2 diabetes. N Engl J Med. 2001;345:870–8.
70. Brenner BM, Cooper ME, de Zeeuw D, Keane WF, Mitch WE, Parving HH, Remuzzi G, Snapinn SM, Zhang Z, Shahinfar S, RENAAL Study Investigators. Effects of losartan on renal and cardiovascular outcomes in patients with type 2 diabetes and nephropathy. N Engl J Med. 2001;345:861–9.
71. Lewis EJ, Hunsicker LG, Clarke WR, Berl T, Pohl MA, Lewis JB, Ritz E, Atkins RC, Rohde R, Raz I, Collaborative Study Group. Renoprotective effect of the angiotensin-receptor antagonist irbesartan in patients with nephropathy due to type 2 diabetes. N Engl J Med. 2001;345:851–60.
72. Chaturvedi N, Sjolie A-K, Stephenson JM, Abrahamian H, Kiepes M, Castellarian A, Rogulja-Pepeonik Z, Fuller JH. Effect of lisinopril on progression of retinopathy in normotensive people with type 1 diabetes. Lancet. 1998;351:28–31.
73. Malik RA, Williamson S, Abbott CA, Carrington AL, Iqbal J, Schady W, Boulton AJM. Effect of angiotensin-converting enzyme (ACE) inhibitor trandolapril on human diabetic neuropathy: randomised double-blind controlled trial. Lancet. 1998;352:1978–81.
74. Agabiti-Rosei E, Heagerty AM, Rizzoni D. Effects of antihypertensive treatment on small artery remodelling. J Hypertens. 2009;27:1107–14.
75. Schiffrin EL, Deng LY, Larochelle P. Effects of a -blocker or a converting enzyme inhibitor on resistance arteries in essential hypertension. Hypertension. 1994;23:83–91.
76. Schiffrin EL, Deng LY, Larochelle P. Progressive improvement in the structure of resistance arteries of hypertensive patients after 2 years of treatment with an angiotensin I-converting enzyme inhibitor: comparison with effects of a -blocker. Am J Hypertens. 1995;8:229–36.
77. Schiffrin EL, Park JB, Intengan HD, Touyz RM. Correction of arterial structure and endothelial dysfunction in human essential hypertension by the angiotensin receptor antagonist losartan. Circulation. 2000;101:1653–9.
78. Schiffrin EL, Park JB, Pu Q. Effect of crossing over hypertensive patients from a -blocker to an angiotensin receptor antagonist on resistance artery structure and endothelial function. J Hypertens. 2002;20:71–8.
79. Park JB, Intengan HD, Schiffrin EL. Reduction of resistance artery stiffness by treatment with the AT1-receptor antagonist losartan in essential hypertension. J Renin Angiotensin Aldosterone Syst. 2000;1:40–5.
80. Schiffrin EL, Pu Q, Park JB. Effect of amlodipine compared to atenolol on small arteries of previously untreated essential hypertensive patients. Am J Hypertens. 2002;15:105–10.
81. Rizzoni D, Porteri E, De Ciuceis C, Sleiman I, Rodella L, Rezzani R, Paiardi S, Bianchi R, Ruggeri G, Boari GE, Muiesan ML, Salvetti M, Zani F, Miclini M, Rosei EA. Effect of treatment with candesartan or enalapril on subcutaneous small artery structure in hypertensive patients with noninsulin-dependent diabetes mellitus. Hypertension. 2005;45:659–65.
82. Savoia C, Touyz RM, Endemann DH, Pu Q, Ko EA, De Ciuceis C, Schiffrin EL. Angiotensin receptor blocker added to previous antihypertensive agents on arteries of diabetic hypertensive patients. Hypertension. 2006;48:271–7.
83. Savoia C, Touyz RM, Volpe M, Schiffrin EL. Angiotensin type 2 receptor in resistance arteries of type 2 diabetic hypertensive patients. Hypertension. 2007;49:341–6.

84. De Ciuceis C, Savoia C, Arrabito E, Porteri E, Mazza M, Rossini C, Duse S, Semeraro F, Agabiti Rosei C, Alonzo A, Sada L, La Boria E, Sarkar A, Petroboni B, Mercantini P, Volpe M, Rizzoni D, Agabiti Rosei E. Effects of a long-term treatment with aliskiren or ramipril on structural alterations of subcutaneous small-resistance arteries of diabetic hypertensive patients. Hypertension. 2014;64(4):717–24.
85. Diep QN, El Mabrouk M, Cohn JS, Endemann D, Amiri F, Virdis A, Fritsch Neves M, Schiffrin EL. Structure, endothelial function, cell growth, and inflammation in blood vessels of angiotensin II-infused rats: role of peroxisome proliferator-activated receptor-gamma. Circulation. 2002;105:2296–302.
86. Diep QN, Amiri F, Touyz RM, Cohn JS, Endemann D, Schiffrin EL. PPAR activator effects on Ang II-induced vascular oxidative stress and inflammation. Hypertension. 2002;40:866–71.
87. Forst T, Michelson G, Ratter F, Weber MM, Anders S, Mitry M, Wilhelm B, Pfützner A. Addition of liraglutide in patients with type 2 diabetes well controlled on metformin monotherapy improves several markers of vascular function. Diabet Med. 2012;29:1115–8.
88. Ott C, Raf U, Schmidt S, Kistner I, Friedrich S, Bramlage P, Harazny JM, Schmieder RE. Effects of saxagliptin on early microvascular changes in patients with type 2 diabetes. Cardiovasc Diabetol. 2014;13:19.
89. Berndt-Zipfel C, Michelson G, Dworak M, Mitry M, Löffler A, Pfützner A, Forst T. Vildagliptin in addition to metformin improves retinal blood flow and erythrocyte deformability in patients with type 2 diabetes mellitus—results from an exploratory study. Cardiovasc Diabetol. 2013;12:59.
90. Ott C, Jumar A, Striepe K, Friedrich S, Karg MV, Bramlage P, Schmieder RE. A randomised study of the impact of the SGLT2 inhibitor dapagliflozin on microvascular and macrovascular circulation. Cardiovasc Diabetol. 2017;16:26.
91. Buus NH, Mathiassen ON, Fenger-Grøn M, Præstholm MN, Sihm I, Thybo NK, Schroeder AP, Thygesen K, Aalkjær C, Pedersen OL, Mulvany MJ, Christensen KL. Small artery structure during antihypertensive therapy is an independent predictor of cardiovascular events in essential hypertension. J Hypertens. 2013;31:791–7.
92. Rizzoni D, Agabiti RE. Small artery remodeling in diabetes mellitus. Nutr Metab Cardiovasc Dis. 2009;19:587–92.

Cardiovascular Effects of Anti-angiogenic Drugs

13

Harry A. J. Struijker-Boudier

13.1 Introduction

Cancer treatment is the cause of an emerging chronic cardiovascular health problem in cancer survivors. Data show that more than 1.5 million people are diagnosed with a malignancy every year in the USA [1, 2], and in 2016, the American Cancer Society reported that there were 15.5 million cancer survivors in the country [1]. The 5-year survival rate for patients treated for cancer is currently 67% [1], but approximately 75% of these cancer survivors have some form of chronic health problems, of which cardiovascular diseases are the leading cause of morbidity and mortality [1]. The risk of cardiovascular disease is 8 times higher in cancer survivors than in the general population, and the relative risks of coronary artery disease and heart failure in cancer survivors are 10 times and 15 times higher, respectively, than in their siblings without cancer [1, 2].

There is a wide range of cardiovascular toxicities associated with anti-cancer therapy. Traditional chemotherapeutics, such as the anthracyclines, have been associated with severe cardiotoxicity, requiring strict control of the dose for a drug like doxorubicin [3]. With the emergence of many novel anti-cancer drugs, new forms of vascular toxicity, including systemic and pulmonary hypertension, stroke, acute coronary syndromes, arterial and arteriolar stenosis, and thrombosis, have been reported [4]. Due to an extensive drug discovery program by pharmaceutical and biotech companies, there has been an explosion of novel anti-cancer agents on the market (Table 13.1). In particular, monoclonal antibodies, VEGF-receptor fusion

H. A. J. Struijker-Boudier (✉)
Department of Pharmacology & Toxicology, Cardiovascular Research Institute Maastricht, Maastricht University, Maastricht, the Netherlands
e-mail: h.struijkerboudier@maastrichtuniversity.nl

E. Agabiti-Rosei et al. (eds.), *Microcirculation in Cardiovascular Diseases*, Updates in Hypertension and Cardiovascular Protection,
https://doi.org/10.1007/978-3-030-47801-8_13

Table 13.1 Major classes of anti-cancer drugs

Major classes of anti-cancer drugs
Antimetabolites
Alkylating agents
Vinca alkaloids
Proteasome inhibitors
Monoclonal antibodies
VEGF receptor fusion molecules
Antimicrotubule agents
Tyrosine kinase inhibitors

Table 13.2 Tyrosine kinase inhibitors

Tyrosine kinase inhibitors
Sorafenib
Sunitinib
Pazopanib
Axitinib
Regorafenib
Cabozantinib
Vandetanib
Lenvatinib
Nilotinib
Ponatinib
Dasatinib

molecules, and tyrosine kinase inhibitors (TKIs; Table 13.2) have been brought to the market rapidly without thorough investigation of their potential cardiovascular side effects.

13.2 Mechanisms of Angiogenesis

A common feature of the three recently introduced classes of anti-cancer drugs (Table 13.2) is that they inhibit the action of vascular growth factors, such as vascular endothelial growth factor (VEGF), platelet-derived growth factor (PDGF), and fibroblast growth factor (FGF) [5]. These growth factors stimulate, via the enzyme tyrosine kinase, a range of signal transduction molecules that cause angiogenesis of the tumor, invasion of tissues, and metastasis. The role of angiogenesis in cancer progression and of angiogenesis inhibition as a form of cancer treatment was introduced several decades ago by Judah Folkman, when he observed an association between solid tumor growth and vascular supply [6, 7]. He showed that a soluble factor isolated from tumor tissue could promote neovascularization of tumors in vivo. A few years later, this substance was identified and sequenced as the vascular endothelial growth factor (VEGF) [8]. In the decades following these early observations, our understanding of the process of angiogenesis has increased substantially, and the reader is referred to recent review articles for more detailed information [9, 10].

The VEGF signaling pathway plays a central role in the development of anti-angiogenic drugs. VEGF is one of the five members of a family of structurally related proteins that are involved in the regulation of vascular endothelium [11]. The VEGF gene undergoes alternative splicing to form these multiple isoforms. The VEGF gene is upregulated by hypoxia, reactive oxygen species, inflammatory cytokines, and other growth factors [12]. VEGF binds three tyrosine kinase receptors leading to a tyrosine kinase signaling cascade that stimulates production of factors that induce vasodilation, cell proliferation migration, and differentiation into mature blood vessels [12, 13]. Major signaling pathways include phosphoinositide 3-kinase/Akt/protein kinase B-mammalian target of rapamycin and activation of endothelial nitric oxide (NO) synthase and inducible NO synthase, leading to downstream release of potent vasodilators, including NO and prostacyclin [12, 13]. Besides promoting angiogenesis, VEGF also increases vascular permeability and is required for the maintenance of a differentiated EC phenotype and EC survival [13]. It thus has many physiological actions, including angiogenesis during embryogenesis, wound healing, and menstruation [12]. In preeclampsia, a placenta-derived tyrosine kinase, a splice variant of the VEGF gene, is markedly increased, thereby contributing to the pathophysiology of this condition, characterized by hypertension, proteinuria, and edema [13, 14].

13.3 Anti-angiogenic Drugs in Cancer Treatment

The VEGF signaling pathway (VSP) has been a primary target for the development of anti-angiogenic drugs. Table 13.2 summarized the major anti-angiogenic drugs from the tyrosine kinase inhibitors group. More than ten of the representatives of this class have been or are currently investigated for their anti-cancer potential. All these agents are small molecule multiple TKIs with varying specificities for VEGF receptor subtypes. Because the kinase domains in the VEGF receptors share structural similarity with the kinase domains in other signaling receptors, these TKIs target multiple pathways [11]. These drugs have been approved for the treatment of different forms of cancer, including gastrointestinal stromal tumors, advanced renal cell carcinoma, advanced pancreatic neuroendocrine tumors, advanced hepatocellular carcinoma, metastatic medullary thyroid cancer, and metastatic colorectal cancer. It is beyond the scope of this chapter to critically review the effects of these drugs in each of these forms of cancer. The reader is referred to recent specialized reviews for this purpose [11, 15–17].

Despite the recent explosion of VSP inhibitors, the benefits of these agents have thus far been modest [11]. There is still a lack of understanding how different cancers become vascularized and how they evade the effects of anti-angiogenic therapy by building up resistance towards VEGF-targeted therapy. In addition, incomplete knowledge on duration and scheduling of therapy as well as toxicities hampers the optimal therapy with anti-angiogenic drugs. The use of predictive biomarkers may be a way to closely monitor their therapeutic efficacy [15].

13.4 Adverse Cardiovascular Effects of Anti-angiogenic Drugs

Hypertension is the most common cardiovascular effect of VSP inhibitors, and its underlying mechanisms will be dealt with in the next section of this chapter. A second important potential toxic target of VSP inhibitors is left ventricular dysfunction and cardiomyopathy. Although it is difficult to judge in cancer patients with complex clinical condition and multiple chemotherapeutic interventions to what degree VSP inhibitors specifically cause left ventricular dysfunction, both retrospective observational data from individual trials with VSP inhibitors and meta-analyses suggest that up to 30% of patients had an absolute decrease in ejection fraction [11, 18]. Similar analyses suggest a significantly increased incidence of cardiomyopathy [11, 18]. One of the most logical explanations for VSP inhibitor-induced cardiomyopathy is a decreased capillary density (rarefaction) in the heart, causing hypoxia and cardiac dysfunction [19].

A further potential cardiovascular toxicity of VSP inhibitors is arterial and venous thromboembolism. Numerous studies have suggested that the incidence of thromboembolic events is increased in patients treated with VSP inhibitors [11, 18]. Meta-analyses have shown an incidence of thromboembolic events in cancer patients treated with VSP inhibitors of around 12%, depending on the type of cancer treated [20]. Mechanisms underlying the thromboembolic events in VSP inhibitor-treated patients include alterations in endothelial cell function and expression of various factors involved in hemostasis and thrombolysis as well as immune complex-mediated platelet activation [11, 18].

13.5 Angiogenesis Inhibition-Related Hypertension: Incidence

Hypertension is a common side effect of cancer treatment with VSP inhibitors. The reported incidence of hypertension shows a wide range, depending on the agents used, their dosing schedule, the type of cancer treated, and the diagnostic criteria for hypertension classification. Several recent review articles summarize these different results [11, 12, 21–23]. A particularly important source of difference in incidence data is the level of blood pressure regarded as hypertensive and the method by which blood pressure is assessed. In view of the recent controversies between hypertension guidelines in different parts of the world as well as the classification systems used by cancer expert organizations to assess chemotherapeutic toxicities, it is difficult to give exact incidence data [12, 21].

Almost 100% of patients treated with VSP inhibitors have an absolute increase in blood pressure, with a subset developing chronic hypertension [12]. Data used from meta-analyses indicate that the overall hypertension incidence during VSP inhibitor treatment ranges from 20% to 90% [12]. The blood pressure increase induced by VSP inhibitors occurs rapidly within hours to days after start of treatment. Ambulatory blood pressure monitoring demonstrated that blood pressure increases over the first 24 h after start of the therapy, with sustained blood pressure increase after 1 week [12, 24].

Hypertension can rapidly lead to levels >150/100 mm Hg which may require adjustment of VSP inhibitors dosing or start antihypertensive treatment. For a detailed overview of the management of VSP inhibitor-induced hypertension, the reader is referred to a recent excellent review article [23].

13.6 Mechanisms of VSP Inhibition-Related Hypertension

Table 13.3 summarizes the most important mechanisms that have been implicated in VSP inhibition-related hypertension. The first suggestion that angiogenesis may cause hypertension stems from the 1990s—before VSP inhibitors were on the market—in several papers reviewing the role of microvascular density in the pathogenesis of hypertension [25, 26]. There is convincing evidence in the meantime that small arteriolar and capillary rarefaction are major hallmarks of hypertension [27, 28]. Such evidence was obtained in various animal models for hypertension but was more difficult to obtain in humans because of lack of methodologies to assess human microcirculation. Recent technological innovations, particularly in thc area of studies of the retina, have confirmed the important role of structural microvascular changes in the development and maintenance of hypertension in humans [29, 30]. In several clinical studies, capillary rarefaction has been shown in response to anti-angiogenic agents [23, 31–33].

Other potential mechanisms of VSP inhibition-related hypertension focus on changes in endothelial cell function. Inhibition of the nitric oxide signaling pathway has been raised on the basis of experimental animal models, but clinical studies have not been conclusive thus far [13, 22]. An interesting alternative endothelium-related mechanism was proposed by Danser and co-workers in a series of both clinical and experimental studies, measuring endothelin metabolism [13, 34]. In these studies, they found that antiangiogenic treatment increases endothelin-1 levels. Besides increased endothelin-1 levels, VEGF inhibition is associated with an increased vasoconstrictor response to endothelin-1. These studies suggest a possible therapeutic solution to VSP inhibition-related hypertension by means of the use of endothelin receptor blockers.

Less well-established potential mechanisms of VSP inhibition-related hypertension involve increased production of reactive oxidative stress molecules and arterial stiffening [12, 22]. Further research is needed to confirm the role of these latter mechanisms in the etiology and treatment of VSP inhibition-related hypertension.

Table 13.3 Potential mechanisms of VSP inhibitor-induced hypertension

Potential mechanisms
Microvascular rarefaction
Reduced NO production
Reduced prostacyclin production
Increased generation of endothelin
Increased production reactive oxygen species
Renal dysfunction
Increased arterial stiffness

13.7 Conclusion

Angiogenesis is a fundamental mechanism in dynamics of the structure of the microcirculation. It contributes both to the maintenance of an optimal perfusion of tissues as to the spread and growth of tumors. Thus, anti-angiogenic drugs can be expected to be double-edged swords, inhibiting tumor growth on the one hand but interfering with cardiovascular control of blood pressure and tissue perfusion on the other hand. In this respect, hypertension is not an unexpected frequent consequence of anti-angiogenic therapy.

References

1. Abe JF. The future of onco-cardiology. Circ Res. 2016;119:896–9.
2. Akam-Venhata J, Franco VI, Lipschultz SE. Late cardiotoxicity issues for childhood cancer survivors. Curr Treat Options Cardiovasc Med. 2016;18:47–57.
3. Zamorano JL, Lancellotti P, Muñoz DR, et al. 2016 ESC position paper on cancer treatment and cardiovascular toxicity. Eur Heart J. 2016;37:2768–801.
4. Herrmann J, Yang EH, Iliescu CA, et al. Vascular toxicities of cancer therapies. Circulation. 2016;133:1272–89.
5. Ancker OV, Wehlad M, Bauer J, Interfanger M, Grimm D. The adverse effect of hypertension in the treatment of thyroid cancer with multi-kinase inhibitors. Int J Mol Sci. 2017;18:625–44.
6. Folkman J. Clinical applications of research on angiogenesis. New Engl J Med. 1995;333:1757–63.
7. Folkman J, Merler E, Abernathy C, Williams G. Isolation of a tumor factor responsible for angiogenesis. J Exp Med. 1971;133:275–88.
8. Leung DW, Cachianes G, Kuang WJ, Goeddel DV, Ferrara N. Vascular endothelial growth factor is a secreted angiogenic mitogen. Science. 1989;246:1306–9.
9. Simons M, Eichmann A. Molecular control of arterial morphogenesis. Circ Res. 2015;116:1712–24.
10. Potente M, Carmeliet P. The link between angiogenesis and endothelial metabolism. Annu Rev Physiol. 2017;79:43–66.
11. Bair SM, Chorieiri TK, Moslehi J. Cardiovascular complications associated with novel angiogenesis inhibitions. Trends Cardiovasc Med. 2013;23:104–13.
12. Small HY, Monterano AC, Rios FJ, Savoia C, Touyz RM. Hypertension due to antiangiogenic cancer therapy with VEGF inhibitors. Can J Cardiol. 2014;30:534–43.
13. Lankhorst F, Daleh L, Danser AH, Meiracker AH. Etiology of angiogenesis-inhibition-related hypertension. Curr Opin Pharma. 2015;21:7–13.
14. Duhig K, Vandermolen B, Sherman A. Recent advances in the diagnosis and management of pre-eclampsia. F1000Res. 2018;7:242.
15. Vasudev NS, Reynolds AR. Anti-angiogenic therapy for cancer: current progress, unresolved questions and future directions. Angiogenesis. 2014;17:471–94.
16. Lee CC, Shiao HY, Wang WC, Hsieh HP. Small molecule EGFR TKI's for the treatment of cancer. Exp Opin Invest Drugs. 2014;23:1333–48.
17. Roskori R. The role of small molecule kit protein TKI's in the treatment of neoplastic disorders. Pharmacol Res. 2018;133:35–52.
18. Rees ML, Khakoo AY. Molecular mechanisms of hypertension and heart failure due to anti angiogenic cancer therapy. Heart Fail Clin. 2011;7:299–311.
19. May D, Gilon D, Djorov V, Itin A, Lazarus A, et al. Transgenic system for conditional induction and rescue of chronic myocardial hibernation provides insight into genomic programs of hibernation. Proc Natl Acad Sci USA. 2008;105:282–7.

20. Nalluri SR, Chu D, Keresztes R, Zhu X, Wu S. Risk of venous thromboembolism with the angiogenesis inhibitor bevacizumab in cancer patients: a meta-analysis. JAMA. 2008;300:2277–85.
21. Li M, Kroetz DL. Bevacizumab-induced hypertension. Pharmacol Ther. 2018;182:152–60.
22. Touyz RM, Herrman SMS, Hermann J. Vascular toxicities with VEGF inhibitor therapies. J Am Soc Hypertens. 2018;12:409–25.
23. Caletti S, Paini A, Coschignano MA, De Ciuceis C, Nardin M, et al. Management of VEGF-targeted therapy induced hypertension. Curr Hypertens Rep. 2018;20:68–76.
24. Dorff TB, Pull SK, Quinn DI. Novel tyrosine kinase inhibitors for renal cell carcinoma. Expert Rev Clin Pharmacol. 2014;7:67–73.
25. Struijker-Boudier HAJ, le Noble JL, Messing MW, et al. The microcirculation and hypertension. J Hypertens. 1992;10(suppl):147–56.
26. le Noble FAC, Stassen FR, Hackeng WJ, Struijker-Boudier H. Angiogenesis and hypertension. J Hypertens. 1998;14:1563–72.
27. Struijker-Boudier HAJ, Agabiti-Rosei E, Bruneval P et al. evaluation of the microcirculation in hypertension and cardiovascular disease. Eur Heart J. 2007;28:2834–40.
28. Mulvany M, Aalkjaer C. Structure and function of small arteries. Physiol Rev. 1990;70:921–61.
29. Leahy MJ, editor. Microcirculation imaging: Wiley-Blackwell; 2012. p. 1–393.
30. Rizzoni D, Agabiti Rosei C, De Ciuceis C, et al. New methods to study the microcirculation. Am J Hypertens. 2018;31:265–73.
31. Steeghs N, Gerldershlom H, Roodt JO, Christensen O, Rajagopalan P, et al. Hypertension and rarefaction during treatment with telatinib, a small molecule angiogenesis inhibitor. Clin Cancer Res. 2008;14:3470–6.
32. Veldt AA, Boer MP, Boven E, Eringa EC, Eertwegh AJ, et al. Reduction in skin microvascular density and changes in vessel morphology in patients treated with sunitinib. Anti-Cancer Drugs. 2010;21:439–46.
33. Mourad JJ, des Guetz G, Debbati H, Levy BI. Blood pressure rise following angiogenesis inhibition by bevacizumab. Ann Oncol. 2008;19:927–34.
34. Kapper MHW, van Esch JHM, Sluiter W, Sleijfer S, Danser AH, Meiracker AH. Hypertension induced by the tyrosine kinase inhibitor sunitinib is associated with increased circulating endothelin-1 levels. Hypertension. 2010;56:675–81.

14 Pathophysiological Mechanisms Implicated in Organ Damage and Cardiovascular Events

Reza Aghamohammadzadeh and Anthony M. Heagerty

Abbreviations

BAT	Brown adipose tissue
CAD	Coronary artery disease
DM	Diabetes mellitus
EC	Endothelial cell
EDCFs	Endothelium-derived constricting factors
EDHF	Endothelium-derived hyperpolarising factor
EDRFs	Endothelium-derived relaxing factors
EFS	Electric field stimulation
MetS	Metabolic syndrome
NO	Nitric oxide
PVAT	Perivascular adipose tissue
RAAS	Renin-angiotensin-aldosterone system
ROS	Reactive oxygen species
SNS	Sympathetic nervous system
T2DM	Type 2 diabetes mellitus
VSMC	Vascular smooth muscle cell
WAT	White adipose tissue

R. Aghamohammadzadeh · A. M. Heagerty (✉)
Division of Cardiovascular Sciences, School of Medical Sciences, Core Technology Facility, The University of Manchester, Manchester, UK
e-mail: reza.zadeh@manchester.ac.uk; tony.heagerty@manchester.ac.uk

E. Agabiti-Rosei et al. (eds.), *Microcirculation in Cardiovascular Diseases*, Updates in Hypertension and Cardiovascular Protection,
https://doi.org/10.1007/978-3-030-47801-8_14

There are numerous instigators and facilitators of damage to the physiological integrity of microcirculation. These include factors in the blood stream, changes in the vessel wall itself, as well as those influencing the microvessels in a paracrine fashion. In particular, the renin-angiotensin-aldosterone system (RAAS), the overactive sympathetic nervous system, metabolic dysregulation including hyperinsulinaemia, adipokine imbalance, and perivascular adipose tissue damage. Clearly, MetS is a chronic disorder with a complex aetiology. The changes in microcirculation in MetS are the driving force behind organ damage, cardiovascular events and the phenotypes that physicians aim to remedy in clinic. In this chapter we explore microcirculation and the pathological processes affecting this organ tissue.

14.1 Introduction

There is a growing consensus that MetS can lead to microvascular damage, and this is the initial pathophysiological stage in development of diabetes as well as hypertension. The histopathological alterations in microcirculation are of considerable interest because once they have developed, they result in an adverse cardiovascular risk profile. Historically, the microcirculation has been studied in the context of hypertension and diabetes mellitus, but more recently, obesity has emerged as a major epidemic, and we have turned our attention to understanding how adipose tissue affects the microcirculation in health and disease.

To explore organ damage and cardiovascular events from the perspective of the microcirculation, we need to examine the link between the microvasculature, perivascular adipose tissue, adipokines and the nervous system and how imbalances in the intricate interplay between these tissues and organ systems can both lead to and be a consequence of disease states such as obesity, diabetes and hypertension.

The microcirculation is a network of blood vessels consisting of arterioles, capillaries and venules. It contributes to systemic blood pressure control as well as perfusing vital organs and tissues. Almost all the microcirculation in the human body is surrounded by adipose tissue which exerts its own effects and controls the adjacent vessels. In health it has important regulatory effects which become distorted in disease states such as obesity. At the level of the smaller arteries, this contributes to vascular resistance and development of hypertension and diabetes. From a therapeutic perspective, it is imperative to understand better the pathological mechanisms behind microvascular damage in an attempt to reverse or prevent alterations to vascular structure and physiology. Our knowledge is fast expanding in this field, and it is important to treat the microcirculation as an organ system in its own right and explore the pathobiological basis of the constituents of the MetS which we frequently encounter in everyday practice and profile our patients in terms of structural and physiological alterations. This may well result in using already available drugs in new settings, such as the possibility of using an antihypertensive agent such as eplerenone in obese populations with the aim of preventing the almost inevitable advent of type 2 DM.

14.2 Microcirculation in Health and Disease

At the cellular level, vascular smooth muscle cells (VSMC) control much of the peripheral systemic blood flow and contribute to the clinic blood pressure. Vessel tone is controlled by a myriad of factors including molecules secreted by endothelial cells, sympathetic innervations of the microvessels and perivascular adipose tissue, as well as circulating hormones produced in distant organs and delivered locally by microcirculation itself. In small resistance arteries that determine peripheral vascular resistance, myogenic tone is the contractile response to pressure that is observed independent of neurohormonal stimuli and the endothelium [1]. This is the body's natural defence against increases in blood pressure in order to limit the downward transmission of this enhanced pressure which can ultimately result in end organ damage.

The contractile apparatus of VSMC is made up of thin actin filaments, intermediate filaments or cytoskeleton of the cell and thick myosin filaments [2, 3]. VSMC contraction is the main contributor to peripheral BP; thus maintaining the healthy workings of the VSMC relaxation pathways to counteract the contracting forces is crucial to prevent adverse development of vessel tone in the microcirculation. VSMC relaxation was thought to be primarily an effect of nitric oxide (NO) and from endothelial cells. Relaxation independent of NO and prostacyclin was later thought to result from an endothelium-derived hyperpolarising factor (EDHF). In fact a number of factors fit this profile and lead to hyperpolarisation of the VSMC including NO, hydrogen sulphide, reactive oxygen species and metabolites of arachidonic acid [4, 5]. A large number of these molecules are synthesised and released in perivascular adipose tissue (PVAT). Thus it becomes evident that alterations in the perivascular adipose tissue environment which surrounds the microcirculation can drastically alter the resting tone of microvessels and their ability to respond to secondary pathophysiological forces.

Endothelial cells (ECs) themselves are affected by a number of factors, the levels of which can be altered in disease states. ECs express receptors for molecules including acetylcholine, catecholamines, angiotensin and serotonin and produce a number of agents including endothelium-derived relaxing factors (EDRFs) as well as endothelium-derived constricting factors (EDCFs). In health, the constricting and relaxing forces exist in an elegant and fine balance which can be altered by physiological stimuli to help control vessel tone, and alterations in any of these have detrimental results and contribute to adverse cardiovascular outcomes [6–9]. Once the message to change vascular tone is 'sensed' from the bloodstream, the endothelium 'transmits' it outward to the VSMCs via EDRFs or EDCFs. When the endothelium receives the signal required for vasodilatation, there is an increase in its $[Ca^{2+}]i$ leading to opening of IKca and SKca channels resulting in efflux of K^+ and its accumulation in the myoendothelial junction. This hyperpolarises the endothelial cell, and subsequently the VSMCs become hyperpolarised by either (1) transmission of the hyperpolarisation from EC to VSMC via gap junctions or (2) induction of hyperpolarisation in VSMC by the K^+ released from the EC into the myoendothelial junction. The K^+ is thought to activate Na^+-K^+-ATPase on VSMC [4, 5, 10]. This explains the 'inside-out' mechanism of VSMC relaxation and the role the ECs play in this process. The 'outside-in' control of VSMCs is both due to sympathetic innervation and the complex PVAT environment surrounding the microvessels.

The changes observed in hypertension involve both vascular and cardiac remodelling where the predominant haemodynamic feature is a high cardiac output [11]. The high cardiac output is accompanied by an increased heart rate and raised oxygen consumption [12]. In the mild early stages of essential hypertension, the peripheral resistance is low, although crucially, it remains inappropriately high for cardiac output. However, over a 20-year period, the high cardiac index and normal total peripheral resistance pattern changes to a low cardiac index and high resistance pattern [13]. This haemodynamic profile could be a consequence of a high sympathetic nervous tone. Studies have demonstrated that autonomic blockade of the heart in mild hypertensive patients restored cardiac output to normal, and there was a combination of increased sympathetic tone and decreased parasympathetic activity [14, 15]. This would suggest that the abnormality in essential hypertension is one of integrated function in the medulla oblongata and not purely a consequence of microvascular changes per se. The contribution of the sympathetic nervous system will be discussed elsewhere in this chapter.

14.3 Structural Changes in the Circulation and Endothelial Dysfunction

In hypertension, the walls of medium-sized arteries are thickened, but their contribution to vascular resistance remains small in comparison to that of microcirculation. The heart and medium-sized arteries demonstrate a hypertrophic response to rising blood pressure. At the level of the resistance arteries, there is evidence of a reduced lumen diameter and increased media thickness: lumen diameter ratio [16, 17].

There is an argument for the existence of excessive sympathetic nervous activity in the early stages of the development of hypertension being the trigger for the majority of this disorder and the small artery morphology changing shape in consequence. However, at least in certain experimental models, structural changes may precede hypertension; hence this topic remains contentious and focus of further research [18].

At normal pressures, the resistance vessels exhibit myogenic tone, which enables arteries to constrict or dilate in response to changes in upstream blood pressure, independent of neurohormonal influences. This myogenic response is only observed in smaller resistance arteries, which facilitate autoregulation of blood flow and stabilise capillary pressure. This ensures a constant flow of oxygenated blood as well as pressure within the target organs downstream. In hypertension, hypertrophy is observed in vessels which do not possess myogenic tone, whereas in smaller resistance arteries, initial increase in pressure will bring about an increased myogenic constriction. If an individual has untreated hypertension, then there will be prolonged myogenic constriction as the resistance vasculature endeavours to protect the target organs downstream from pressure-induced damage brought about by an increase in blood flow. Prolonged vasoconstriction will lead to inward eutrophic remodelling and/or reduced arterial distensibility.

Inward eutrophic remodelling is an adaptation observed after prolonged vasoconstriction and is thought to preserve a lumen diameter for long periods. This process is also the mechanism by which wall stress can be normalised whilst maintaining vasomotor tone. Remodelling following prolonged vasoconstriction involves a migratory process whereby existing vascular smooth muscle cells reposition themselves in the vascular wall and resulting in a narrow lumen. Thus, the breakdown of the autoregulatory mechanism of myogenic tone is the trigger for the vessel wall hypertrophic response to offset the increased wall stress.

Interestingly, vessels from patients with type 2 diabetes mellitus (T2DM) demonstrate hypertrophy, increase in distensibility and a highly significant loss of myogenic responsiveness compared with patients with essential hypertension and control patients. The study demonstrated that vasoconstrictor function remains normal in patients with T2DM and T2DM+ hypertension and essential hypertension. In fact, there is little evidence to suggest that hypertension is associated with abnormalities of contractile function at all. Both in vitro and in vivo studies have suggested that contraction is normal. There is some controversy with regard to whether the structural alterations in the vascular wall lead to exaggerated constriction and vascular amplification in hypertension. Whilst endothelium-dependent dilation is normal in patients with essential hypertension, it is abnormal in patients with T2DM and T2DM+ hypertension with a significant correlation between dilator impairment and the degree of dyslipidemia recorded in all groups [19]. This demonstrates the complex interplay between the three elements of the MetS as well as the closely related pathobiological processes behind hypertension, DM and obesity/dyslipidaemia.

In terms of cardiovascular risk, recent data from Italy have demonstrated that there is an increased risk of development of cardiovascular events in patients whose small arteries demonstrate hypertrophy rather than eutrophic inward remodelling [20].

What we have learned from physiological and anatomical experimental protocols is that it is simply impossible to be categorically definitive and certain about our observations, as the ultimate experimental design is one that accounts for all the confounders and examines the field of interest without removing it from its natural environment. This is why ex vivo and in vitro experiments don't always closely resemble the natural pathobiology observed in vivo. In support of this, some work in intact animals seems to suggest that vascular amplification seen in isolated vascular beds in hypertension is not observed when the circulation is examined as a whole. The same is true for potential vascular relaxation abnormalities in hypertension. When studying hypertension, there is often associated dyslipidaemia, and we know that oxidised LDL can reduce the bioavailability of nitric oxide which contributes to endothelium-dependent dilator function which has been reported to be abnormal in patients with high blood pressure and dyslipidaemia, patients with dyslipidaemia and coronary artery disease or patients with hypercholesterolaemia [21].

It is possible that the level of oxidised LDL in the bloodstream of individuals with hypertension is a significant contributor to endothelial dysfunction rather than rising blood pressure itself. In support of this, the use of statins has been demonstrated to restore endothelial integrity to near normal as soon as the cholesterol levels are improved [21]. It becomes clear that studying small vessel function in hypertension is not entirely accurate without accounting or adjusting for any variations in lipid levels. Moreover, endothelial function declines with age, thus highlighting the need to at least attempt to account for age as a variable in our protocols.

Hypercholesterolaemia is associated with an increased expression of a type 1 angiotensin receptor (ATl), and binding of angiotensin II to the ATl receptor is associated with an increase in oxidative stress and a reduced bioavailability of nitric oxide. It follows that, in animal studies, ACE inhibitors or angiotensin receptor blockers can ameliorate endothelial dysfunction and reduce plaque load independent of their antihypertensive properties.

Inflammation, oxidative stress and monocyte recruitment all play their part in initiating endothelial dysfunction in obesity. There is also disruption to the fine balance between the vasoconstrictor action of endothelin-1 and the vasodilator effect of NO in endothelial cells. In health, insulin activates phosphoinositide 3-kinase leading to an increased NO production secondary to eNOS phosphorylation [22]. Postprandial physiological surge in insulin concentrations leads to dilatation of precapillary arterioles, thus improving blood flow and delivery of nutrients to tissues, a process known as nutritive flow [23]. In obesity, NO-mediated vasorelaxation is impaired, leading to vasoconstriction via unopposed endothelin-1 action [22, 23]. Reduced endothelial nitric oxide bioavailability in obesity is a significant consequence of the reactions between free radicals and NO. Reactive oxygen species such as the superoxide anion react with nitric oxide to produce peroxynitrite and deplete endothelial NO levels.

Changes are not limited to the microcirculation. Vessel stiffness is another important contributing factor in the pathophysiology of obesity-related hypertension. The association of vessel stiffness is strongest for waist circumference and visceral adiposity, rather than global obesity as measured by BMI [24]. Obesity is a complex multifaceted disorder, and dysregulation of any number of factors can affect vascular stiffness. The adipokine leptin has been linked with impairment of arterial distensibility, and its raised levels in obesity may well be a contributing factor in arterial stiffness [25].

14.4 Perivascular Adipose Tissue and the Microcirculation

Perivascular adipose tissue is certainly one factor which links obesity and hypertension. Adipocytes are ubiquitously found throughout the body and traditionally regarded as energy stores; however, there is now ample evidence suggesting that adipocytes produce a large number of metabolically active substances with both endocrine and paracrine properties. The term perivascular adipose tissue (PVAT) is used to describe the adipocytes surrounding almost every blood vessel in the human body. It provides mechanical support for the microvessels and also secretes vasoactive and metabolically essential cytokines known as adipokines which can regulate vascular function and affect vessel tone. PVAT function has been investigated in dog, pig and rat models as well as some ex vivo and in vitro studies of human blood vessels. The functional and structural properties of PVAT seem to vary between species and anatomical site being examined. Thus the site of tissue harvest for experiments is of vital importance in order to compare like-with-like when scrutinising the findings from different studies. Intra-abdominal and visceral fat depots are linked with an adverse cardiometabolic profile and higher mortality associated with obesity [26, 27]. Increased gluteofemoral fat mass negatively correlates with levels of inflammatory cytokines and is positively linked to raise concentrations of adipokines linked with lower metabolic and cardiovascular risk [28]. Intuitively, this is most likely due to the unique properties of different adipose depots and thus unique PVAT constituents.

A number of investigations have reported that healthy human PVAT can exert a local anticontractile effect on adjacent blood vessels, and elegant pharmacological assessments have identified adiponectin, angiotensin 1–7, hydrogen sulphide, hydrogen peroxide and palmitic acid as some of the adipokines playing this role. The exact mechanisms at work are still requiring some clarification. We have shown that adiponectin works via the large calcium-sensitive potassium (BKCa) channel which if absent leads to a loss of normal PVAT relaxing function. Microelectrode studies of de-endothelialised rat mesenteric vessels in constricted arteries have shown that the hyperpolarisation to exogenous adiponectin is inhibited by selective blockade of BKCa. There is also evidence that stimulation of the ß3 adrenoreceptor releases a factor which indirectly activates myocyte BKCa channels. This has been shown to be partly simulated by release of adiponectin from adipocytes after sympathetic nerve stimulation of B3 receptors on adipocytes [29]. PKG also plays a pivotal role in adiponectin release and has provided further clues to decipher the mechanism of action of the vasorelaxant factors released from PVAT [30].

In obesity there is inflammation, increased oxidative stress and loss of normal PVAT anticontractile activity. Hypoxia is proposed as a major trigger for the chronic low-grade inflammation observed in obesity. Adipocytes store fat and undergo significant hypertrophy in obesity. The cross-sectional area of adipocytes is up to 1000 μm^2 larger in obese individuals with metabolic syndrome compared with that of healthy individuals [31]. The diffusion limit of oxygen is around 100 μm [32], and the hypertrophied adipocytes are subject to a decreased oxygen tension. In obesity, there is no increase in angiogenesis to match the increase in adipocyte size; moreover, the postprandial increase of blood flow that occurs in lean subjects is absent [33, 34]. Thus the hypertrophied adipocytes exist in a state of relative hypoxia as confirmed by pimonidazole staining of adipocytes taken from obese mice [32], as well as measurements of partial pressures of oxygen in abdominal subcutaneous adipose tissue of obese humans [35].

The inflammatory state in obesity is marked by an increase in circulating levels of reactive oxygen species (ROS) and pro-inflammatory cytokines such as CRP, IL-6 and TNF-α [36]. In tissue taken from healthy individuals, incubation with

TNF attenuates the PVAT vasorelaxant effect, suggesting that higher TNF levels in obesity may contribute to the reduced vasorelaxant function of PVAT. Similarly, incubation with TNF and IL6 results in a reduction of the PVAT vasorelaxant effect in healthy rat tissue [31]. The evidence for PVAT inflammation in humans is not limited to the peripheral microvasculature. Proteomic analysis of both epicardial and subcutaneous adipose tissue obtained from patients with coronary artery disease (CAD) has demonstrated higher levels of ROS, and mRNA analysis has revealed lower levels of the antioxidant enzyme catalase [37]. There are also reports of a link between high levels of IL6 and increased risk of CAD [38]. Epicardial adipose tissue IL6 mRNA levels are higher in CAD than non-CAD patients, with higher levels correlating with greater degrees of angiographically defined vascular disease [39].

Central to the inflammation concept is the role of the cells from the immune system in facilitating the establishment of an inflammatory environment. Increased macrophage numbers in adipose tissue of both obese animals [40] and humans [41] also support the hypoxia theory, and hypoxic cells secrete chemokines which attract macrophages [35]. When mouse PVAT is made hypoxic, the presence and activation of macrophages are the key modulator of increased vascular contractility. Moreover, following the conditional ablation of macrophages, hypoxia has no effect on vessel contractility [42].

Macrophage recruitment from blood to PVAT is of paramount importance in the loss of PVAT vasorelaxant effect. Monocyte chemotactic protein-1 (MCP-1) levels are increased in adipose tissue of genetically obese and diet-induced obese mice [43], as well in obese humans [44]. Moreover, insulin increases MCP-1 secretion from insulin resistant 3 T3-L1 adipocytes and in ob/ob mice [45]. This explains how the hyperinsulinaemic state in obesity leads to macrophage recruitment in PVAT and subsequent release of cytokines which reduce or abolish attenuates its vasorelaxant effect.

The involvement of macrophages in establishing an inflammatory milieu is not unique to obesity alone. Fractalkine or CX3CL1 is a protein secreted from adipocytes that promotes monocyte adhesion to human adipocytes [46]. Fractalkine levels are increased in diabetes as well as in obesity and hypertension [47]. There is a significant increase in expression of CX3CL1 receptor gene in blood leukocytes from patients with arterial hypertension [48, 49]. It is thought that this protein may be a significant contributor in bring about the macrophage-induced loss of PVAT vasorelaxant function.

Macrophages aren't the only immune cells implicated in the PVAT damage and inflammation theory. There is also evidence for the role of eosinophils in this regard [50]. PVAT vasorelaxant effect is lost in mice deficient in eosinophils and restored following eosinophil reconstitution. This loss of PVAT function was shown to be due to a reduction in bioavailability of adiponectin and adipocyte-derived nitric oxide, which was restored post eosinophil reconstitution. The mechanistic link is thought to be adiponectin and nitric oxide released after activation of adipocyte-expressed β3 adrenoceptors by catecholamines, with eosinophils being a novel source of these mediators.

Sustained weight loss in obese animals leads to improvement in the PVAT vasorelaxant function with attendant reductions in inflammatory markers [51]. Weight-reducing surgery is associated with a restoration of normal anticontractile function despite patients remaining morbidly obese [52]. A direct link between PVAT function and blood pressure is difficult to establish; however, we have reported such a correlation in an animal model of obesity where weight gain was shown to correlate with damage to PVAT vasorelaxant function and an increase in blood pressure [53]. When considering therapeutic options in MetS, one can count on potential adiponectin analogues as anti-diabetic and vasorelaxant agents as well as free radical scavengers [31] and eplerenone [42] as anti-inflammatory agents used to rescue the damaged PVAT or even used pre-emptively in the obese and normotensive and normoglycaemic individuals in order to prevent the onset of DM and hypertension and resultant cardiovascular events [54, 55].

Not all adipokines are vasorelaxant. Some have vasoconstrictor properties and may play a significant role in controlling microvessel tone by counteracting the effect of the vasorelaxant adipokines.

Chemerin is one such adipokine that may be a potentially significant player in the loss of PVAT vasorelaxant function by behaving as a possible link between obesity, BMI [56], PVAT, diabetes [57] and hypertension both in adults and children [58, 59]. Adipose tissue from obese patients exhibits higher chemerin levels compared with lean controls, and higher chemerin release is associated with insulin resistance. Chemerin stimulates VSMC proliferation and migration via a ROS-dependent signalling pathway, which can contribute to raising blood pressure [60]. Furthermore, chemerin has a direct vasoconstriction, as well as enhancing agonist-induced contractions through G_i proteins, resulting in the activation of L-type Ca^{2+} channels, as well as Src and Rho kinase [61]. It has been linked with increasing BP in mice, and importantly, its levels fall with loss of adipocyte mass following exercise or bariatric surgery [62].

Importantly, chemerin induces insulin resistance in human skeletal muscle cells at the level of insulin receptor substrate 1, Akt and glycogen synthase kinase 3 phosphorylation and glucose uptake, and ERK inhibition prevents chemerin-induced insulin resistance [56]. Following weight loss, the significant decrease in chemerin levels in the 3 months after bariatric surgery is associated with a decrease in HOMA-IR and blood glucose [63]. There is also evidence of chemerin inducing ICAM-1 and E-selectin expression in endothelial cells [64]. Given that it plays a role in monocytes recruitment, insulin resistance and vasoconstriction and its levels correlate with weight gain and drop following weight loss, chemerin is one of the major adipokines that could be targeted in therapeutic strategies to treat MetS.

14.5 The RAAS and ROS within PVAT

In obesity, there are raised circulating levels of the components of the renin-angiotensin-aldosterone system (RAAS). Adipocytes have an intrinsic RAAS system including ACE, angiotensin type 1 and type 2 receptors, and they secrete angiotensinogen, the levels of which are raised in obesity [65]. The source of the adipocyte renin activity remains controversial and unclear [66]. The raised circulating aldosterone levels in obesity correlate with the degree of visceral adiposity and waist-hip ratio [67–69] and have a twofold effect: firstly contribute to increased blood volume by increasing sodium reabsorption and secondly ROS generation. Aldosterone increases the expression of TNF-α from macrophages, and activation of the mineralocorticoid receptor results in generation of reactive oxygen species (ROS). Accordingly, blockade of the mineralocorticoid receptor, using eplerenone, leads to a reduction of ROS and increased levels of adiponectin in obese and diabetic mice [70]. The superoxide anions generated directly by the macrophages and by MR stimulation further contribute to the PVAT damage. Aldosterone activates NADPH oxidase thus increasing ROS levels leading to oxidative post-translational changes to guanylyl cyclase rendering it NO-insensitive [71]. ROS can also reduce NO bioavailability by forming molecules such as peroxynitrite, thus contributing to endothelial dysfunction. ROS can also stimulate the mineralocorticoid receptor (MR) [72], thereby theoretically contributing to further elevations in ROS levels and forming a vicious circle. Superoxide generated by NADPH oxidase in response to electric field stimulation enhances the contractile response of adjacent small arteries [73], and angiotensin II type 1 receptor blockade using candesartan reduces this PVAT-mediated potentiation of EFS-induced contractile response. It is important to note that in obesity there is both sympathetic nerve overactivity and increased angiotensin II levels, providing a further potential explanation for the increased vascular resistance in obesity [74].

At the endothelial level, aldosterone decreases glucose-6-phosphate dehydrogenase (G6PD) activity. G6PD is a cytosolic enzyme and the main source of intracellular NADPH which functions to limit ROS activity [75]. There are two aldosterone receptor antagonists in clinical use; spironolactone is a nonselective aldosterone receptor antagonist, whereas eplerenone is a selective aldosterone receptor antagonist which has a lower degree of cross reactivity with sex-steroid hormones and a longer half-life than spironolactone [76]. Spironolactone increases the expression of G6PD and its activity, as well as raising NADPH levels leading to a reduction in ROS generation in aortas of aldosterone-treated mice [75].

It is not clear to what extent the blood pressure reduction is a result of blood volume and cardiac output reduction secondary to reduced sodium reabsorption, or due to a reduction in sympathetic activity through the direct CNS effect of aldosterone [77, 78]. Certainly, a reduction in ROS generation within PVAT would partly restore the favourable vasorelaxant profile lost, in part, following hypoxia-induced inflammatory damage in obesity.

The ROS-induced PVAT damage in obesity would suggest that antioxidants and free radical scavengers could be therapeutic agents to reverse this damage and possibly lower blood pressure in obesity. We've shown in ex vivo experiments that SOD and catalase can restore the PVAT vasorelaxant property in both human and murine models of obesity [52, 53]. A 3-week administration of desmethyl tirilazad (lazaroid), a potent antioxidant, significantly ameliorates blood pressure in SHR rats [79].

Intuitively, it has been proposed that prevention of ROS generation using NADPH oxidase inhibitors may be a better way of tackling oxidative stress than scavenging the free radicals once they have been generated, although clinical studies need to assess the feasibility of this theory [80].

14.6 Sympathetic Nervous System

Sympathetic nerves don't just innervate the vessels; they also innervate the adipose tissue that surrounds the microvasculature, and secretion of catecholamines from these nerves modulates the contractile state of VSMCs. Sympathetic nerve-derived noradrenaline induces vasoconstriction via activation of alpha 1 - adrenoceptors, on VSMCs, and white adipose tissue itself responds to the sympathetic nerve-derived catecholamines which play key role in lipolysis. Sympathetic denervation of adipose tissue in vivo leads to an increase in lipid deposits [81, 82]. Electrical stimulation of sympathetic nerves in epididymal adipose tissue in vitro increases lipolysis, a process which is attenuated by inhibition of β-adrenoceptors [83]. Under certain stress stimuli such as food deprivation and cold exposure, sympathetic activity varies between white (WAT) and brown adipose tissue (BAT) [84]. In fasting conditions, a reduction in sympathetic activity in BAT leads to a reduction in thermogenesis and ATP utilisation in BAT [85]. In contrast, in WAT, reduction in blood glucose levels lead to increased sympathetic activity, thus increasing lipolysis and releasing of free fatty acids which are to be metabolised into ATP [86].

Fluorescence histochemical studies have shown that nerve fibres make direct contact with adipocytes in brown adipose tissue [87], and using electron microscopy, catecholaminergic nerve plexuses in the extracellular space between brown adipocytes from human perirenal brown adipose tissue have been visualised [88]. There is further evidence indicating that the autonomic nerves present in BAT are sympathetic in origin [89]. This field of research remains contentious, and there are reports that only as little as 2–3% of adipocytes are innervated by the sympathetic nerves [90].

Further research is necessary to determine whether autonomic nerves are directly innervating the adipocytes in PVAT; however, existing evidence suggests that PVAT has predominantly sympathetic nerve fibre innervations, but the degree to which these fibres influence the anti-contractile function of PVAT is not entirely clear and seems to vary depending on a number of factors, including the PVAT location. This is partly explained by the fact that PVAT from different anatomical locations may possess different qualities as a consequence of its unique constituent parts. For example, PVAT around the abdominal aorta is predominantly made up of brown adipocytes, and PVAT around mesenteric vessels is mainly white adipocytes [91]. Electrical stimulation of the PVAT surrounding the superior mesenteric artery demonstrates the fat's anti-contractile effect on the vessel, but stimulation of the abdominal aorta demonstrates the pro-contractile effect of PVAT. This adds to the complexity of elucidating the effect of PVAT on microcirculation. Even slight permutations in experimental design such as the voltage for electric field stimulation of PVAT tissue seem to result in drastically conflicting results [92].

In obesity, an imbalance in the hypothalamic-pituitary axis [93] results in an overactive sympathetic nervous system [94, 95]. The mechanisms underlying this are manyfold [95, 96] and include changes in the levels of adipokines such as leptin, which has a stimulatory effect on SNS, and adiponectin with an inhibitory effect on SNS. In obesity, levels of leptin are increased, and adiponectin are reduced, thus highlighting the effect of adipokine derangements on the SNS and the consequent increase in microvessel tone contributing to raised systemic blood pressures.

In obese hypertensive patients, there is an increase in SNS activity in cardiac and renal nerves [96]. The degree of overactivity varies according to the location of increased adiposity with abdominal and visceral fat depots being associated with the greatest increases in SNS activity [97]. Alarmingly, this increase in SNS activity seems to be independent of actual weight gain and observed only after 12 days of high-fat feeding in rats [98] and similarly in nonobese humans with small increases in body weight [99]. Once again, this shines a light on the complexity of the pathophysiology leading to microvessel disorders and end organ damage.

Obesity has a differential effect on local SNS activity. A significant increase in renal sympathetic activity is observed in obese individuals, but compared with lean normotensive subjects, the obese normotensive individuals exhibit suppression of their cardiac sympathetic activity, whilst the hypertensive obese individuals show an increased sympathetic activity in both cardiac and renal nerves [100]. The degree of renal sympathetic stimulation in obesity is similar in both normotensive and hypertensive cohorts, thus emphasising the significance of other contributory factors such as the suppression of cardiac sympathetic drive in the normotensive individuals [101]. A lower cardiac sympathetic tone in obesity can be viewed as a protective factor, and it may be it is overactivity that tips the balance in favour of the development of hypertension, but it is not clear which factors dictate the differential activity of local SNS. Also there is evidence of central stimulation of the SNS by reactive oxygen species in obesity. There are raised levels of oxidative stress markers such as the superoxide anion, and animal studies suggest that NADPH oxidase-dependent oxidative stress in the brain may be a cause of increased sympathetic tone leading to hypertension in high-fat fed animals [102].

We are not even close to understanding the full scope of pathological changes seen in cardiovascular disease. Some changes seem to be almost immediate and transient thus antagonising the notions of physiological adaptations which take a longer time course to establish. For example, Gosmanov et al. have reported the acute effects of high-fat ingestion by normotensive obese individuals [103]. Both bolus oral ingestion and the intravenous infusion of fat resulted in a significant rise in systolic BP, attenuated endothelial function, increased oxidative stress and activation of the sympathetic nervous system at measured by heart rate variability. Clearly, this is body's immediate reaction to fat ingestion, and the longer-term pathobiological changes may well be as least in part as a direct result of frequent ingestion of high-fat meals. This elegant experiment also highlights the need for public education and promotion of healthy eating.

Describing all the pathophysiological changes in microcirculation that lead to organ damage and cardiovascular events is beyond the scope of one chapter. It requires careful exploration of the normal physiology of the microvessels and the multitude of factors that interact with the microvasculature in order to maintain homeostasis in health. In this chapter, we have made a humble attempt at identifying and presenting a few of the factors involved in this complex process in health and disease.

References

1. Khavandi K, Greenstein AS, Sonoyama K, Withers S, Price A, Malik RA, et al. Myogenic tone and small artery remodelling: insight into diabetic nephropathy. Nephrol Dial Transplant. 2009;24(2):361–9.
2. Chang JB, editor. Textbook of angiology. New York: Springer; 2000.
3. Beverly J, Hunt LP, Schachter M, Halliday A, editors. An introduction to vascular biology. 2nd ed. Cambridge: Cambridge University Press; 2002.
4. Feletou M, Vanhoutte PM. EDHF: an update. Clin Sci (Lond). 2009;117(4):139–55.
5. Bellien J, Thuillez C, Joannides R. Contribution of endothelium-derived hyperpolarizing factors to the regulation of vascular tone in humans. Fundam Clin Pharmacol. 2008;22(4):363–77.
6. Furchgott RF. Role of endothelium in responses of vascular smooth muscle. Circ Res. 1983;53(5):557–73.
7. Furchgott RF, Cherry PD, Zawadzki JV, Jothianandan D. Endothelial cells as mediators of vasodilation of arteries. J Cardiovasc Pharmacol. 1984;6(Suppl 2):S336–43.
8. Sumpio BE, Riley JT, Dardik A. Cells in focus: endothelial cell. Int J Biochem Cell Biol. 2002;34(12):1508–12.
9. Vallance P. Endothelial regulation of vascular tone. Postgrad Med J. 1992;68(803):697–701.
10. Busse R, Edwards G, Feletou M, Fleming I, Vanhoutte PM, Weston AH. EDHF: bringing the concepts together. Trends Pharmacol Sci. 2002;23(8):374–80.
11. Widimsky J, Fejfarova MH, Fejfar Z. Changes of cardiac output in hypertensive disease. Cardiologia. 1957;31(5):381–9.
12. SPJ J. Autonomic nervous and behavioural factors in hypertension. In: JHBB L, editor. Hypertension: pathophysiology diagnosis and management. New York: Raven Press; 1990. p. 2083–90.
13. Lund-Johansen P. Twenty-year follow-up of hemodynamics in essential hypertension during rest and exercise. Hypertension. 1991;18(5 Suppl):III54–61.
14. Julius S, Pascual AV, Sannerstedt R, Mitchell C. Relationship between cardiac output and peripheral resistance in borderline hypertension. Circulation. 1971;43(3):382–90.
15. Esler M, Julius S, Zweifler A, Randall O, Harburg E, Gardiner H, et al. Mild high-renin essential hypertension. Neurogenic human hypertension? N Engl J Med. 1977;296(8):405–11.
16. Bright R. Tabular view of the morbid appearances in 100 cases connected with albuminous urine with observations. Guy's Hosp Rep. 1836;1:380–400.
17. Heagerty AM, Aalkjaer C, Bund SJ, Korsgaard N, Mulvany MJ. Small artery structure in hypertension. Dual processes of remodeling and growth. Hypertension. 1993;21(4):391–7.
18. Zacchigna L, Vecchione C, Notte A, Cordenonsi M, Dupont S, Maretto S, et al. Emilin1 links TGF-beta maturation to blood pressure homeostasis. Cell. 2006;124(5):929–42.
19. Schofield I, Malik R, Izzard A, Austin C, Heagerty A. Vascular structural and functional changes in type 2 diabetes mellitus: evidence for the roles of abnormal myogenic responsiveness and dyslipidemia. Circulation. 2002;106(24):3037–43.
20. Izzard AS, Rizzoni D, Agabiti-Rosei E, Heagerty AM. Small artery structure and hypertension: adaptive changes and target organ damage. J Hypertens. 2005;23(2):247–50.

21. Goode GK, Heagerty AM. In vitro responses of human peripheral small arteries in hypercholesterolemia and effects of therapy. Circulation. 1995;91(12):2898–903.
22. Kotsis V, Stabouli S, Papakatsika S, Rizos Z, Parati G. Mechanisms of obesity-induced hypertension. Hypertens Res. 2010;33(5):386–93.
23. Yudkin JS, Eringa E, Stehouwer CD. "Vasocrine" signalling from perivascular fat: a mechanism linking insulin resistance to vascular disease. Lancet. 2005;365(9473):1817–20.
24. Safar ME, Czernichow S, Blacher J. Obesity, arterial stiffness, and cardiovascular risk. J Am Soc Nephrol. 2006;17(4 Suppl 2):S109–11.
25. Singhal A, Farooqi IS, Cole TJ, O'Rahilly S, Fewtrell M, Kattenhorn M, et al. Influence of leptin on arterial distensibility: a novel link between obesity and cardiovascular disease? Circulation. 2002;106(15):1919–24.
26. Fox CS, Massaro JM, Hoffmann U, Pou KM, Maurovich-Horvat P, Liu CY, et al. Abdominal visceral and subcutaneous adipose tissue compartments: association with metabolic risk factors in the Framingham Heart Study. Circulation. 2007;116(1):39–48.
27. Gesta S, Tseng YH, Kahn CR. Developmental origin of fat: tracking obesity to its source. Cell. 2007;131(2):242–56.
28. Manolopoulos KN, Karpe F, Frayn KN. Gluteofemoral body fat as a determinant of metabolic health. Int J Obes. 2010;34(6):949–59.
29. Saxton SN, Ryding KE, Aldous RG, Withers SB, Ohanian J, Heagerty AM. Role of sympathetic nerves and adipocyte catecholamine uptake in the Vasorelaxant function of perivascular adipose tissue. Arterioscler Thromb Vasc Biol. 2018;38(4):880–91.
30. Withers SB, Simpson L, Fattah S, Werner ME, Heagerty AM. cGMP-dependent protein kinase (PKG) mediates the anticontractile capacity of perivascular adipose tissue. Cardiovasc Res. 2013;101(1):130–7.
31. Greenstein AS, Khavandi K, Withers SB, Sonoyama K, Clancy O, Jeziorska M, et al. Local inflammation and hypoxia abolish the protective anticontractile properties of perivascular fat in obese patients. Circulation. 2009;119(12):1661–70.
32. Hosogai N, Fukuhara A, Oshima K, Miyata Y, Tanaka S, Segawa K, et al. Adipose tissue hypoxia in obesity and its impact on adipocytokine dysregulation. Diabetes. 2007;56(4):901–11.
33. Bulow J, Astrup A, Christensen NJ, Kastrup J. Blood flow in skin, subcutaneous adipose tissue and skeletal muscle in the forearm of normal man during an oral glucose load. Acta Physiol Scand. 1987;130(4):657–61.
34. Coppack SW, Evans RD, Fisher RM, Frayn KN, Gibbons GF, Humphreys SM, et al. Adipose tissue metabolism in obesity: lipase action in vivo before and after a mixed meal. Metabolism. 1992;41(3):264–72.
35. Pasarica M, Sereda OR, Redman LM, Albarado DC, Hymel DT, Roan LE, et al. Reduced adipose tissue oxygenation in human obesity: evidence for rarefaction, macrophage chemotaxis, and inflammation without an angiogenic response. Diabetes. 2009;58(3):718–25.
36. Trayhurn P, Wang B, Wood IS. Hypoxia in adipose tissue: a basis for the dysregulation of tissue function in obesity? Br J Nutr. 2008;100(2):227–35.
37. Salgado-Somoza A, Teijeira-Fernandez E, Fernandez AL, Gonzalez-Juanatey JR, Eiras S. Proteomic analysis of epicardial and subcutaneous adipose tissue reveals differences in proteins involved in oxidative stress. Am J Physiol Heart Circ Physiol. 2010;299(1):H202–9.
38. Pai JK, Pischon T, Ma J, Manson JE, Hankinson SE, Joshipura K, et al. Inflammatory markers and the risk of coronary heart disease in men and women. N Engl J Med. 2004;351(25):2599–610.
39. Eiras S, Teijeira-Fernandez E, Shamagian LG, Fernandez AL, Vazquez-Boquete A, Gonzalez-Juanatey JR. Extension of coronary artery disease is associated with increased IL-6 and decreased adiponectin gene expression in epicardial adipose tissue. Cytokine. 2008;43(2):174–80.
40. Rausch ME, Weisberg S, Vardhana P, Tortoriello DV. Obesity in C57BL/6J mice is characterized by adipose tissue hypoxia and cytotoxic T-cell infiltration. Int J Obes. 2008;32(3):451–63.

41. Weisberg SP, McCann D, Desai M, Rosenbaum M, Leibel RL, Ferrante AW Jr. Obesity is associated with macrophage accumulation in adipose tissue. J Clin Invest. 2003;112(12):1796–808.
42. Withers BS, Agabiti-Rosei C, Linvingstone DM, Little MC, Aslam R, Malik RA, et al. Macrophage activation is responsible for loss of anticontractile function in inflamed perivascular fat. Arterioscler Thromb Vasc Biol. 2011;31:908.
43. Kanda H, Tateya S, Tamori Y, Kotani K, Hiasa K, Kitazawa R, et al. MCP-1 contributes to macrophage infiltration into adipose tissue, insulin resistance, and hepatic steatosis in obesity. J Clin Invest. 2006;116(6):1494–505.
44. Kim CS, Park HS, Kawada T, Kim JH, Lim D, Hubbard NE, et al. Circulating levels of MCP-1 and IL-8 are elevated in human obese subjects and associated with obesity-related parameters. Int J Obes. 2006;30(9):1347–55.
45. Sartipy P, Loskutoff DJ. Monocyte chemoattractant protein 1 in obesity and insulin resistance. Proc Natl Acad Sci U S A. 2003;100(12):7265–70.
46. Shah R, Hinkle CC, Ferguson JF, Mehta NN, Li M, Qu L, et al. Fractalkine is a novel human adipochemokine associated with type 2 diabetes. Diabetes. 2011;60(5):1512–8.
47. Sirois-Gagnon D, Chamberland A, Perron S, Brisson D, Gaudet D, Laprise C. Association of common polymorphisms in the fractalkine receptor (CX3CR1) with obesity. Obesity (Silver Spring). 2010;19(1):222–7.
48. Timofeeva AV, Goryunova LE, Khaspekov GL, Kovalevskii DA, Scamrov AV, Bulkina OS, et al. Altered gene expression pattern in peripheral blood leukocytes from patients with arterial hypertension. Ann N Y Acad Sci. 2006;1091:319–35.
49. Flierl U, Bauersachs J, Schafer A. Modulation of platelet and monocyte function by the chemokine fractalkine (CX3 CL1) in cardiovascular disease. Eur J Clin Investig. 2015;45(6):624–33.
50. Withers SB, Forman R, Meza-Perez S, Sorobetea D, Sitnik K, Hopwood T, et al. Eosinophils are key regulators of perivascular adipose tissue and vascular functionality. Sci Rep. 2017;7:44571.
51. Bussey CE, Withers SB, Aldous RG, Edwards G, Heagerty AM. Obesity-related perivascular adipose tissue damage is reversed by sustained weight loss in the rat. Arterioscler Thromb Vasc Biol. 2016;36(7):1377–85.
52. Aghamohammadzadeh R, Greenstein AS, Yadav R, Jeziorska M, Hama S, Soltani F, et al. The effects of bariatric surgery on human small artery function: evidence for reduction in perivascular adipocyte inflammation, and the restoration of normal anticontractile activity despite persistent obesity. J Am Coll Cardiol. 2013;62(2):128–35.
53. Aghamohammadzadeh R, Unwin RD, Greenstein AS, Heagerty AM. Effects of obesity on perivascular adipose tissue vasorelaxant function: nitric oxide, inflammation and elevated systemic blood pressure. J Vasc Res. 2015;52(5):299–305.
54. Aghamohammadzadeh R, Withers S, Lynch F, Greenstein A, Malik R, Heagerty A. Perivascular adipose tissue from human systemic and coronary vessels: the emergence of a new pharmacotherapeutic target. Br J Pharmacol. 2012;165(3):670–82.
55. Aghamohammadzadeh R, Heagerty AM. Obesity-related hypertension: epidemiology, pathophysiology, treatments, and the contribution of perivascular adipose tissue. Ann Med. 2012;44(Suppl 1):S74–84.
56. Sell H, Laurencikiene J, Taube A, Eckardt K, Cramer A, Horrighs A, et al. Chemerin is a novel adipocyte-derived factor inducing insulin resistance in primary human skeletal muscle cells. Diabetes. 2009;58(12):2731–40.
57. Ouwens DM, Bekaert M, Lapauw B, Van Nieuwenhove Y, Lehr S, Hartwig S, et al. Chemerin as biomarker for insulin sensitivity in males without typical characteristics of metabolic syndrome. Arch Physiol Biochem. 2012;118(3):135–8.
58. Schipper HS, Nuboer R, Prop S, van den Ham HJ, de Boer FK, Kesmir C, et al. Systemic inflammation in childhood obesity: circulating inflammatory mediators and activated CD14++ monocytes. Diabetologia. 2012;55(10):2800–10.

59. Verrijn Stuart AA, Schipper HS, Tasdelen I, Egan DA, Prakken BJ, Kalkhoven E, et al. Altered plasma adipokine levels and in vitro adipocyte differentiation in pediatric type 1 diabetes. J Clin Endocrinol Metab. 2011;97(2):463–72.
60. Kunimoto H, Kazama K, Takai M, Oda M, Okada M, Yamawaki H. Chemerin promotes the proliferation and migration of vascular smooth muscle and increases mouse blood pressure. Am J Physiol Heart Circ Physiol. 2015;309(5):H1017–28.
61. Ferland DJ, Darios ES, Neubig RR, Sjogren B, Truong N, Torres R, et al. Chemerin-induced arterial contraction is Gi- and calcium-dependent. Vasc Pharmacol. 2017;88:30–41.
62. Chakaroun R, Raschpichler M, Kloting N, Oberbach A, Flehmig G, Kern M, et al. Effects of weight loss and exercise on chemerin serum concentrations and adipose tissue expression in human obesity. Metabolism. 2011;61(5):706–14.
63. Sell H, Divoux A, Poitou C, Basdevant A, Bouillot JL, Bedossa P, et al. Chemerin correlates with markers for fatty liver in morbidly obese patients and strongly decreases after weight loss induced by bariatric surgery. J Clin Endocrinol Metab. 2010;95(6):2892–6.
64. Landgraf K, Friebe D, Ullrich T, Kratzsch J, Dittrich K, Herberth G, et al. Chemerin as a mediator between obesity and vascular inflammation in children. J Clin Endocrinol Metab. 2012;97(4):E556–64.
65. Van Harmelen V, Ariapart P, Hoffstedt J, Lundkvist I, Bringman S, Arner P. Increased adipose angiotensinogen gene expression in human obesity. Obesity. 2000;8(4):337–41.
66. Engeli S, Negrel R, Sharma AM. Physiology and pathophysiology of the adipose tissue renin-angiotensin system. Hypertension. 2000;35(6):1270–7.
67. Goodfriend TL, Calhoun DA. Resistant hypertension, obesity, sleep apnea, and aldosterone: theory and therapy. Hypertension. 2004;43(3):518–24.
68. Goodfriend TL, Egan BM, Kelley DE. Aldosterone in obesity. Endocr Res. 1998;24(3–4):789–96.
69. Goodfriend TL, Kelley DE, Goodpaster BH, Winters SJ. Visceral obesity and insulin resistance are associated with plasma aldosterone levels in women. Obes Res. 1999;7(4):355–62.
70. Guo C, Ricchiuti V, Lian BQ, Yao TM, Coutinho P, Romero JR, et al. Mineralocorticoid receptor blockade reverses obesity-related changes in expression of adiponectin, peroxisome proliferator-activated receptor-gamma, and proinflammatory adipokines. Circulation. 2008;117(17):2253–61.
71. Maron BA, Zhang YY, Handy DE, Beuve A, Tang SS, Loscalzo J, et al. Aldosterone increases oxidant stress to impair guanylyl cyclase activity by cysteinyl thiol oxidation in vascular smooth muscle cells. J Biol Chem. 2009;284(12):7665–72.
72. Wang H, Shimosawa T, Matsui H, Kaneko T, Ogura S, Uetake Y, et al. Paradoxical mineralocorticoid receptor activation and left ventricular diastolic dysfunction under high oxidative stress conditions. J Hypertens. 2008;26(7):1453–62.
73. Gao YJ, Takemori K, Su LY, An WS, Lu C, Sharma AM, et al. Perivascular adipose tissue promotes vasoconstriction: the role of superoxide anion. Cardiovasc Res. 2006;71(2):363–73.
74. Lu C, Su LY, Lee RM, Gao YJ. Mechanisms for perivascular adipose tissue-mediated potentiation of vascular contraction to perivascular neuronal stimulation: the role of adipocyte-derived angiotensin II. Eur J Pharmacol. 2011;634(1–3):107–12.
75. Leopold JA, Dam A, Maron BA, Scribner AW, Liao R, Handy DE, et al. Aldosterone impairs vascular reactivity by decreasing glucose-6-phosphate dehydrogenase activity. Nat Med. 2007;13(2):189–97.
76. Maron BA, Leopold JA. Aldosterone receptor antagonists: effective but often forgotten. Circulation. 2010;121(7):934–9.
77. de Paula RB, da Silva AA, Hall JE. Aldosterone antagonism attenuates obesity-induced hypertension and glomerular hyperfiltration. Hypertension. 2004;43(1):41–7.
78. Rahmouni K, Barthelmebs M, Grima M, Imbs JL, De Jong W. Involvement of brain mineralocorticoid receptor in salt-enhanced hypertension in spontaneously hypertensive rats. Hypertension. 2001;38(4):902–6.
79. Vaziri ND, Ni Z, Oveisi F, Trnavsky-Hobbs DL. Effect of antioxidant therapy on blood pressure and NO synthase expression in hypertensive rats. Hypertension. 2000;36(6):957–64.

80. Drummond GR, Selemidis S, Griendling KK, Sobey CG. Combating oxidative stress in vascular disease: NADPH oxidases as therapeutic targets. Nat Rev Drug Discov. 2011;10(6):453–71.
81. Maryam B, Victor TA, Adkison MG, Wade SW, Timothy JB. Central nervous system origins of the sympathetic nervous system outflow to white adipose tissue. Am J Phys Regul Integr Comp Phys. 1998;275(1):R291–R9.
82. Michelle TF, Timothy JB. Sympathetic but not sensory denervation stimulates white adipocyte proliferation. Am J Phys Regul Integr Comp Phys. 2006;291(6):R1630–R7.
83. Correll JW. Adipose tissue: ability to respond to nerve stimulation in vitro. Science. 1963;140(3565):387–8.
84. Nilton AB, Marcia NB, Timothy JB. Differential sympathetic drive to adipose tissues after food deprivation, cold exposure or glucoprivation. Am J Phys Regul Integr Comp Phys. 2008;294(5):R1445–R52.
85. Egawa M, Yoshimatsu H, Bray GA. Effects of 2-deoxy-D-glucose on sympathetic nerve activity to interscapular brown adipose tissue. Am J Phys Regul Integr Comp Phys. 1989;257(6):R1377–R85.
86. Niijima A. Nervous regulation of metabolism. Prog Neurobiol. 1989;33(2):135–47.
87. Wirsen C. Adrenergic innervation of adipose tissue examined by fluorescence microscopy. Nature. 1964;202(4935):913.
88. Lever JD, Jung RT, Nnodim JO, Leslie PJ, Symons D. Demonstration of a catecholaminergic innervation in human perirenal brown adipose tissue at various ages in the adult. Anat Rec. 1986;215(3):251–5.
89. Cannon B, Nedergaard J, Lundberg JM, Hokfelt T, Terenius L, Goldstein M. Neuropeptide tyrosine (NPY) is co-stored with noradrenaline in vascular but not in parenchymal sympathetic nerves of brown adipose tissue. Exp Cell Res. 1986;164(2):546–50.
90. Slavin BG, Ballard KW. Morphological studies on the adrenergic innervation of white adipose tissue. Anat Rec. 1978;191(3):377–89.
91. Torok J, Zemancikova A, Kocianova Z. Interaction of perivascular adipose tissue and sympathetic nerves in arteries from normotensive and hypertensive rats. Physiol Res. 2016;65(Supplementum 3):S391–S9.
92. Saxton SN, Clark BJ, Withers SB, Eringa EC, Heagerty AM. Mechanistic links between obesity, diabetes, and blood pressure: role of perivascular adipose tissue. Physiol Rev. 2019;99(4):1701–63.
93. Chrousos GP. The role of stress and the hypothalamic–pituitary–adrenal axis in the pathogenesis of the metabolic syndrome: neuro-endocrine and target tissue-related causes. Int J Obes. 2000;24(2):S50–S5.
94. Manolis AJ, Poulimenos LE, Kallistratos MS, Gavras I, Gavras H. Sympathetic overactivity in hypertension and cardiovascular disease. Curr Vasc Pharmacol. 2014;12(1):4–15.
95. Smith MM, Minson CT. Obesity and adipokines: effects on sympathetic overactivity. J Physiol. 2019;590(8):1787–801.
96. Lemche E, Chaban OS, Lemche AV. Neuroendorine and epigentic mechanisms subserving autonomic imbalance and HPA dysfunction in the metabolic syndrome. Front Neurosci. [Review]. 2016;10:142.
97. Tentolouris N, Liatis S, Katsilambros N. Sympathetic system activity in obesity and metabolic syndrome. Ann N Y Acad Sci. 2006;1083(1):129–52.
98. Muntzel Martin S, Al-Naimi Omar Ali S, Barclay A, Ajasin D. Cafeteria diet increases fat mass and chronically elevates lumbar sympathetic nerve activity in rats. Hypertension. 2019;60(6):1498–502.
99. Davy KP, Orr JS. Sympathetic nervous system behavior in human obesity. Neurosci Biobehav Rev. 2009;33(2):116–24.
100. Rumantir MS, Vaz M, Jennings GL, Collier G, Kaye DM, Seals DR, et al. Neural mechanisms in human obesity-related hypertension. J Hypertens. 1999;17(8):1125–33.
101. Esler M, Straznicky N, Eikelis N, Masuo K, Lambert G, Lambert E. Mechanisms of sympathetic activation in obesity-related hypertension. Hypertension. 2006;48(5):787–96.

102. Nagae A, Fujita M, Kawarazaki H, Matsui H, Ando K, Fujita T. Sympathoexcitation by oxidative stress in the brain mediates arterial pressure elevation in obesity-induced hypertension. Circulation. 2009;119(7):978–86.
103. Gosmanov AR, Smiley DD, Robalino G, Siquiera J, Khan B, Le NA, et al. Effects of oral and intravenous fat load on blood pressure, endothelial function, sympathetic activity, and oxidative stress in obese healthy subjects. Am J Physiol Endocrinol Metab. 2010;299(6):E953–8.

The Role of Perivascular Adipose Tissue in Arterial Function in Health and Disease

15

Claudia Agabiti-Rosei, Clarissa Barp, Sophie N. Saxton, and Anthony M. Heagerty

15.1 PVAT Structure

The PVAT that surrounds small blood vessels comprises white adipocytes, inflammatory immune cells, stem cells, nerves and stromal tissue plus microfeeding vessels [1–5]. In the aorta whilst many of the cellular constituents will be similar, the adipocytes are brown in nature [6, 7]. Such adipocytes have a role in thermogenesis, whereas white adipocytes are regarded as energy stores [8, 9]. For many years physiological studies of isolated blood vessels in vitro skeletonised the artery in advance of pharmacological testing. The removal of the PVAT meant that any putative functional contribution to arterial tone was lost. It was not until 1991 that Soltis and Cassis tested the hypothesis that PVAT might play a role in vascular tonal regulation [10]. Using isolated segments of rat aorta, they demonstrated the presence of a diminished response to the exogenous application of noradrenaline when the PVAT was retained, in comparison with arteries that had been denuded of fat tissue. This effect was attributed to the dense sympathetic innervation in PVAT. Also, there was an assumption that externally applied vasoconstrictor substances might have more difficulty in gaining access to the outside of the blood vessel because of the barrier posed by the PVAT [10, 11]. Ten years later Lohn and colleagues demonstrated that the vasoconstrictor responses to angiotensin II (AngII), serotonin and phenylephrine were 95%, 80% and 30% lower, respectively, when PVAT was retained [12].

C. Agabiti-Rosei (✉)
Clinica Medica, Department of Clinical and Experimental Sciences, University of Brescia, Brescia, Italy

C. Barp
Department of Pharmacology, Biological Sciences Centre – Block D, Universidade Federal de Santa Catarina, Florianopolis, Brazil

S. N. Saxton · A. M. Heagerty
Division of Cardiovascular Sciences, The University of Manchester, Core Technology Facility, Manchester, UK
e-mail: sophie.saxton@manchester.ac.uk; tony.heagerty@manchester.ac.uk

E. Agabiti-Rosei et al. (eds.), *Microcirculation in Cardiovascular Diseases*, Updates in Hypertension and Cardiovascular Protection,
https://doi.org/10.1007/978-3-030-47801-8_15

Furthermore this effect was found to be independent of nitric oxide (NO) formation. It appeared to be mediated by a substance released from the fat, as transfer of the organ bath solution from vessels with PVAT or cultured rat adipocytes, on to arteries without PVAT, resulted in a rapid reduction in the constrictor response [13, 14]. Further data suggested that PVAT released a soluble factor that induced vasodilatation, which was calcium dependent and regulated by tyrosine kinase acting via ATP-dependent potassium channels [15, 16]. The exact identity was left uncertain at this time. It seems clear now that there is a variety of vasodilator adipokines responsible for relaxation which may also be dependent upon the spasmogen used. Several candidates have been suggested including angiotensin 1–7, free fatty acids such as methyl palmitate and hydrogen sulphide, as well as inflammatory cytokines such as interleukin-6 and tumour necrosis factor [1]. There is also an emerging body of evidence that the predominating factor is adiponectin [13, 17]. In this regard it is clear that adiponectin can act as a vasodilator by activating endothelial nitric oxide synthase (NOS) with the consequent production of NO. In addition, adiponectin has anti-inflammatory properties and can inhibit macrophage activation and reduce the proliferation of vascular smooth muscle cells [18, 19]. Adiponectin gene knockout mice exhibit a loss of anticontractile function from PVAT, and their phenotype demonstrates hypertension and glucose intolerance [13]. Pathological states, such as obesity, hypertension, atherosclerosis, type 2 diabetes and myocardial infarction, are associated with endothelial dysfunction and a reduction in plasma adiponectin levels [20–22]. Recent studies have demonstrated adiponectin-mediated anticontractile activity in human small arteries [3]; however, the exact mechanism by which healthy PVAT induces vasodilation is difficult to identify, as there appears to be variation dependent upon anatomical location and animal species [2, 3, 5, 17]. For example, in the rat aorta, PVAT appears to cause vasorelaxation by opening ATP-dependent K^+ channels in vascular smooth muscle cells, whereas in rat mesenteric arteries voltage-dependent potassium channels seem to be more important [14–16]. There is evidence also of modulation of venous function by PVAT and a recent report suggesting that the contractile responses of the inferior vena cava to various agonists can be diminished by activation of K_V channels by adipokines released from PVAT [23]. Clearly PVAT can play an important role as a dual modulator of vascular tone. The fine balance between PVAT-derived vasodilator and vasoconstrictor mediators may be crucial for the maintenance of local vascular tone.

15.2 The Role of Inflammation and Oxidative Stress on PVAT Phenotype

Recently, a review from Huang Cao and colleagues discussed the importance of PVAT regulation of vascular tone and underlined the significance of the functional integrity of PVAT, rather than the amount of PVAT itself [24]. In the healthy lean state, PVAT-derived adipokines and cytokines can act on endothelial cells through a NO-dependent mechanism and directly on smooth muscle cells to induce vasorelaxation of the vessel wall [3, 16]. In contrast, in obesity, PVAT shifts its phenotype, demonstrating an increase in oxidative stress, pro-inflammatory cytokines/adipokines release, immune cells migration and the secretory profile of PVAT changes with a reduction in the expression of vasorelaxing factors and an increase in vasoconstrictor factors with, in particular, macrophage infiltration [24]. There is compelling evidence for an increased level of adipokines in the inflammatory state, such as in obesity, and of a decreased cardiometabolic protective role of adiponectin. These obesity-related changes in adipose tissue lead to a significant decrease in the anticontractile effect of PVAT even though its mass is increased. This shift is probably linked to the development of insulin resistance, type 2 diabetes mellitus and hypertension [25–27]. In summary, in obesity, the anticontractile effect is significantly decreased, despite an increase in PVAT mass [3].

The mechanism by which adipose inflammation is related to obesity remains unknown, although hypoxia may be key: adipocytes become hypertrophic leading to inadequate perfusion and consequent local hypoxia that is mediated mainly by hypoxia-inducible factor alpha (HIF-1 α) [28–30]. HIF-1 α levels are higher in the adipose tissue of obese subjects, and it reverts to normal after weight loss [31]. HIF-1 α is a main hypoxia-inducible transcription factor linked to increases in inflammatory cytokines like TNF-α and IL-6 and reduces the bioavailability of adiponectin [32]. Greenstein et al. have shown that the anticontractile function of PVAT is lost in obese patients and that the experimental application of TNF-α and IL-6 to intact PVAT around healthy rat blood vessels decreased the dilator effect of PVAT [3]. Similarly, it has been shown that the induction of experimental hypoxia for 2.5 h causes PVAT inflammation and loss of its anticontractile function. These changes were reversed, and the anticontractile property rescued by catalase and superoxide dismutase [3]. In addition, using preincubation with the cytokine antagonist infliximab in the organ bath can restore normal PVAT function. Furthermore, after induced hypoxia, the incubation of the vessels with aldosterone can reproduce the inflammatory phenotype in adipose tissue, with loss of anticontractile property, causing PVAT resident macrophage activation, and the uses of free radical scavengers are able to rescue the anticontractile function [33].

In macrophage deficient CD11b-DTR mice during hypoxic and inflammatory conditions [33, 34], macrophages are a key modulator of oxidative stress and systemic inflammation in adipose tissue. The important role of IL-6 and MCP-1 in the recruitment of inflammatory cells, in particular monocytes and macrophages during obesity, is described in a recent review [24]. In particular, it has been demonstrated that 2 weeks of high-fat diet feeding caused an increase in MCP-1 and IL-6 expression in PVAT which was not seen with other adipose tissue depots [35]. Furthermore, animal models with MCP-1 gene deletion show a significantly reduced infiltration of inflammatory cells, and lipid deposition in the arterial wall and IL-6 inhibition reduce intrinsic mechanical stiffness in low-density lipoprotein receptor knockout mice [36, 37]. From such studies we can conclude that both these cytokines are directly involved in the development of arterial stiffness and increased cardiovascular disease risk.

In addition to involvement of the macrophages in PVAT function, it has been demonstrated that eosinophils also play an important role in the loss of PVAT anticontractile function [38]. Withers et al. have shown that the anticontractile effect of PVAT was lost in eosinophil-deficient mice and could be rescued after eosinophil reconstitution with an increase of adiponectin and adipocyte-derived NO bioavailability. In addition, these eosinophil-deficient mice exhibited hypertension and hyperglycaemia [38]. Mechanistically, it appears that eosinophils are a source of catecholamines and activate adipocyte β_3-adrenoceptors to release the vasodilators adiponectin and NO. This study and others [39] suggest that targeting eosinophil number in obesity may be a useful therapeutic target. Wu et al. have demonstrated that hypereosinophilia induced by helminth infection improved glucose and insulin tolerance in obesity. Further work is needed to fully elucidate the mechanisms by which eosinophils are involved in PVAT and vascular functionality and their potential as a therapeutic target in obesity.

15.3 The Release of Vasoactive Factors from PVAT

Whilst there was a school of thought that PVAT was only providing mechanical support for the vasculature, there is an accumulating body of evidence that it is highly metabolically active and can regulate vascular tone in a paracrine and endocrine fashion [1–4]. The exact nature of this regulation depends on the species being tested and the vascular beds that are studied. In addition, the distinctive profile of PVAT activity is dependent certainly in vitro on the spasmogen used to stimulate the arterial tissue. Against this background it is clear that in a variety of vascular beds including skeletal muscle and mesenteric small arteries, as well as larger conduit arteries such as the superior mesenteric and the thoracic aorta, there is a significant contribution to an anticontractile effect in a response to a variety of vasoconstrictor stimulae. Locally, this could play a role in peripheral vascular resistance and in consequence blood pressure as well as nutrient delivery to skeletal muscle. The loss of this latter property could lead to a decrease in glucose uptake and the development of type 2 diabetes. The vasorelaxant activity can be ascribed to a number of adipokines that can be produced and secreted from PVAT, as well as a contribution from PVAT in terms of catecholamine reuptake into the adipocytes. Below is a summary of some of the principle dilator adipokines that may be responsible for the anticontractile function in healthy lean individuals.

1. *Adiponectin.* This adipokine is secreted in a number of polymeric forms with varying molecular weights [40–42]. The ligand binds to two types of adiponectin receptor (type 1 and type 2) [43–45]. Type 1 (adiopoR1) is located on both endothelial and vascular smooth muscle cells, and it is activation of this receptor which will stimulate a number of pathways including smooth cell muscle differentiation and growth. With regard to vasorelaxation, activation of the receptor stimulates the production of AMPK which in turn will increase the phosphorylation of endothelial NOS leading to an increased bioavailability of NO in both the endothelium and smooth muscle cells [46]. In addition, in vascular smooth cells, AMPK regulates the opening of large conductance Ca^{2+}-activated K^+ channels

[47, 48]. In hypertension circulating levels of adiponectin are decreased, and there is restoration to normality when blood pressure is lowered [49]. Low plasma adiponectin is also observed in obesity [20]. The adiponectin gene knockout mouse develops a hypertensive and diabetic phenotype, and the noradrenaline and electrical field stimulation-induced anticontractile effects of PVAT are lost [13].

2. *Leptin*. This adipokine is recognised to play a vital role in regulating satiety and body weight [50]. Whilst it has central actions, leptin can induce vasodilatation by both endothelium-dependent and endothelium-independent mechanisms [51]. It is of interest that in the spontaneously rat where there is loss of PVAT anticontractile activity, there is reduced expression of leptin in the PVAT with reduced eNOS activation [52].
3. *Nitric oxide*. The ubiquitous nature of NO means that it is widely recognised as an important vasodilator [53]. It is synthesised by three NOS enzymes: endothelial (eNOS), neuronal (nNOS) and inducible (iNOS). We have demonstrated that the anticontractile effect of PVAT stimulated by noradrenaline in small mesenteric arteries is NOS-dependent [3, 16]. When the PVAT is incubated with a nonspecific NOS inhibitor, the anticontractile effect is completely abolished. It is thought that early in obesity, there is an adaptive increase in NO which may preserve vascular function [54]. In a high-fat fed mouse model of obesity, insulin and leptin which are elevated stimulate NO production from mesenteric PVAT. However, in chronic obesity increased superoxide production results in a reduction in NO bioavailability [55].
4. *Hydrogen sulphide*. There is evidence in aortic PVAT that hydrogen sulphide (H_2S) is produced and that it can induce vasodilation of vascular smooth muscle cells by opening of ATP-sensitive K^+ channels or by activation of endothelial intermediate conductance Ca^{2+}-activated K^+ channels [15, 56]. A number of vasoconstrictors including phenylephrine, 5HT and angiotensin II stimulate the production of H_2S from aortic PVAT, and plasma H_2S is reduced in the spontaneously hypertensive rat and in a pharmacologically reduced mouse model of hypertension using oral administration of a NOS inhibitor [57, 58].
5. *Hydrogen peroxide*. Hydrogen peroxide (H_20_2) is a known vasodilator in endothelium-denuded vessels by the activation of soluble guanylate cyclase in vascular smooth muscle [59]. Some researchers have advanced the hypothesis that H_20_2 plays an important role in this pathway, and Gao et al. found that phenylephrine-induced PVAT anticontractile activity could be diminished using an H_20_2 scavenger.
6. *Palmitic acid methyl ester*. Palmitic acid methyl ester (PAME) is released from PVAT in the presence of Ca^{2+} and induces vasodilation of the aorta via K_v channels in a NO-independent and endothelial-dependent manner [60]. In the spontaneously hypertensive rat, PAME secretion from PVAT is reduced, and the vasodilator effect of exogenous PAME in these animals is significantly less than in wild-type normotensive rats.
7. *Angiotensin 1–7*. PVAT is able to produce angiotensin 1–7, and antagonists ameliorate the PVAT anticontractile effect in aorta [61]. Its physiological role is

uncertain, and its contribution to hypertension and metabolic disorders such as type 2 diabetes is still not fully elucidated.

15.4 Autonomic Innervation of Adipose Tissue

There is clear evidence that adipocytes contain a complete adrenergic system including β_3-adrenergic receptors, uptake transporters and metabolic enzymes, and this system may modulate the anticontractile function of PVAT [5, 13, 62, 63]. It is still uncertain whether there is precise autonomic innervation of PVAT. Certainly there is clear histopathological proof that sympathetic nerve fibres run through PVAT and innervate the vasculature [5, 64, 65]. It is attractive to postulate that direct stimulation of sympathetic nerve fibres could cause vasoconstriction by activation of α_1-adrenoreceptors on the vasculature and that this can be modulated by anticontractile function as a result of the same nerve fibre causing discharge of catecholamines that will bind to adipocyte β_3-adrenergic receptors. There are reports that the use of β-adrenoreceptor agonists causes hypotension in rodents and dogs and that this has been shown to induce PVAT-dependent relaxation in pre-constricted mesenteric resistance arteries [66]. Our most recent studies have demonstrated that adipocyte located β_3-adrenoreceptors do play a vital role in the anticontractile effect of PVAT, and activation of these receptors by sympathetic nerve-derived noradrenaline triggers the release of the vasodilator adiponectin [13]. In addition, we have shown that the anticontractile function is, at least in part, a result of PVAT sequestering sympathetic nerve-derived noradrenaline via organic cation transporter 3, thereby preventing the noradrenaline from reaching the α1-adrenoreceptors on the vasculature.

15.5 Autonomic Dysfunction in Obesity

It is clear that the sympathetic nervous system becomes pathologically overactive in obesity, and increased neural activity has been shown to correlate with increases in body mass index and the percentage of body fat [67, 68]. Sympathetic over-activity is observed after only 12 days of high-fat feeding in rats and is independent of weight gain [69]. Again our studies indicate that in obesity there is a loss of PVAT anticontractile function [3, 55], which may be a result of adrenoceptor desensitisation similar to that which occurs in heart failure [70]. Inevitably, as described above, this will lead to an increase in peripheral vascular resistance and a decrease in the tissue uptake of glucose. The obese phenotype is associated with adipocyte hypertrophy in PVAT depots with intense inflammation, and the activation of macrophages with a decreased in PVAT located eosinophils [3]. We have demonstrated that the macrophage is central to PVAT function, and without macrophages, even in acute hypoxic states which are known to reduce anticontractile activity of PVAT, there is no loss of the anticontractile effect which is mediated by the preservation of adiponectin bioavailability [33]. Furthermore, the increase in PVAT eosinophils correlates with the loss of PVAT anticontractile function, and the restoration of eosinophil donation is associated with a rescuing of the PVAT function, accompanied by a normalisation of the obesity-associated hypertension and diabetes [38]. The mechanism is still unclear although originally it was assumed that the eosinophil was operating by the transformation of macrophages to a classically active phenotype which is pro-inflammatory, and in doing so this reduced the bioavailability of adiponectin. However, our detailed studies suggest that this is not the case, and it may well be that the catecholamines such as noradrenaline contained within eosinophils are locally released thereby stimulating β_3-adrenergic receptors, and promoting the secretion of adiponectin is the most likely explanation [38].

15.6 Involvement of Renin-Angiotensin-Aldosterone System on PVAT Function

The renin-angiotensin-aldosterone system (RAAS) is involved in systemic blood pressure regulation and in renal electrolyte homeostasis. Several studies have demonstrated the presence of local RAS activity in both white and brown perivascular adipose tissue [71]. Angiotensin II (Ang II) is the major component of RAAS with a variety of physiological actions [72]; in the adipose tissue, Ang II may play a role in adipocyte growth and differentiation, stimulating lipogenesis, pre-adipocytes recruitment and their differentiation in mature adipocytes [73]. The effects of Ang II are mediated by two main membrane receptors: the angiotensin type 1 receptor (AT1R) and the angiotensin type 2 receptor (AT2R). The AT1R is responsible for most biological effects of Ang II, including blood pressure control, trophic and pro-inflammatory effects, whereas the AT2R antagonises several AT1R-mediated effects. It has been demonstrated that Ang II through AT1R stimulates transcription factor expression, which may promote adipocytes differentiation and increase of triglycerides content in human adipocytes culture and also in 3 T3-L1 cells, thus leading to adipocyte hypertrophy [74]. The crucial role of AT1R is confirmed by the demonstration that in essential hypertensive patients, in adipocytes in culture and in a rat model of type 2 diabetes, AT1R antagonist is able to increase adiponectin concentrations and to improve insulin sensitivity [75]. Conversely, the binding between Ang II and its AT2R may determine tissue regeneration [76, 77] and upregulation of adiponectin production in neonatal rat ventricular myocytes [77].

Oxidative stress is closely correlated to RAAS; Ang II is a potent inductor of reactive oxygen substances (ROS), and during obesity systemic and related adipose tissue, Ang II levels are increased. In addition to Ang II, aldosterone is also an important mediator of RAAS effects. A review of 2011 discusses the known interaction between adiponectin and aldosterone, demonstrating an inverse relationship [78]. Previous data demonstrate the presence of crosstalk between aldosterone and Ang II which modulates and stimulates the signalling transduction such as the Ang II effects on PVAT [79].

The local RAAS system of PVAT may operate in combination or independently of circulating RAAS components. The precise function of RAAS in perivascular adipose tissue is still controversial but may contribute to vascular tone, leading to structural and functional alterations, especially in pathological conditions such as hypertension and obesity, amplifying the effect of systemic RAAS. It has been hypothesised a crosstalk between RAAS of smooth muscle cells and of endothelial cells which could explain at least in part the implication of tissue RAAS in the perivascular adipose tissue on structural alterations of small resistance arteries. In view of the relevant role of the RAAS system and PVAT on vascular function, it is interesting to understand the possible relationship between them.

In 2006, Gálvez and colleagues observed several changes in in the function and in the mass of PVAT in spontaneously hypertensive rats (SHR) and suggested an interrelation with increased vascular resistance [80]. In the SHR model, they observed a reduced anticontractile effect and mass of mesenteric PVAT compared with control mice, presenting smaller adipocytes and lower leptin content. Subsequently, Lee demonstrated that the changes on PVAT structure and function were observed also in Ang II-induced hypertension in adult male Wistar rats [61]. In addition, it has been shown that Ang II is able to mediate PVAT-associated increase of contractile response to perivascular neuronal excitation by electrical field stimulation, possibly through superoxide production [81]. Recently, it has been sown that angiotensin [1–7] could mediates the reduced anticontractile function in aorta PVAT of SHR rats), as angiotensin [1–7] receptor blockade was able to inhibit the anticontractile effect of PVAT [82, 83]. Additionally, angiotensin 1–7 has been also implicated in PVAT anticontractile function of inferior vena cava [23].

Using a mouse model, we have demonstrated that small mesenteric arteries PVAT loses its anticontractile effect after two different inflammatory stimulation: aldosterone and hypoxia [33]. Among multiple factors that may be involved in the control of obesity and hypertension vascular consequences, we have shown that eplerenone reduces the inflammatory effects caused by both aldosterone and hypoxia, and this property of mineralocorticoid-receptors blockers may be of potential therapeutic interest [33]. Further, in a recent study, we demonstrated that blocking ACE-inhibitor and also AT1R have similar effects improving the anticontractile properties of the PVAT following hypoxia. In our study in vitro induction of a hypoxic environment could stimulate the loss of anticontractile perivascular adipose tissue function seen in obese patients that could be prevented using inhibitors or the renin-angiotensin cascade [84].

In conclusion, RAAS system seems to be important in modulating PVAT effects in vascular physiology, and changes in its components production can lead to vascular impairment. Therefore RAAS blockade may protect the vasculature and may reverse the lack of PVAT anticontractile function in vascular diseases.

15.7 Clinical Translation: Melatonin

Several evidences have demonstrated that obesity is involved in pathogenesis and progression of cardiovascular disease (Fig. 15.1). The weight loss represents the obvious treatment for obesity through the lifestyle changes as a first-line treatment and the surgical intervention (as bariatric surgery). Recently it has been shown that

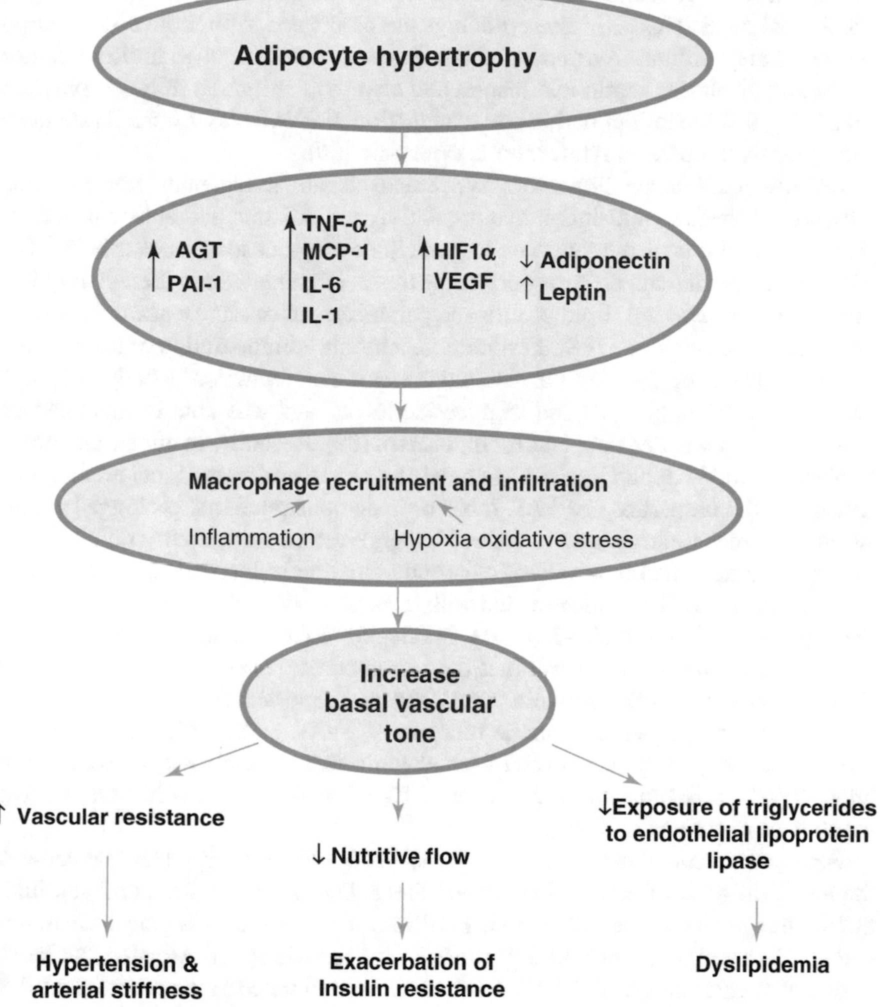

Fig. 15.1 Hypothetical link between adipose tissue inflammation in obesity, oxidative stress and vascular disease

there is reduction of adipose tissue inflammation and increase in local adiponectin, and NO bioavailability in obese subjects 6 months after bariatric surgery [85]. In this study, although patients remained morbidly obese, it was observed a significant reduction in body mass index and blood pressure, an improvement in lipid profile and blood glucose concentration and a restoration of anticontractile function of PVAT. These findings are very interesting because they could explain the cases of obese people that do not present complications as hypertension or type 2 diabetes mellitus. More recently, it has been demonstrated that in diet-induced obese rats, sustained weight loss improves PVAT function with restoration of its anticontractile property through reduction in local TNFα and increased NO production mainly by endothelial NOS (NOS-3). These findings are associated with reduction of adipocytes size and infiltration of eosinophils, decrease in macrophage infiltration, normalisation of plasma leptin and insulin and also with reduction in blood pressure, suggesting that the inflammation and dysfunction in PVAT play a crucial role in the pathogenesis of obesity and metabolic syndrome [86].

Melatonin (N-acetyl-5-methoxytryptamine) is an endogenous hormone that presents antioxidant, anti-inflammatory, anti-hyperlipidemic and anti-hypertensive properties, and it is also being seen to participate on glucose homeostasis [87, 88]. All these properties can be corroborated by the fact that melatonin therapy is able to improve blood pressure, lipid profile and parameters of oxidative stress in patients with metabolic syndrome [88]. Furthermore, chronic administration of melatonin in rats fed a normal or high-fat diet resulted in significant reduction in body weight, in circulating insulin, glucose and triglyceride levels, and was able to modulate the normal circadian pattern of plasma adiponectin [89]. Recently, in our animal model of obesity (B6.V-Lepob/OlaHsd), we have observed the effects of melatonin on the anticontractile properties of PVAT. It has been demonstrated that prolonged administration of melatonin is able to reduce hyperglycemia associated with obesity in hyperphagic mice and to ameliorate the inflammation in the perivascular environment. We saw that ob/ob mice treated with melatonin showed a marked reduction in the expression of endothelin-1 (ET-1), interleukin-6 (IL-6) and metalloproteases 2 and 9. In addition, we observed that the increased expression of both TNF-α and CD68 in visceral fat sections from ob/ob mice was significantly reduced after melatonin treatment. The anticontractile function of PVAT, partially lost in our animal model of obesity, may be restored by melatonin treatment only in presence of an intact PVAT, indicating the importance of PVAT oxidative stress in vascular dysfunction of obese animals [90].

Aging represents another condition associated with a progressively decrease of the nighttime peak of melatonin concentrations. During aging, structural and functional changes have been observed. Particularly, the effect of aging on vascular endothelium and small muscle cells (SMC) has been widely investigated, but less is known about the changes of PVAT. A senescence-accelerated prone mouse (SAMP8) is a model of age-related vascular dysfunction with associated increase in blood pressure and cognitive decline. It has been demonstrated that there was overexpression of endothelin-1 (ET-1), inducible NOS (iNOS) and cyclooxygenase-2 (COX-2) in the vasculature of these animals. All of these markers associated with oxidative stress showed a reduction in the level of vascular eNOS, cyclooxygenase-1 (COX-1) and adiponectin. In addition, in SAMP8 mice, the PVAT had lost its protective anticontractile effect, but the long-term treatment with melatonin was able to increase some vasculoprotective markers, to decrease oxidative stress and inflammation and to restore the anticontractile effect of perivascular adipose tissue. Decreased expression of adiponectin and adiponectin receptor 1 was also observed in visceral fat of untreated aging mice, whereas a significant increase was observed after melatonin treatment. The increased production/activity of adiponectin might contribute to explain the improvement of the anticontractile action of perivascular fat observed in the mesenteric small resistance arteries of SAMP8 mice after chronic treatment with melatonin [91].

Besides the use of melatonin, the blockade of RAAS system has been seen as a potential target to treat vascular diseases. It has been demonstrated that the use of mineralocorticoid receptor (MR) antagonist, as spironolactone, prevents vascular remodelling of a mice model of type 2 diabetes mellitus [92]. Interestingly, Briones et al. showed that PVAT adipocytes present functional aldosterone synthase which is able to generate aldosterone in a process regulated by Ang II through AT1R [93]. They also showed that the use of eplerenone, another MR antagonist, could improve the acetylcholine-induced relaxation in obese diabetic (db/db) mice, without effect in a control group. Further, in a mice model of specifically adipocyte-MR overexpression (adipo-MROE), animals presented features of metabolic syndrome and impaired vasoconstriction to phenylephrine with preserved endothelial function, showing that activation of adipose-MR could change functional properties of arteries [93]. These finds demonstrate a link between PVAT-RAAS system and vascular dysfunction, mainly through aldosterone action. Although it is necessary to further investigate to understand the effect of chronical treatment with MR antagonists over PVAT, this seems to be a potential target to the treatment for vascular complications of diabetes and metabolic syndrome.

15.8 Conclusions

PVAT depots are highly metabolically active and contribute to a physiologically important paracrine activity of vasodilation by the release of a broad spectrum of vasorelaxing adipokines. Perhaps the most prominent of these is adiponectin. In obesity with the development of localised tissue inflammation, the bioavailability of these PVAT-derived vasorelaxing adipokines is reduced, and blood pressure rises as a result of vasoconstriction with diabetes ensuing as a result of a decrease in glucose uptake. Manipulating the inflammasome PVAT is the next challenge in terms of producing novel ways to prevent the complications of obesity which cause so much mortality and morbidity in a culturated society.

References

1. Aghamohammadzadeh R, Withers S, Lynch F, Greenstein A, Malik R, Heagerty A. Perivascular adipose tissue from human systemic and coronary vessels: the emergence of a new pharmacotherapeutic target. Br J Pharmacol. 2012;165:670–82.
2. Szasz T, Webb RC. Perivascular adipose tissue: more than just structural support. Clin Sci (Lond). 2012;122:1–12.
3. Greenstein AS, Khavandi K, Withers SB, Sonoyama K, Clancy O, Jeziorska M, Laing I, Yates AP, Pemberton PW, Malik RA, Heagerty AM. Local inflammation and hypoxia abolish the protective anticontractile properties of perivascular fat in obese patients. Circulation. 2009;119:1661–70.

4. Galvez-Prieto B, Dubrovska G, Cano MV, Delgado M, Aranguez I, Gonzalez MC, Ruiz-Gayo M, Gollasch M, Fernandez-Alfonso MS. A reduction in the amount and anti-contractile effect of periadventitial mesenteric adipose tissue precedes hypertension development in spontaneously hypertensive rats. Hypertens Res. 2008;31:1415–23.
5. Bulloch JM, Daly CJ. Autonomic nerves and perivascular fat: interactive mechanisms. Pharmacol Ther. 2014;143:61–73.
6. Padilla J, Jenkins NT, Vieira-Potter VJ, Laughlin MH. Divergent phenotype of rat thoracic and abdominal perivascular adipose tissues. Am J Physiol Regul Integr Comp Physiol. 2013;304:R543–52.
7. Galvez-Prieto B, Bolbrinker J, Stucchi P, de Las Heras AI, Merino B, Arribas S, Ruiz-Gayo M, Huber M, Wehland M, Kreutz R, Fernandez-Alfonso MS. Comparative expression analysis of the renin-angiotensin system components between white and brown perivascular adipose tissue. J Endocrinol. 2008;197:55–64.
8. Coelho M, Oliveira T, Fernandes R. Biochemistry of adipose tissue: an endocrine organ. Arch Med Sci. 2013;9:191–200.
9. Florian W Kiefer, Paul Cohen, Jorge Plutzky. Fifty Shades of Brown: Perivascular Fat, Thermogenesis, and Atherosclerosis. Circulation. 2012;126(9):1012–5.
10. Soltis EE, Cassis LA. Influence of perivascular adipose tissue on rat aortic smooth muscle responsiveness. Clin Exp Hypertens A. 1991;13:277–96.
11. Weston AH, Egner I, Dong Y, Porter EL, Heagerty AM, Edwards G. Stimulated release of a hyperpolarizing factor (ADHF) from mesenteric artery perivascular adipose tissue: involvement of myocyte BKCa channels and adiponectin. Br J Pharmacol. 2013;169:1500–9.
12. Lohn M, Dubrovska G, Lauterbach B, Luft FC, Gollasch M, Sharma AM. Periadventitial fat releases a vascular relaxing factor. FASEB J. 2002;16:1057–63.
13. Saxton SN, Ryding KE, Aldous RG, Withers SB, Ohanian J, Heagerty AM. Role of sympathetic nerves and adipocyte catecholamine uptake in the vasorelaxant function of perivascular adipose tissue. Arterioscler Thromb Vasc Biol. 2018;38:880–91.
14. Lynch FM, Withers SB, Yao Z, Werner ME, Edwards G, Weston AH, Heagerty AM. Perivascular adipose tissue-derived adiponectin activates BK(Ca) channels to induce anticontractile responses. Am J Physiol Heart Circ Physiol. 2013;304:H786–95.
15. Fang L, Zhao J, Chen Y, Ma T, Xu G, Tang C, Liu X, Geng B. Hydrogen sulfide derived from periadventitial adipose tissue is a vasodilator. J Hypertens. 2009;27:2174–85.
16. Bussey CE, Withers SB, Saxton SN, Bodagh N, Aldous RG, Heagerty AM. β3-Adrenoceptor stimulation of perivascular adipocytes leads to increased fat cell-derived NO and vascular relaxation in small arteries. Br J Pharmacol. 2018;175:3685–98.
17. Withers SB, Bussey CE, Saxton SN, Melrose HM, Watkins AE, Heagerty AM. Mechanisms of Adiponectin-Associated Perivascular Function in Vascular Disease. Arterioscler Thromb Vasc Biol. 2014;34(8):1637–42.
18. Yokota T, Oritani K, Takahashi I, Ishikawa J, Matsuyama A, Ouchi N, Kihara S, Funahashi T, Tenner AJ, Tomiyama Y, Matsuzawa Y. Adiponectin, a new member of the family of soluble defense collagens, negatively regulates the growth of myelomonocytic progenitors and the functions of macrophages. Blood. 2000;96:1723–32.
19. Tsao TS, Lodish HF, Fruebis J. ACRP30, a new hormone controlling fat and glucose metabolism. Eur J Pharmacol. 2002;440:213–21.
20. Arita Y, Kihara S, Ouchi N, Takahashi M, Maeda K, Miyagawa J, Hotta K, Shimomura I, Nakamura T, Miyaoka K, Kuriyama H, Nishida M, Yamashita S, Okubo K, Matsubara K, Muraguchi M, Ohmoto Y, Funahashi T, Matsuzawa Y. Paradoxical decrease of an adipose-specific protein, adiponectin, in obesity. Biochem Biophys Res Commun. 1999;257:79–83.
21. Imatoh T, Miyazaki M, Momose Y, Tanihara S, Une H. Adiponectin levels associated with the development of hypertension: a prospective study. Hypertens Res. 2008;31:229–33.
22. Nayak S, Soon SQ, Kunjal R, Ramadoo R, Baptiste O, Persad J, Temull V, Diptee L, Balgobin S. Relationship between adiponectin, inflammatory markers and obesity in type 2 diabetic and non-diabetic Trinidadians. Arch Physiol Biochem. 2009;115:28–33.

23. Lu C, Zhao AX, Gao YJ, Lee RM. Modulation of vein function by perivascular adipose tissue. Eur J Pharmacol. 2011;657:111–6.
24. Huang Cao ZF, Stoffel E, Cohen P. Role of perivascular adipose tissue in vascular physiology and pathology. Hypertension. 2017;69:770–7.
25. Berg AH, Scherer PE. Adipose tissue, inflammation, and cardiovascular disease. Circ Res. 2005;96:939–49.
26. Lazar MA. How obesity causes diabetes: not a tall tale. Science. 2005;307:373–5.
27. Gao YJ, Holloway AC, Zeng ZH, Lim GE, Petrik JJ, Foster WG, Lee RM. Prenatal exposure to nicotine causes postnatal obesity and altered perivascular adipose tissue function. Obes Res. 2005;13:687–92.
28. Clambey ET, McNamee EN, Westrich JA, Glover LE, Campbell EL, Jedlicka P, de Zoeten EF, Cambier JC, Stenmark KR, Colgan SP, Eltzschig HK. Hypoxia-inducible factor-1 alpha-dependent induction of FoxP3 drives regulatory T-cell abundance and function during inflammatory hypoxia of the mucosa. Proc Natl Acad Sci U S A. 2012;109:E2784–93.
29. Halberg N, Khan T, Trujillo ME, Wernstedt-Asterholm I, Attie AD, Sherwani S, Wang ZV, Landskroner-Eiger S, Dineen S, Magalang UJ, Brekken RA, Scherer PE. Hypoxia-inducible factor 1alpha induces fibrosis and insulin resistance in white adipose tissue. Mol Cell Biol. 2009;29:4467–83.
30. Lolmede K, Durand de Saint Front V, Galitzky J, Lafontan M, Bouloumie A. Effects of hypoxia on the expression of proangiogenic factors in differentiated 3T3-F442A adipocytes. Int J Obes Relat Metab Disord. 2003;27:1187–95.
31. Cancello R, Henegar C, Viguerie N, Taleb S, Poitou C, Rouault C, Coupaye M, Pelloux V, Hugol D, Bouillot JL, Bouloumie A, Barbatelli G, Cinti S, Svensson PA, Barsh GS, Zucker JD, Basdevant A, Langin D, Clement K. Reduction of macrophage infiltration and chemoattractant gene expression changes in white adipose tissue of morbidly obese subjects after surgery-induced weight loss. Diabetes. 2005;54:2277–86.
32. Chen B, Lam KS, Wang Y, Wu D, Lam MC, Shen J, Wong L, Hoo RL, Zhang J, Xu A. Hypoxia dysregulates the production of adiponectin and plasminogen activator inhibitor-1 independent of reactive oxygen species in adipocytes. Biochem Biophys Res Commun. 2006;341:549–56.
33. Withers SB, Agabiti-Rosei C, Livingstone DM, Little MC, Aslam R, Malik RA, Heagerty AM. Macrophage activation is responsible for loss of anticontractile function in inflamed perivascular fat. Arterioscler Thromb Vasc Biol. 2011;31:908–13.
34. Furukawa S, Fujita T, Shimabukuro M, Iwaki M, Yamada Y, Nakajima Y, Nakayama O, Makishima M, Matsuda M, Shimomura I. Increased oxidative stress in obesity and its impact on metabolic syndrome. J Clin Invest. 2004;114:1752–61.
35. Chatterjee TK, Stoll LL, Denning GM, Harrelson A, Blomkalns AL, Idelman G, Rothenberg FG, Neltner B, Romig-Martin SA, Dickson EW, Rudich S, Weintraub NL. Proinflammatory phenotype of perivascular adipocytes: influence of high-fat feeding. Circ Res. 2009;104:541–9.
36. Deshmane SL, Kremlev S, Amini S, Sawaya BE. Monocyte chemoattractant protein-1 (MCP-1): an overview. J Interf Cytokine Res. 2009;29:313–26.
37. Du B, Ouyang A, Eng JS, Fleenor BS. Aortic perivascular adipose-derived interleukin-6 contributes to arterial stiffness in low-density lipoprotein receptor deficient mice. Am J Physiol Heart Circ Physiol. 2015;308:H1382–90.
38. Withers SB, Forman R, Meza-Perez S, Sorobetea D, Sitnik K, Hopwood T, Lawrence CB, Agace WW, Else KJ, Heagerty AM, Svensson-Frej M, Cruickshank SM. Eosinophils are key regulators of perivascular adipose tissue and vascular functionality. Sci Rep. 2017;7:44571.
39. Wu D, Molofsky AB, Liang HE, Ricardo-Gonzalez RR, Jouihan HA, Bando JK, Chawla A, Locksley RM. Eosinophils sustain adipose alternatively activated macrophages associated with glucose homeostasis. Science. 2011;332:243–7.
40. Maeda N, Shimomura I, Kishida K, Nishizawa H, Matsuda M, Nagaretani H, Furuyama N, Kondo H, Takahashi M, Arita Y, Komuro R, Ouchi N, Kihara S, Tochino Y, Okutomi K, Horie M, Takeda S, Aoyama T, Funahashi T, Matsuzawa Y. Diet-induced insulin resistance in mice lacking adiponectin/ACRP30. Nat Med. 2002;8:731–7.

41. Fruebis J, Tsao TS, Javorschi S, Ebbets-Reed D, Erickson MR, Yen FT, Bihain BE, Lodish HF. Proteolytic cleavage product of 30-kDa adipocyte complement-related protein increases fatty acid oxidation in muscle and causes weight loss in mice. Proc Natl Acad Sci U S A. 2001;98:2005–10.
42. Kovacova Z, Tencerova M, Roussel B, Wedellova Z, Rossmeislova L, Langin D, Polak J, Stich V. The impact of obesity on secretion of adiponectin multimeric isoforms differs in visceral and subcutaneous adipose tissue. Int J Obes. 2012;36:1360–5.
43. Lee MH, Klein RL, El-Shewy HM, Luttrell DK, Luttrell LM. The adiponectin receptors AdipoR1 and AdipoR2 activate ERK1/2 through a Src/Ras-dependent pathway and stimulate cell growth. Biochemistry. 2008;47:11682–92.
44. Ding M, Xie Y, Wagner RJ, Jin Y, Carrao AC, Liu LS, Guzman AK, Powell RJ, Hwa J, Rzucidlo EM, Martin KA. Adiponectin induces vascular smooth muscle cell differentiation via repression of mammalian target of rapamycin complex 1 and FoxO4. Arterioscler Thromb Vasc Biol. 2011;31:1403–10.
45. Ding M, Carrao AC, Wagner RJ, Xie Y, Jin Y, Rzucidlo EM, Yu J, Li W, Tellides G, Hwa J, Aprahamian TR, Martin KA. Vascular smooth muscle cell-derived adiponectin: a paracrine regulator of contractile phenotype. J Mol Cell Cardiol. 2012;52:474–84.
46. Almabrouk TA, Ugusman AB, Katwan OJ, Salt IP, Kennedy S. Deletion of AMPKalpha1 attenuates the anticontractile effect of perivascular adipose tissue (PVAT) and reduces adiponectin release. Br J Pharmacol. 2017;174(20):3398–3410.
47. Fukao M, Mason HS, Britton FC, Kenyon JL, Horowitz B, Keef KD. Cyclic GMP-dependent protein kinase activates cloned BKCa channels expressed in mammalian cells by direct phosphorylation at serine 1072. J Biol Chem. 1999;274:10927–35.
48. Foller M, Jaumann M, Dettling J, Saxena A, Pakladok T, Munoz C, Ruth P, Sopjani M, Seebohm G, Ruttiger L, Knipper M, Lang F. AMP-activated protein kinase in BK-channel regulation and protection against hearing loss following acoustic overstimulation. FASEB J. 2012;26:4243–53.
49. Yilmaz MI, Sonmez A, Caglar K, Celik T, Yenicesu M, Eyileten T, Acikel C, Oguz Y, Yavuz I, Vural A. Effect of antihypertensive agents on plasma adiponectin levels in hypertensive patients with metabolic syndrome. Nephrology (Carlton). 2007;12:147–53.
50. Brennan AM, Mantzoros CS. Drug insight: the role of leptin in human physiology and pathophysiology—emerging clinical applications. Nat Clin Pract Endocrinol Metab. 2006;2:318–27.
51. Leung YM, Kwan CY. Dual vascular effects of leptin via endothelium: hypothesis and perspective. Chin J Physiol. 2008;51:1–6.
52. Galvez-Prieto B, Somoza B, Gil-Ortega M, Garcia-Prieto CF, de Las Heras AI, Gonzalez MC, Arribas S, Aranguez I, Bolbrinker J, Kreutz R, Ruiz-Gayo M, Fernandez-Alfonso MS. Anticontractile effect of perivascular adipose tissue and Leptin are reduced in hypertension. Front Pharmacol. 2012;3:103.
53. Michel T, Feron O. Nitric oxide synthases: which, where, how, and why? J Clin Invest. 1997;100:2146–52.
54. Gil-Ortega M, Stucchi P, Guzman-Ruiz R, Cano V, Arribas S, Gonzalez MC, Ruiz-Gayo M, Fernandez-Alfonso MS, Somoza B. Adaptative nitric oxide overproduction in perivascular adipose tissue during early diet-induced obesity. Endocrinology. 2010;151:3299–306.
55. Aghamohammadzadeh R, Unwin RD, Greenstein AS, Heagerty AM. Effects of obesity on perivascular adipose tissue vasorelaxant function: nitric oxide, inflammation and elevated systemic blood pressure. J Vasc Res. 2015;52:299–305.
56. Mustafa AK, Sikka G, Gazi SK, Steppan J, Jung SM, Bhunia AK, Barodka VM, Gazi FK, Barrow RK, Wang R, Amzel LM, Berkowitz DE, Snyder SH. Hydrogen sulfide as endothelium-derived hyperpolarizing factor sulfhydrates potassium channels. Circ Res. 2011;109:1259–68.
57. Yan H, Du J, Tang C. The possible role of hydrogen sulfide on the pathogenesis of spontaneous hypertension in rats. Biochem Biophys Res Commun. 2004;313:22–7.
58. Zhong G, Chen F, Cheng Y, Tang C, Du J. The role of hydrogen sulfide generation in the pathogenesis of hypertension in rats induced by inhibition of nitric oxide synthase. J Hypertens. 2003;21:1879–85.

59. Gao YJ, Lu C, Su LY, Sharma AM, Lee RM. Modulation of vascular function by perivascular adipose tissue: the role of endothelium and hydrogen peroxide. Br J Pharmacol. 2007;151:323–31.
60. Lee YC, Chang HH, Chiang CL, Liu CH, Yeh JI, Chen MF, Chen PY, Kuo JS, Lee TJ. Role of perivascular adipose tissue-derived methyl palmitate in vascular tone regulation and pathogenesis of hypertension. Circulation. 2011;124:1160–71.
61. Lee RM, Ding L, Lu C, Su LY, Gao YJ. Alteration of perivascular adipose tissue function in angiotensin II-induced hypertension. Can J Physiol Pharmacol. 2009;87:944–53.
62. Pizzinat N, Marti L, Remaury A, Leger F, Langin D, Lafontan M, Carpene C, Parini A. High expression of monoamine oxidases in human white adipose tissue: evidence for their involvement in noradrenaline clearance. Biochem Pharmacol. 1999;58:1735–42.
63. Ayala-Lopez N, Martini M, Jackson WF, Darios E, Burnett R, Seitz B, Fink GD, Watts SW. Perivascular adipose tissue contains functional catecholamines. Pharmacol Res Perspect. 2014;2:e00041.
64. Bamshad M, Aoki VT, Adkison MG, Warren WS, Bartness TJ. Central nervous system origins of the sympathetic nervous system outflow to white adipose tissue. Am J Phys. 1998;275:R291–9.
65. Foster MT, Bartness TJ. Sympathetic but not sensory denervation stimulates white adipocyte proliferation. Am J Physiol Regul Integr Comp Physiol. 2006;291:R1630–7.
66. Shen YT, Cervoni P, Claus T, Vatner SF. Differences in beta 3-adrenergic receptor cardiovascular regulation in conscious primates, rats and dogs. J Pharmacol Exp Ther. 1996;278:1435–43.
67. Smith MM, Minson CT. Obesity and adipokines: effects on sympathetic overactivity. J Physiol. 2012;590:1787–801.
68. Manolis AJ, Poulimenos LE, Kallistratos MS, Gavras I, Gavras H. Sympathetic overactivity in hypertension and cardiovascular disease. Curr Vasc Pharmacol. 2014;12:4–15.
69. Muntzel MS, Al-Naimi OA, Barclay A, Ajasin D. Cafeteria diet increases fat mass and chronically elevates lumbar sympathetic nerve activity in rats. Hypertension. United States. 2012;60:1498–502.
70. Post SR, Hammond HK, Insel PA. Beta-adrenergic receptors and receptor signaling in heart failure. Annu Rev Pharmacol Toxicol. 1999;39:343–60.
71. Engeli S, Gorzelniak K, Kreutz R, Runkel N, Distler A, Sharma AM. Co-expression of renin-angiotensin system genes in human adipose tissue. J Hypertens. 1999;17:555–60.
72. Touyz RM, Schiffrin EL. Signal transduction mechanisms mediating the physiological and pathophysiological actions of angiotensin II in vascular smooth muscle cells. Pharmacol Rev. 2000;52:639–72.
73. Cassis LA, Police SB, Yiannikouris F, Thatcher SE. Local adipose tissue renin-angiotensin system. Curr Hypertens Rep. 2008;10:93–8.
74. Jones BH, Standridge MK, Moustaid N. Angiotensin II increases lipogenesis in 3T3-L1 and human adipose cells. Endocrinology. 1997;138:1512–9.
75. Furuhashi M, Ura N, Takizawa H, Yoshida D, Moniwa N, Murakami H, Higashiura K, Shimamoto K. Blockade of the renin-angiotensin system decreases adipocyte size with improvement in insulin sensitivity. J Hypertens. 2004;22:1977–82.
76. Sowers JR. Endocrine functions of adipose tissue: focus on adiponectin. Clin Cornerstone. 2008;9:32–8. discussion 39-40
77. Guo B, Li Y, Han R, Zhou H, Wang M. Angiotensin II upregulation of cardiomyocyte adiponectin production is nitric oxide/cyclic GMP dependent. Am J Med Sci. 2011;341:350–5.
78. Flynn C, Bakris GL. Interaction between adiponectin and aldosterone. Cardiorenal Med. 2011;1:96–101.
79. Lemarie CA, Paradis P, Schiffrin EL. New insights on signaling cascades induced by cross-talk between angiotensin II and aldosterone. J Mol Med (Berl). 2008;86:673–8.
80. Galvez B, de Castro J, Herold D, Dubrovska G, Arribas S, Gonzalez MC, Aranguez I, Luft FC, Ramos MP, Gollasch M, Fernandez Alfonso MS. Perivascular adipose tissue and mesenteric vascular function in spontaneously hypertensive rats. Arterioscler Thromb Vasc Biol. 2006;26:1297–302.

81. Gao YJ, Takemori K, Su LY, An WS, Lu C, Sharma AM, Lee RM. Perivascular adipose tissue promotes vasoconstriction: the role of superoxide anion. Cardiovasc Res. 2006;71:363–73.
82. Lu C, Su LY, Lee RM, Gao YJ. Alterations in perivascular adipose tissue structure and function in hypertension. Eur J Pharmacol. 2011;656:68–73.
83. Lee RM, Lu C, Su LY, Gao YJ. Endothelium-dependent relaxation factor released by perivascular adipose tissue. J Hypertens. 2009;27:782–90.
84. Rosei CA, Withers SB, Belcaid L, De Ciuceis C, Rizzoni D, Heagerty AM. Blockade of the renin-angiotensin system in small arteries and anticontractile function of perivascular adipose tissue. J Hypertens. 2015;33:1039–45.
85. Aghamohammadzadeh R, Greenstein AS, Yadav R, Jeziorska M, Hama S, Soltani F, Pemberton PW, Ammori B, Malik RA, Soran H, Heagerty AM. Effects of bariatric surgery on human small artery function: evidence for reduction in perivascular adipocyte inflammation, and the restoration of normal anticontractile activity despite persistent obesity. J Am Coll Cardiol. 2013;62:128–35.
86. Bussey CE, Withers SB, Aldous RG, Edwards G, Heagerty AM. Obesity-related perivascular adipose tissue damage is reversed by sustained weight loss in the rat. Arterioscler Thromb Vasc Biol. 2016;36:1377–85.
87. Bonnefont-Rousselot D, Collin F, Jore D, Gardes-Albert M. Reaction mechanism of melatonin oxidation by reactive oxygen species in vitro. J Pineal Res. 2011;50:328–35.
88. Prunet-Marcassus B, Desbazeille M, Bros A, Louche K, Delagrange P, Renard P, Casteilla L, Penicaud L. Melatonin reduces body weight gain in Sprague Dawley rats with diet-induced obesity. Endocrinology. 2003;144:5347–52.
89. Kozirog M, Poliwczak AR, Duchnowicz P, Koter-Michalak M, Sikora J, Broncel M. Melatonin treatment improves blood pressure, lipid profile, and parameters of oxidative stress in patients with metabolic syndrome. J Pineal Res. 2011;50:261–6.
90. Agabiti-Rosei C, De Ciuceis C, Rossini C, Porteri E, Rodella LF, Withers SB, Heagerty AM, Favero G, Agabiti-Rosei E, Rizzoni D, Rezzani R. Anticontractile activity of perivascular fat in obese mice and the effect of long-term treatment with melatonin. J Hypertens. 2014;32:1264–74.
91. Rios-Lugo MJ, Cano P, Jimenez-Ortega V, Fernandez-Mateos MP, Scacchi PA, Cardinali DP, Esquifino AI. Melatonin effect on plasma adiponectin, leptin, insulin, glucose, triglycerides and cholesterol in normal and high fat-fed rats. J Pineal Res. 2010;49:342–8.
92. Agabiti-Rosei C, Favero G, De Ciuceis C, Rossini C, Porteri E, Rodella LF, Franceschetti L, Maria Sarkar A, Agabiti-Rosei E, Rizzoni D, Rezzani R. Effect of long-term treatment with melatonin on vascular markers of oxidative stress/inflammation and on the anticontractile activity of perivascular fat in aging mice. Hypertens Res. 2017;40:41–50.
93. Briones AM, Nguyen Dinh Cat A, Callera GE, Yogi A, Burger D, He Y, Correa JW, Gagnon AM, Gomez-Sanchez CE, Gomez-Sanchez EP, Sorisky A, Ooi TC, Ruzicka M, Burns KD, Touyz RM. Adipocytes produce aldosterone through calcineurin-dependent signaling pathways: implications in diabetes mellitus-associated obesity and vascular dysfunction. Hypertension. 2012;59:1069–78.

Prognostic Role of Microvascular Damage and Effect of Treatment

16

Enrico Agabiti-Rosei, Claudia Agabiti-Rosei, and Damiano Rizzoni

16.1 Introduction

The microcirculation includes vessels, from small resistance arteries to capillaries, and postcapillary venules, with an internal lumen diameter below 300 μm. Alterations of the microcirculation definitely contribute to the increase of peripheral resistance, which is the hemodynamic hallmark of established hypertension.

In fact, essential hypertension is associated with a narrowing of the internal lumen and with a greater media wall thickness, with consequent increase in the media to lumen ratio [1]. This increase in the media to lumen ratio may be the consequence of an inward eutrophic remodeling (rearrangement of otherwise normal material around a narrowed lumen) or of an inward hypertrophic remodeling (vascular smooth muscle cell hypertrophy or hyperplasia) [2]. Eutrophic remodeling of subcutaneous small arteries is commonly seen in essential hypertension; on the contrary a hypertrophic remodeling, with evident smooth muscle cell growth, has been shown in patients with type II diabetes mellitus [3, 4], obesity [5, 6] and also metabolic syndrome [7] (even regardless of the presence or absence of elevated or normal blood pressure levels), as well as in some forms of secondary hypertension, including renovascular hypertension [8], Cushing's syndrome [9], acromegaly [10] and, possibly, primary aldosteronism [8]. Structural remodeling of the microvascular networks includes also rarefaction in the most distal part, with potential consequences for tissue perfusion and exchange or transport of nutrients [11, 12].

Not only the structure of the microcirculation may be altered in cardiovascular and/or metabolic diseases, but also the functional characteristics of small vessels may be impaired. In fact, an impairment of the endothelial function, as evaluated by the vasodilator response to acetylcholine, has been observed in human small arteries in essential hypertension [3, 8] as well as in type 2 diabetes mellitus [3, 4] and obesity [5, 6].

E. Agabiti-Rosei (✉) · C. Agabiti-Rosei · D. Rizzoni
Department of Clinical and Experimental Sciences, University of Brescia, Brescia, Italy
e-mail: enrico.agabitirosei@unibs.it; damiano.rizzoni@unibs.it

E. Agabiti-Rosei et al. (eds.), *Microcirculation in Cardiovascular Diseases*, Updates in Hypertension and Cardiovascular Protection,
https://doi.org/10.1007/978-3-030-47801-8_16

16.2 Pathophysiological Consequences of Alteration of Microcirculation

An important consequence of the presence of increased media to lumen ratio may be an impairment of vasodilator reserve. In fact, remodeling of small resistance arteries is characterized by a narrowing of the lumen, which leads to an increase of resistance to flow even at maximal dilatation, i.e. in the absence of vascular tone. A close correlation was observed, in hypertensive patients, between media to lumen ratio of subcutaneous small resistance arteries and coronary flow reserve [13, 14] or minimum vascular resistance in the forearm, as evaluated from the maximum post-ischemic flow [15], suggesting that structural alterations in small resistance arteries may be present at the same time in different vascular districts, including the coronary circulation. Alterations in the microcirculation may therefore play an important role in the development of organ damage in hypertension.

According to Park and Schiffrin, small artery remodeling is the most prevalent and probably the earliest form of target organ damage in human mild essential hypertension [16]. In a review Schiffrin [17] addressed the topic of the time course of changes in morphologic and mechanical aspects of resistance arteries as hypertension evolves with time. He suggested that an increase in the media to lumen ratio of small resistance arteries might be present very early, but its severity parallels the increase in blood pressure values [17].

Recently, interest has been focused on the possible relationships between alterations in the microcirculation and in the macrocirculation (i.e. large arteries stiffness and relative changes in the mechanical properties). The media to lumen ratio of subcutaneous small resistance arteries was significantly related to indices of large artery stiffness as well as with central systolic and pulse pressure [18]. Thus, the crosstalk between the small and large artery exaggerates arterial damage, following a vicious circle with a relevant role probably played by an enhancement of pulse wave reflection from distal reflection sites [19, 20].

16.3 Prognostic Significance of Changes of Microcirculation

An increase in the media to lumen ratio of subcutaneous small arteries was proved to be a powerful predictor of cardiovascular events in hypertension [21] (Fig. 16.1). Hypertrophic remodeling of small vessels was associated with an even higher incidence of events, compared with eutrophic remodeling [22] (Fig. 16.2).

In more details, a direct demonstration of a prognostic role of microvascular structural alterations, independently of blood pressure values, has made available in 2003 [21] (Fig. 16.1). In that study, 151 hypertensive and/or diabetic patients, together with a group of normotensives, were evaluated. In all subjects the tunica media to internal lumen ratio of subcutaneous small resistance arteries was assessed. The subjects were re-evaluated after an average follow-up time of 5.6 years.

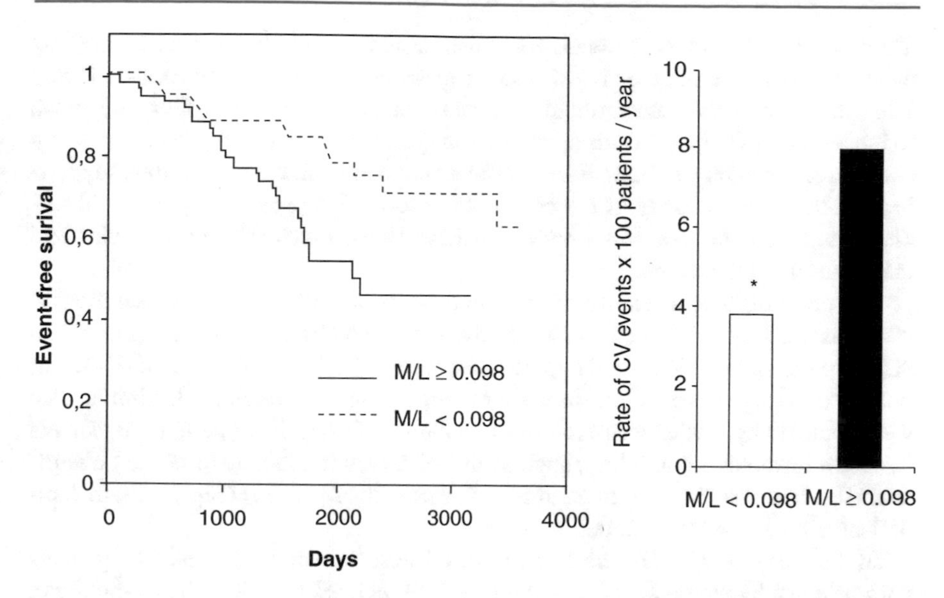

Fig. 16.1 Left: event-free survival (Kaplan-Meier method) in group of patients with a media to lumen ratio of subcutaneous small arteries ≥0.098 (mean and median values observed in the whole population) ($n = 64$, solid line) or <0.098 ($n = 64$, dotted line). Mantel-Cox test between curves, $p = 0.015$; Breslow test between curves, $p = 0.036$. Right, incidence of cardiovascular (CV) events in the subgroups of patients. From reference 21

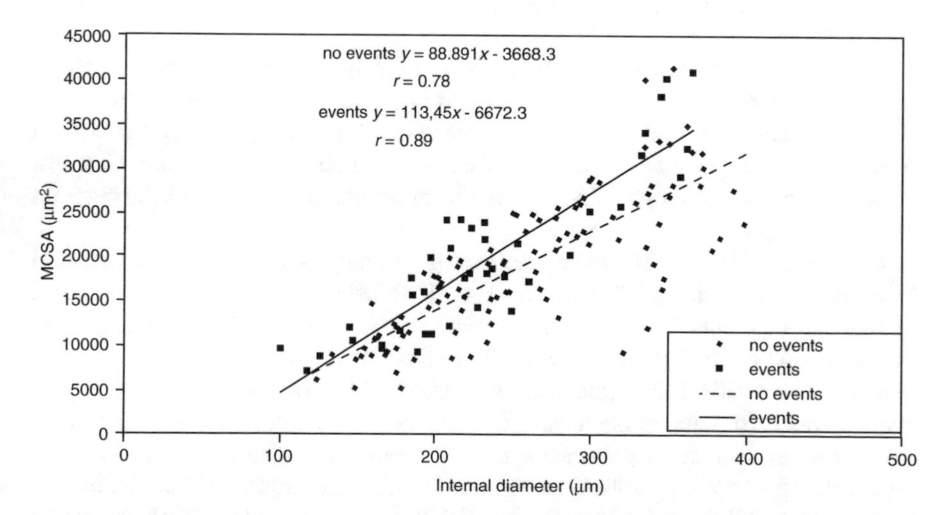

Fig. 16.2 Medial cross-sectional area (MCSA; mm²) plotted against lumen diameter of the small arteries of patients with hypertension who have subsequently had a cardiovascular event or who were event-free at the time of follow-up. The slopes of the two regression lines are significantly different ($p = 0.00009$). Data are from reference 22

Thirty-seven subjects had a documented fatal or nonfatal cardiovascular event (5.32 events/100 pts./year). A Cox multivariate regression analysis was performed, taking into consideration the conventional cardiovascular risk factors were. Only the media to lumen ratio of subcutaneous small resistance arteries and pulse pressure (usually considered a marker of large arteries stiffness) remained in the model, thus suggesting that the great majority of the prognostic information provided by classical cardiovascular risk factors was present in these two indices of microvascular and macrovascular alterations.

The prognostic significance of structural alterations of small resistance arteries was subsequently confirmed by De Ciuceis et al. in a larger population (more than 300 normotensive subjects and hypertensive patients) at lower global cardiovascular risk, also taking into account only major cardiovascular events [23]. Similar data were obtained by Mathiassen et al. in a population of essential hypertensive patients [24]. Hypertrophic remodeling, such as that observed in diabetic or obese patients, again resulted associated with an even worse prognosis [22, 25] (Fig. 16.2), in comparison with eutrophic remodeling.

In the study by De Ciuceis et al. [23], 65 normotensive subjects, 111 patients with primary hypertension (37 of them with associated type 2 diabetes mellitus), 109 patients with secondary forms of hypertension and 18 normotensive diabetic patients, for a total number of 303 subjects and patients, were included. Subjects were re-evaluated after an average follow-up time of 6.9 years in order to assess the occurrence of cardio-cerebrovascular events. Eleven subjects died of a fatal cardio-cerebrovascular event, 14 had a major, nonfatal cardiovascular event (stroke or myocardial infarction), 23 had a minor cardiovascular event and 255 had no cardiovascular event.

A significant difference was observed in event-free survival between subjects and patients with a media to lumen ratio above or below two different cutpoints: (a) 0.11 (two standard deviations above the mean of normal control subjects) and (b) 0.098 (mean and median value of the entire population). The Mantel-Cox test between survival curves gave highly significant results ($p = 0.0001$ for the cutpoint of 0.098).

Similar results were obtained by restricting the analysis to patients with essential hypertension. Similarly, Mathiassen et al. [24] have investigated 159 essential hypertensive patients. Thirty patients suffered a cardiovascular event during a follow-up period of more than 10 years. The authors tested two different cutpoints of media to lumen ratio: 0.083 (mean value of the hypertensive cohort) and 0.098 (two standard deviations above the mean of normal control subjects). Also in this study, event-free survival was significantly different between patients with low or high media to lumen ratio of subcutaneous small resistance arteries (Mantel-Cox test between cumulative survival curves: $p = 0.010$ for the cutpoint of 0.098, $p = 0.022$ for the cutpoint of 0.083) [26].

We have pooled together the two studies [23, 24], using Comprehensive Meta-Analysis Software (Englewood, NJ, USA) [27]. The results are reported in Fig. 16.3. A total of 270 patients with essential hypertension were included. The cutpoint of 0.098 for media to lumen ratio of subcutaneous small resistance arteries was used,

Meta Analysis

Study name	Odds ratio	Lower limit	Upper limit	Z-Value	p-Value
Mathiassen	3,448	1,509	7,880	2,936	0,003
De Ciuceis	3,067	1,048	8,975	2,045	0,041
	3,301	1,715	6,355	3,574	0,000

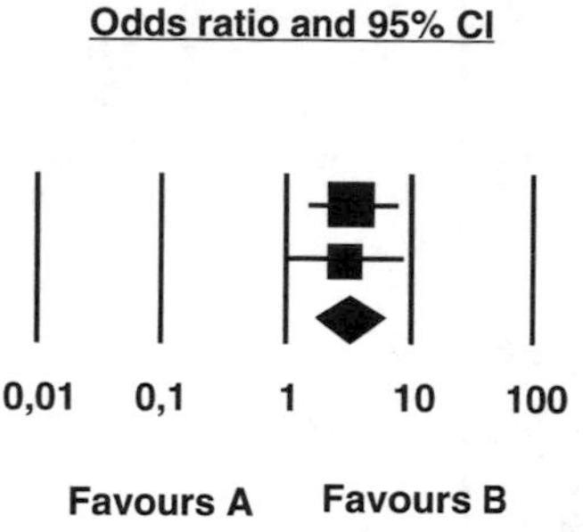

Fig. 16.3 Meta-analysis of two studies [23, 24]: only patients with essential hypertension are included ($n = 270$). Odds ratio for cardiovascular events in favour of patients with a media-to-lumen ratio of subcutaneous small arteries below 0.098 ($p < 0.0001$). CI: confidence interval. From reference 27

since it was the same in the two studies. The results clearly indicate a worse prognosis for those hypertensive patients with a media to lumen ratio above 0.098, with a p value >0.0001.

Moreover, we have performed an additional analysis in our database of subjects and patients in whom an evaluation of small resistance artery structure was performed [27] (part of them were included in the previously mentioned study) [23], restricting the analysis to 119 patients with diabetes mellitus, using the two cutpoints (0.098 and 0.11) previously identified [21, 23]. The results for cutpoint 0.098 are reported in Fig. 16.4 [27]. A significant difference was observed for cumulative survival, in favour of those with a lower media to lumen ratio ($p < 0.05$ at least).

As previously mentioned, it is not presently known whether capillary rarefaction may possess a prognostic significance. However, a preliminary study suggests that microvascular rarefaction might be correlated with media/lumen ratio of small arteries [28]; hence, it is possible that also vascular changes observed at a more distal level might contribute to the higher incidence of cardiovascular events observed in hypertensive and/or diabetic patients.

16.4 Effect of Treatment

Several intervention studies with specific antihypertensive drugs have demonstrated an improvement or even an almost complete normalization of the structure of subcutaneous small resistance arteries, in particular using angiotensin-converting enzyme (ACE) inhibitors (cilazapril, perindopril, lisinopril), calcium channel blockers (nifedipine, amlodipine, isradipine) and angiotensin II receptor blockers (losartan, irbesartan, candesartan, olmesartan and valsartan) [17, 29]. On the contrary, the β-blocker atenolol and the diuretic hydrochlorothiazide had limited effects on resistance vessels,

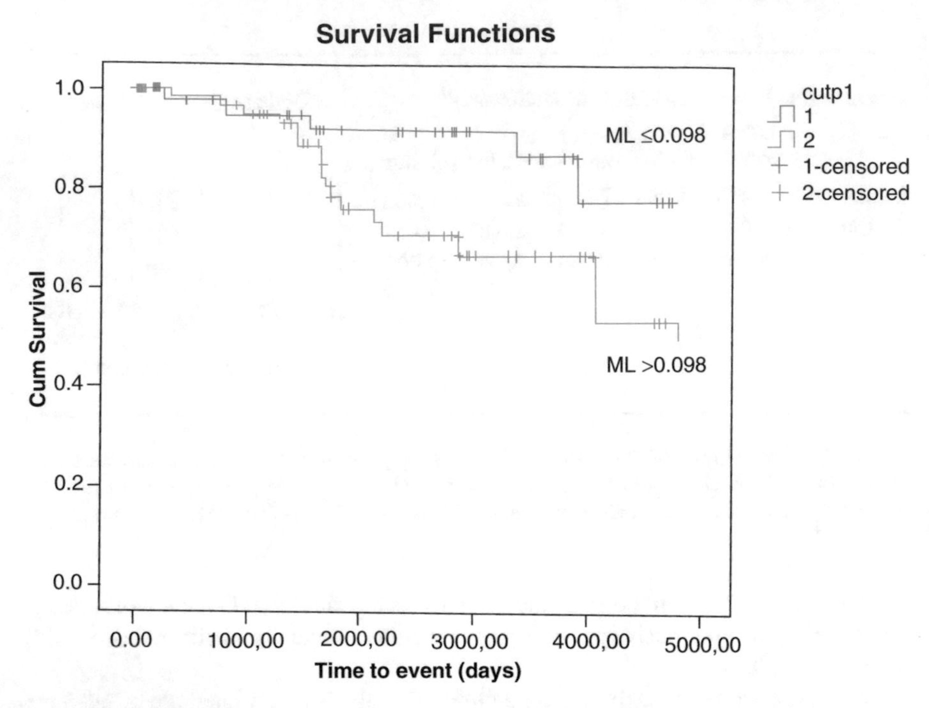

Fig. 16.4 Kaplan-Meier cum survival curves between normotensive or hypertensive diabetic patients with an M/L of subcutaneous small arteries below or above 0.098 (cutp1): log rank (Mantel-Cox test $p = 0.049$, Breslow-Wilcoxon test: $p = 0.081$, Tarone-Ware test: $p = 0.058$. *Cum* cumulative, *M/L* media-to-lumen ratio. From reference 27

despite a similar blood pressure reduction [17, 29]. More than 300 patients were investigated in these intervention studies, using a reliable and precise micromyographic approach [30–32]. ACE inhibitors proved to be significantly more effective than the β-blocker atenolol in terms of changes in media to lumen ratio [29]. The same result was obtained comparing dihydropyridinic calcium channel blockers and atenolol, or angiotensin receptor blockers and atenolol [29].

Several reasons may be advocated to explain the disparate effects of different drug classes on small artery structure. It is possible that ACE inhibitors and angiotensin receptor blockers possess growth inhibiting and antioxidant properties that may be responsible for their beneficial effect on the microcirculation [2]. Additional possible reasons are the lack of vasodilator properties of atenolol [33] or a parallel improvement of alterations in the macrovasculature observed with other drugs [34, 35]. In fact, drugs that improve microvascular structure are also particularly effective in reducing central blood pressure, thus providing a possible additional benefit, probably by a reduction of the reflection waves from peripheral sites [19, 20, 34, 35]. Also in patients with type 2 diabetes mellitus, drugs that block the renin-angiotensin-aldosterone system seem to be more effective than atenolol in terms of effects on microvascular structure [11, 36–38]. However, in patients with more severe hypertension and high global cardiovascular risk (i.e. patients with left ventricular hypertrophy or concomitant diabetes mellitus), only a reduction, but not a full normalization of the media to lumen ratio of subcutaneous small arteries, was obtained [3, 29]. In fact, media to lumen ratio remained significantly higher in respect to what observed in normotensive controls.

Few data are presently available about patients with type 1 diabetes mellitus. A study from Greenstein et al. [39] suggests that, with poor metabolic control, small arteries from patients with type 1 diabetes mellitus show hypertrophic growth in response to elevated blood pressure, similar to that seen in type 2 diabetes mellitus. However, metabolic improvements enable eutrophic remodeling to occur in response to an increase in blood pressure.

After surgical correction of obesity and consistent weight loss, a significant improvement of microvascular structure and of some oxidative stress/inflammation markers was observed [5]. Therefore, consistent weight loss obtained by bariatric surgery may substantially improve microvascular structure [8]. It should also be noted that, during antihypertensive treatment, the regression of microvascular structural alterations in the subcutaneous small arteries of hypertensive patients is paralleled by an improvement of coronary flow reserve [40].

In addition, it was also demonstrated that the severity of structural alterations in subcutaneous small resistance arteries might predict the hemodynamic and clinical outcome of patients with primary aldosteronism after adrenalectomy, since the presence of vascular remodeling implies lower chances of blood pressure normalization at long-term follow-up [26].

In general, there is evidence that the pathophysiological consequences of the regression of small artery remodeling might be as follows:

- A better control of blood pressure with reduced vascular reactivity [26]
- An increased organ flow reserve, especially in the heart [40]
- A reduction of central blood pressure [19, 20, 34, 35]

Buus NH et al. [41] demonstrated a prognostic role of changes of microvascular structure, as evaluated by the media to lumen ratio of subcutaneous small resistance arteries during antihypertensive treatment, since a normalization or a reduction of the altered structure was associated with less cardiovascular events, independently from arterial pressure changes. This demonstration of the prognostic role of changes in microvascular structure during treatment, independently of the extent of blood pressure reduction, could substantially support the idea to consider microvascular structure as an intermediate endpoint in the evaluation of the benefits of antihypertensive treatment.

16.5 Conclusion

In hypertension, the increase in peripheral resistance occurs at the level of microcirculation. It was clearly demonstrated that wall thickness is increased in relation to internal lumen and that this alteration significantly contributes to peripheral resistance. The increased media/lumen ratio may impair organ flow reserve [13]. This may be important in the maintenance and, probably, also in the progressive worsening of hypertensive disease. The presence of structural alterations represents a prognostically relevant factor, in terms of development of target organ damage and incidence of cardiovascular events, thus allowing a prediction of hypertension complications [22, 25].

However, new, noninvasive techniques are needed before suggesting extensive application of the evaluation of microvascular morphology for the cardiovascular risk stratification in hypertensive patients. Some new techniques for evaluation of microvascular morphology in the retina, presently under clinical investigation, seem to represent a promising and interesting future perspective.

At present time, we may safely state that the evaluation of microvascular structure is progressively moving from bench to bedside, and it could represent, in the near future, an evaluation to be performed in all hypertensive patients, in order to obtain a better stratification of cardiovascular risk. In addition, it could be probably considered as an intermediate endpoint for the evaluation of the effects of antihypertensive therapy, provided that a demonstration of a prognostic value of noninvasive measures of microvascular structure is made available.

References

1. Mulvany MJ, Aalkjaer C. Structure and function of small arteries. Physiol Rev. 1990;70:921–71.
2. Heagerty AM, Aalkjaaer C, Bund SJ, Korsgaard N, Mulvany MJ. Small artery structure in hypertension. Dual process of remodeling and growth. Hypertension. 1993;21:391–7.
3. Rizzoni D, Porteri E, Guelfi D, Muiesan ML, Valentini U, Cimino A, Girelli A, Rodella L, Bianchi R, Sleiman I, Agabiti-Rosei E. Structural alterations in subcutaneous small arteries of normotensive and hypertensive patients with non-insulin dependent diabetes mellitus. Circulation. 2001;103:1238–44.
4. Schofield I, Malik R, Izzard A, Austin C, Heagerty AM. Vascular structural and functional changes in type 2 diabetes mellitus. Evidence for the role of abnormal myogenic responsiveness and dyslipidemia. Circulation. 2002;106:3037–43.
5. De Ciuceis C, Porteri E, Rizzoni D, Corbellini C, La Boria E, Boari GE, Pilu A, Mittempergher F, Di Betta E, Casella C, Nascimbeni R, Agabiti-Rosei C, Ruggeri G, Caimi L, Agabiti-Rosei E. Effects of weight loss on structural and functional alterations of subcutaneous small arteries in obese patients. Hypertension. 2011;58:29–36.
6. Grassi G, Seravalle G, Scopelliti F, Dell'Oro R, Fattori L, Quarti-Trevano F, Brambilla G, Schiffrin EL, Mancia G. Structural and functional alterations of subcutaneous small resistance arteries in severe human obesity. Obesity. 2010;18:92–8.
7. Grassi G, Seravalle G, Brambilla G, Facchetti R, Bolla G, Mozzi E, Mancia G. Impact of the metabolic syndrome on subcutaneous microcirculation in obese patients. J Hypertens. 2010;28:1708–14.
8. Rizzoni D, Porteri E, Castellano M, Bettoni G, Muiesan ML, Muiesan P, Giulini SM, Agabiti-Rosei E. Vascular hypertrophy and remodeling in secondary hypertension. Hypertension. 1996;28:785–90.
9. Rizzoni D, Porteri E, De Ciuceis C, Rodella LF, Paiardi S, Rizzardi N, Platto C, Boari GE, Pilu A, Tiberio GA, Giulini SM, Favero G, Rezzani R, Agabiti-Rosei C, Bulgari G, Avanzi D, Agabiti-Rosei E. Hypertrophic remodeling of subcutaneous small resistance arteries in patients with Cushing's syndrome. Clin Endocrinol Metab. 2009;94:5010–8.
10. Rizzoni D, Porteri E, Giustina A, De Ciuceis C, Sleiman I, Boari GE, Castellano M, Muiesan ML, Bonadonna S, Burattin A, Cerudelli B, Agabiti-Rosei E. Acromegalic patients show the

presence of hypertrophic remodeling of subcutaneous small resistance arteries. Hypertension. 2004;43:561–5.
11. Rizzoni D, Agabiti-Rosei E. Small artery remodeling in diabetes mellitus. Nutr Metab Cardiovasc Dis. 2009;19:587–92.
12. Levy BI, Schiffrin EL, Mourad JJ, Agostini D, Vicaut E, Safar ME, Struijker-Boudier HA. Impaired tissue perfusion: a pathology common to hypertension, obesity, and diabetes mellitus. Circulation. 2008;118:968–76.
13. Rizzoni D, Palombo C, Porteri E, Muiesan ML, Kozàkovà M, La Canna G, Nardi M, Guelfi D, Salvetti M, Morizzo C, Vittone F, Agabiti-Rosei E. Relationships between coronary vasodilator capacity and small artery remodeling in hypertensive patients. J Hypertens. 2003;21:625–32.
14. Rizzoni D, Palombo C, Porteri E, Muiesan ML, Kozakova M, Salvetti M, Morizzo C, De Ciuceis C, Belotti E, Agabiti-Rosei E. Coronary flow reserve and small artery remodelling in hypertensive patients: re-analysis of data and review of the literature. High Blood Press Cardiovasc Prev. 2008;15:127–34.
15. Agabiti Rosei E, Rizzoni D, Castellano M, Porteri E, Zulli R, Muiesan ML, Bettoni G, Salvetti M, Muiesan P, Giulini SM. Media: lumen ratio in human small resistance arteries is related to forearm minimal vascular resistance. J Hypertens. 1995;13:341–7.
16. Park JB, Schiffrin EL. Small artery remodeling is the most prevalent (earliest?) form of target organ damage in mild essential hypertension. J Hypertens. 2001;19:921–30.
17. Schiffrin EL. Remodeling of resistance arteries in essential hypertension and effects of antihypertensive treatment. Am J Hypertens. 2004;17(12 Pt 1):1192–200.
18. Muiesan ML, Salvetti M, Rizzoni D, Paini A, Agabiti-Rosei C, Aggiusti C, Bertacchini F, Stassaldi D, Gavazzi A, Porteri E, De Ciuceis C, Agabiti-Rosei E. Pulsatile hemodynamics and microcirculation: evidence for a close relationship in hypertensive patients. Hypertension. 2013;61:130–6.
19. Laurent S, Briet M, Boutouyrie P. Large and small artery cross-talk and recent morbidity-mortality trials in hypertension. Hypertension. 2009;54:388–92.
20. Rizzoni D, Muiesan ML, Porteri E, De Ciuceis C, Boari GE, Salvetti M, Paini A, Agabiti- Rosei E. Vascular remodeling, macro- and microvessels: therapeutic implications. Blood Press. 2009;18:242–6.
21. Rizzoni D, Porteri E, Boari GEM, De Ciuceis C, Sleiman I, Muiesan ML, Castellano M, Miclini M, Agabiti-Rosei E. Prognostic significance of small artery structure in hypertension. Circulation. 2003;108:2230–5.
22. Izzard AS, Rizzoni D, Agabiti-Rosei E, Heagerty AM. Small artery structure and hypertension: adaptive changes and target organ damage. J Hypertens. 2005;23:247–50.
23. De Ciuceis C, Porteri E, Rizzoni D, Rizzardi N, Paiardi S, Boari GEM, Miclini M, Zani F, Muiesan ML, Donato F, Salvetti M, Castellano M, Tiberio GA, Giulini SM, Agabiti-Rosei E. Structural alterations of subcutaneous small arteries may predict major cardiovascular events in hypertensive patients. Am J Hypertens. 2007;20:846–52.
24. Mathiassen ON, Buus NH, Sihm I, Thybo NK, Mørn B, Schroeder AP, Thygesen K, Aalkjaer C, Lederballe O, Mulvany MJ, Christensen KL. Small artery structure is an independent predictor of cardiovascular events in essential hypertension. J Hypertens. 2007;25:1021–6.
25. Heagerty AM. Predicting hypertension complications from small artery structure. J Hypertens. 2007;25:939–40.
26. Rossi GP, Bolognesi M, Rizzoni D, Seccia TM, Piva A, Porteri E, Tiberio GA, Giulini SM, Agabiti-Rosei E, Pessina AC. Vascular remodeling and duration of hypertension predict outcome of adrenalectomy in primary aldosteronism patients. Hypertension. 2008;51:1366–71.
27. Agabiti-Rosei E, Rizzoni D. Microvascular structure as a prognostically relevant endpoint. J Hypertens. 2017;35:914–21.
28. Paiardi S, Rodella LF, De Ciuceis C, Porteri E, Boari GE, Rezzani R, Rizzardi N, Platto C, Tiberio GA, Giulini SM, Rizzoni D, Agabiti-Rosei E. Immunohistochemical evaluation of microvascular rarefaction in hypertensive humans and in spontaneously hypertensive rats. Clin Hemorheol Microcirc. 2009;42:259–68.

29. Agabiti-Rosei E, Heagerty AM, Rizzoni D. Effects of antihypertensive treatment on small artery remodelling. J Hypertens. 2009;27:1107–14.
30. Virdis A, Savoia C, Grassi G, Lembo G, Vecchione C, Seravalle G, Taddei S, Volpe M, Agabiti-Rosei E, Rizzoni D. Evaluation of microvascular structure in humans: a 'state-of-the-art' document of the Working Group on Macrovascular and Microvascular Alterations of the Italian Society of Arterial Hypertension. J Hypertens. 2014;32:2120–9.
31. Rizzoni D, Aalkjaer C, De Ciuceis C, Porteri E, Rossini C, Agabiti-Rosei C, Sarkar A, Agabiti-Rosei E. How to assess microvascular structure in humans. High Blood Press Cardiovasc Prev. 2011;18:169–77.
32. Rizzoni D, Agabiti-Rosei E. Structural abnormalities of small resistance arteries in essential hypertension. Intern Emerg Med. 2012;7:205–12.
33. Mathiassen ON, Buus NH, Larsen ML, Mulvany MJ, Christensen KL. Small artery structure adapts to vasodilatation rather than to blood pressure during antihypertensive treatment. J Hypertens. 2007;25:1027–34.
34. Smith RD, Yokoyama H, Averill DB, Schiffrin EL, Ferrario CM. Reversal of vascular hypertrophy in hypertensive patients through blockade of angiotensin II receptors. J Am Soc Hypertens. 2008;2:165–72.
35. De Ciuceis C, Salvetti M, Rossini C, Muiesan ML, Paini A, Duse S, La Boria E, Semeraro F, Cancarini A, Agabiti-Rosei C, Sarkar A, Ruggeri G, Caimi L, Ricotta D, Rizzoni D, Agabiti-Rosei E. Effect of antihypertensive treatment on microvascular structure, central blood pressure and oxidative stress in patients with mild essential hypertension. J Hypertens. 2014;32:565–74.
36. Savoia C, Touyz RM, Endemann DH, Pu Q, Ko EA, De Ciuceis C, Schiffrin EL. Angiotensin receptor blocker added to previous antihypertensive agents on arteries of diabetic hypertensive patients. Hypertension. 2006;48:271–7.
37. Rizzoni D, Porteri E, De Ciuceis C, Sleiman I, Rodella L, Rezzani R, Paiardi S, Bianchi R, Ruggeri G, Boari GEM, Muiesan ML, Salvetti M, Zani F, Miclini M, Agabiti-Rosei E. Effects of treatment with candesartan or enalapril on subcutaneous small resistance artery structure in hypertensive patients with NIDDM. Hypertension. 2005;45:659–65.
38. Agabiti Rosei E, Rizzoni D. Small artery remodelling in diabetes. J Cell Mol Med. 2010;14:1030–6.
39. Greenstein AS, Price A, Sonoyama K, Paisley A, Khavandi K, Withers S, Shaw L, Paniagua O, Malik RA, Heagerty AM. Eutrophic remodeling of small arteries in type 1 diabetes mellitus is enabled by metabolic control: a 10-year follow-up study. Hypertension. 2009;54:134–41.
40. Buus NH, Bøttcher M, Jørgensen CG, Christensen KL, Thygesen K, Nielsen TT, Mulvany MJ. Myocardial perfusion during long-term angiotensin-converting enzyme inhibition or beta-blockade in patients with essential hypertension. Hypertension. 2004;44:465–70.
41. Buus NH, Mathiassen ON, Fenger-Grøn M, Præstholm MN, Sihm I, Thybo NK, Schroeder AP, Thygesen K, Aalkjær C, Pedersen OL, Mulvany MJ, Christensen KL. Small artery structure during antihypertensive therapy is an independent predictor of cardiovascular events in essential hypertension. J Hypertens. 2013;31:791–7.

Printed in the United States
by Baker & Taylor Publisher Services